Microbial Contamination Control in the Pharmaceutical Industry

DRUGS AND THE PHARMACEUTICAL SCIENCES
A Series of Textbooks and Monographs

1. Pharmacokinetics, *Milo Gibaldi and Donald Perrier*
2. Good Manufacturing Practices for Pharmaceuticals: A Plan for Total Quality Control, *Sidney H. Willig, Murray M. Tuckerman, and William S. Hitchings IV*
3. Microencapsulation, *edited by J. R. Nixon*
4. Drug Metabolism: Chemical and Biochemical Aspects, *Bernard Testa and Peter Jenner*
5. New Drugs: Discovery and Development, *edited by Alan A. Rubin*
6. Sustained and Controlled Release Drug Delivery Systems, *edited by Joseph R. Robinson*
7. Modern Pharmaceutics, *edited by Gilbert S. Banker and Christopher T. Rhodes*
8. Prescription Drugs in Short Supply: Case Histories, *Michael A. Schwartz*
9. Activated Charcoal: Antidotal and Other Medical Uses, *David O. Cooney*
10. Concepts in Drug Metabolism (in two parts), *edited by Peter Jenner and Bernard Testa*
11. Pharmaceutical Analysis: Modern Methods (in two parts), *edited by James W. Munson*
12. Techniques of Solubilization of Drugs, *edited by Samuel H. Yalkowsky*
13. Orphan Drugs, *edited by Fred E. Karch*
14. Novel Drug Delivery Systems: Fundamentals, Developmental Concepts, Biomedical Assessments, *Yie W. Chien*
15. Pharmacokinetics: Second Edition, Revised and Expanded, *Milo Gibaldi and Donald Perrier*
16. Good Manufacturing Practices for Pharmaceuticals: A Plan for Total Quality Control, Second Edition, Revised and Expanded, *Sidney H. Willig, Murray M. Tuckerman, and William S. Hitchings IV*
17. Formulation of Veterinary Dosage Forms, *edited by Jack Blodinger*
18. Dermatological Formulations: Percutaneous Absorption, *Brian W. Barry*
19. The Clinical Research Process in the Pharmaceutical Industry, *edited by Gary M. Matoren*
20. Microencapsulation and Related Drug Processes, *Patrick B. Deasy*
21. Drugs and Nutrients: The Interactive Effects, *edited by Daphne A. Roe and T. Colin Campbell*
22. Biotechnology of Industrial Antibiotics, *Erick J. Vandamme*

23. Pharmaceutical Process Validation, *edited by Bernard T. Loftus and Robert A. Nash*
24. Anticancer and Interferon Agents: Synthesis and Properties, *edited by Raphael M. Ottenbrite and George B. Butler*
25. Pharmaceutical Statistics: Practical and Clinical Applications, *Sanford Bolton*
26. Drug Dynamics for Analytical, Clinical, and Biological Chemists, *Benjamin J. Gudzinowicz, Burrows T. Younkin, Jr., and Michael J. Gudzinowicz*
27. Modern Analysis of Antibiotics, *edited by Adjoran Aszalos*
28. Solubility and Related Properties, *Kenneth C. James*
29. Controlled Drug Delivery: Fundamentals and Applications, Second Edition, Revised and Expanded, *edited by Joseph R. Robinson and Vincent H. Lee*
30. New Drug Approval Process: Clinical and Regulatory Management, *edited by Richard A. Guarino*
31. Transdermal Controlled Systemic Medications, *edited by Yie W. Chien*
32. Drug Delivery Devices: Fundamentals and Applications, *edited by Praveen Tyle*
33. Pharmacokinetics: Regulatory • Industrial • Academic Perspectives, *edited by Peter G. Welling and Francis L. S. Tse*
34. Clinical Drug Trials and Tribulations, *edited by Allen E. Cato*
35. Transdermal Drug Delivery: Developmental Issues and Research Initiatives, *edited by Jonathan Hadgraft and Richard H. Guy*
36. Aqueous Polymeric Coatings for Pharmaceutical Dosage Forms, *edited by James W. McGinity*
37. Pharmaceutical Pelletization Technology, *edited by Isaac Ghebre-Sellassie*
38. Good Laboratory Practice Regulations, *edited by Allen F. Hirsch*
39. Nasal Systemic Drug Delivery, *Yie W. Chien, Kenneth S. E. Su, and Shyi-Feu Chang*
40. Modern Pharmaceutics: Second Edition, Revised and Expanded, *edited by Gilbert S. Banker and Christopher T. Rhodes*
41. Specialized Drug Delivery Systems: Manufacturing and Production Technology, *edited by Praveen Tyle*
42. Topical Drug Delivery Formulations, *edited by David W. Osborne and Anton H. Amann*
43. Drug Stability: Principles and Practices, *Jens T. Carstensen*

44. Pharmaceutical Statistics: Practical and Clinical Applications, Second Edition, Revised and Expanded, *Sanford Bolton*
45. Biodegradable Polymers as Drug Delivery Systems, *edited by Mark Chasin and Robert Langer*
46. Preclinical Drug Disposition: A Laboratory Handbook, *Francis L. S. Tse and James J. Jaffe*
47. HPLC in the Pharmaceutical Industry, *edited by Godwin W. Fong and Stanley K. Lam*
48. Pharmaceutical Bioequivalence, *edited by Peter G. Welling, Francis L. S. Tse, and Shrikant V. Dinghe*
49. Pharmaceutical Dissolution Testing, *Umesh V. Banakar*
50. Novel Drug Delivery Systems: Second Edition, Revised and Expanded, *Yie W. Chien*
51. Managing the Clinical Drug Development Process, *David M. Cocchetto and Ronald V. Nardi*
52. Good Manufacturing Practices for Pharmaceuticals: A Plan for Total Quality Control, Third Edition, *edited by Sidney H. Willig and James R. Stoker*
53. Prodrugs: Topical and Ocular Drug Delivery, *edited by Kenneth B. Sloan*
54. Pharmaceutical Inhalation Aerosol Technology, *edited by Anthony J. Hickey*
55. Radiopharmaceuticals: Chemistry and Pharmacology, *edited by Adrian D. Nunn*
56. New Drug Approval Process: Second Edition, Revised and Expanded, *edited by Richard A. Guarino*
57. Pharmaceutical Process Validation: Second Edition, Revised and Expanded, *edited by Ira R. Berry and Robert A. Nash*
58. Ophthalmic Drug Delivery Systems, *edited by Ashim K. Mitra*
59. Pharmaceutical Skin Penetration Enhancement, *edited by Kenneth A. Walters and Jonathan Hadgraft*
60. Colonic Drug Absorption and Metabolism, *edited by Peter R. Bieck*
61. Pharmaceutical Particulate Carriers: Therapeutic Applications, *edited by Alain Rolland*
62. Drug Permeation Enhancement: Theory and Applications, *edited by Dean S. Hsieh*
63. Glycopeptide Antibiotics, *edited by Ramakrishnan Nagarajan*
64. Achieving Sterility in Medical and Pharmaceutical Products, *Nigel A. Halls*
65. Multiparticulate Oral Drug Delivery, *edited by Isaac Ghebre-Sellassie*
66. Colloidal Drug Delivery Systems, *edited by Jörg Kreuter*

67. Pharmacokinetics: Regulatory • Industrial • Academic Perspectives, Second Edition, *edited by Peter G. Welling and Francis L. S. Tse*

68. Drug Stability: Principles and Practices, Second Edition, Revised and Expanded, *Jens T. Carstensen*

69. Good Laboratory Practice Regulations: Second Edition, Revised and Expanded, *edited by Sandy Weinberg*

70. Physical Characterization of Pharmaceutical Solids, *edited by Harry G. Brittain*

71. Pharmaceutical Powder Compaction Technology, *edited by Göran Alderborn and Christer Nyström*

72. Modern Pharmaceutics: Third Edition, Revised and Expanded, *edited by Gilbert S. Banker and Christopher T. Rhodes*

73. Microencapsulation: Methods and Industrial Applications, *edited by Simon Benita*

74. Oral Mucosal Drug Delivery, *edited by Michael J. Rathbone*

75. Clinical Research in Pharmaceutical Development, *edited by Barry Bleidt and Michael Montagne*

76. The Drug Development Process: Increasing Efficiency and Cost Effectiveness, *edited by Peter G. Welling, Louis Lasagna, and Umesh V. Banakar*

77. Microparticulate Systems for the Delivery of Proteins and Vaccines, *edited by Smadar Cohen and Howard Bernstein*

78. Good Manufacturing Practices for Pharmaceuticals: A Plan for Total Quality Control, Fourth Edition, Revised and Expanded, *Sidney H. Willig and James R. Stoker*

79. Aqueous Polymeric Coatings for Pharmaceutical Dosage Forms: Second Edition, Revised and Expanded, *edited by James W. McGinity*

80. Pharmaceutical Statistics: Practical and Clinical Applications, Third Edition, *Sanford Bolton*

81. Handbook of Pharmaceutical Granulation Technology, *edited by Dilip M. Parikh*

82. Biotechnology of Antibiotics: Second Edition, Revised and Expanded, *edited by William R. Strohl*

83. Mechanisms of Transdermal Drug Delivery, *edited by Russell O. Potts and Richard H. Guy*

84. Pharmaceutical Enzymes, *edited by Albert Lauwers and Simon Scharpé*

85. Development of Biopharmaceutical Parenteral Dosage Forms, *edited by John A. Bontempo*

86. Pharmaceutical Project Management, *edited by Tony Kennedy*

87. Drug Products for Clinical Trials: An International Guide to Formulation • Production • Quality Control, *edited by Donald C. Monkhouse and Christopher T. Rhodes*

88. Development and Formulation of Veterinary Dosage Forms: Second Edition, Revised and Expanded, *edited by Gregory E. Hardee and J. Desmond Baggot*

89. Receptor-Based Drug Design, *edited by Paul Leff*

90. Automation and Validation of Information in Pharmaceutical Processing, *edited by Joseph F. deSpautz*

91. Dermal Absorption and Toxicity Assessment, *edited by Michael S. Roberts and Kenneth A. Walters*

92. Pharmaceutical Experimental Design, *Gareth A. Lewis, Didier Mathieu, and Roger Phan-Tan-Luu*

93. Preparing for FDA Pre-Approval Inspections, *edited by Martin D. Hynes III*

94. Pharmaceutical Excipients: Characterization by IR, Raman, and NMR Spectroscopy, *David E. Bugay and W. Paul Findlay*

95. Polymorphism in Pharmaceutical Solids, *edited by Harry G. Brittain*

96. Freeze-Drying/Lyophilization of Pharmaceutical and Biological Products, *edited by Louis Rey and Joan C. May*

97. Percutaneous Absorption: Drugs–Cosmetics–Mechanisms–Methodology, Third Edition, Revised and Expanded, *edited by Robert L. Bronaugh and Howard I. Maibach*

98. Bioadhesive Drug Delivery Systems: Fundamentals, Novel Approaches, and Development, *edited by Edith Mathiowitz, Donald E. Chickering III, and Claus-Michael Lehr*

99. Protein Formulation and Delivery, *edited by Eugene J. McNally*

100. New Drug Approval Process: Third Edition, The Global Challenge, *edited by Richard A. Guarino*

101. Peptide and Protein Drug Analysis, *edited by Ronald E. Reid*

102. Transport Processes in Pharmaceutical Systems, *edited by Gordon L. Amidon, Ping I. Lee, and Elizabeth M. Topp*

103. Excipient Toxicity and Safety, *edited by Myra L. Weiner and Lois A. Kotkoskie*

104. The Clinical Audit in Pharmaceutical Development, *edited by Michael R. Hamrell*

105. Pharmaceutical Emulsions and Suspensions, *edited by Francoise Nielloud and Gilberte Marti-Mestres*

106. Oral Drug Absorption: Prediction and Assessment, *edited by Jennifer B. Dressman and Hans Lennernäs*

107. Drug Stability: Principles and Practices, Third Edition, Revised and Expanded, *edited by Jens T. Carstensen and C. T. Rhodes*

108. Containment in the Pharmaceutical Industry, *edited by James P. Wood*

109. Good Manufacturing Practices for Pharmaceuticals: A Plan for Total Quality Control from Manufacturer to Consumer, Fifth Edition, Revised and Expanded, *Sidney H. Willig*

110. Advanced Pharmaceutical Solids, *Jens T. Carstensen*

111. Endotoxins: Pyrogens, LAL Testing, and Depyrogenation, Second Edition, Revised and Expanded, *Kevin L. Williams*

112. Pharmaceutical Process Engineering, *Anthony J. Hickey and David Ganderton*

113. Pharmacogenomics, *edited by Werner Kalow, Urs A. Meyer, and Rachel F. Tyndale*

114. Handbook of Drug Screening, *edited by Ramakrishna Seethala and Prabhavathi B. Fernandes*

115. Drug Targeting Technology: Physical • Chemical • Biological Methods, *edited by Hans Schreier*

116. Drug–Drug Interactions, *edited by A. David Rodrigues*

117. Handbook of Pharmaceutical Analysis, *edited by Lena Ohannesian and Anthony J. Streeter*

118. Pharmaceutical Process Scale-Up, *edited by Michael Levin*

119. Dermatological and Transdermal Formulations, *edited by Kenneth A. Walters*

120. Clinical Drug Trials and Tribulations: Second Edition, Revised and Expanded, *edited by Allen Cato, Lynda Sutton, and Allen Cato III*

121. Modern Pharmaceutics: Fourth Edition, Revised and Expanded, *edited by Gilbert S. Banker and Christopher T. Rhodes*

122. Surfactants and Polymers in Drug Delivery, *Martin Malmsten*

123. Transdermal Drug Delivery: Second Edition, Revised and Expanded, *edited by Richard H. Guy and Jonathan Hadgraft*

124. Good Laboratory Practice Regulations: Second Edition, Revised and Expanded, *edited by Sandy Weinberg*

125. Parenteral Quality Control: Sterility, Pyrogen, Particulate, and Package Integrity Testing: Third Edition, Revised and Expanded, *Michael J. Akers, Daniel S. Larrimore, and Dana Morton Guazzo*

126. Modified-Release Drug Delivery Technology, *edited by Michael J. Rathbone, Jonathan Hadgraft, and Michael S. Roberts*

127. Simulation for Designing Clinical Trials: A Pharmacokinetic-Pharmacodynamic Modeling Perspective, *edited by Hui C. Kimko and Stephen B. Duffull*

128. Affinity Capillary Electrophoresis in Pharmaceutics and Biopharmaceutics, *edited by Reinhard H. H. Neubert and Hans-Hermann Rüttinger*

129. Pharmaceutical Process Validation: An International Third Edition, Revised and Expanded, *edited by Robert A. Nash and Alfred H. Wachter*

130. Ophthalmic Drug Delivery Systems: Second Edition, Revised and Expanded, *edited by Ashim K. Mitra*

131. Pharmaceutical Gene Delivery Systems, *edited by Alain Rolland and Sean M. Sullivan*

132. Biomarkers in Clinical Drug Development, *edited by John C. Bloom and Robert A. Dean*

133. Pharmaceutical Extrusion Technology, *edited by Isaac Ghebre-Sellassie and Charles Martin*

134. Pharmaceutical Inhalation Aerosol Technology: Second Edition, Revised and Expanded, *edited by Anthony J. Hickey*

135. Pharmaceutical Statistics: Practical and Clinical Applications, Fourth Edition, *Sanford Bolton and Charles Bon*

136. Compliance Handbook for Pharmaceuticals, Medical Devices, and Biologics, *edited by Carmen Medina*

137. Freeze-Drying/Lyophilization of Pharmaceutical and Biological Products: Second Edition, Revised and Expanded, *edited by Louis Rey and Joan C. May*

138. Supercritical Fluid Technology for Drug Product Development, *edited by Peter York, Uday B. Kompella, and Boris Y. Shekunov*

139. New Drug Approval Process: Fourth Edition, Accelerating Global Registrations, *edited by Richard A. Guarino*

140. Microbial Contamination Control in Parenteral Manufacturing, *edited by Kevin L. Williams*

141. New Drug Development: Regulatory Paradigms for Clinical Pharmacology and Biopharmaceutics, *edited by Chandrahas G. Sahajwalla*

142. Microbial Contamination Control in the Pharmaceutical Industry, *edited by Luis Jimenez*

143. Generic Drug Development: Solid Oral Dosage Forms, *edited by Leon Shargel and Izzy Kanfer*

144. Introduction to the Pharmaceutical Regulatory Process, *edited by Ira R. Berry*

ADDITIONAL VOLUMES IN PREPARATION

Drug Delivery to the Oral Cavity: Molecules to Market, *edited by Tapash Ghosh and William R. Pfister*

Microbial Contamination Control in the Pharmaceutical Industry

edited by
Luis Jimenez
Genomic Profiling Systems, Inc.
Bedford, Massachusetts, U.S.A.

CRC Press
Taylor & Francis Group
Boca Raton London New York

CRC Press is an imprint of the
Taylor & Francis Group, an **informa** business

CRC Press
Taylor & Francis Group
6000 Broken Sound Parkway NW, Suite 300
Boca Raton, FL 33487-2742

First issued in paperback 2019

ISBN-13: 978-0-8247-5753-3 (hbk)
ISBN-13: 978-0-367-39394-6 (pbk)

Library of Congress Cataloging-in-Publication Data
A catalog record for this book is available from the Library of Congress.

Visit the Taylor & Francis Web site at
http://www.taylorandfrancis.com

and the CRC Press Web site at
http://www.crcpress.com

Preface

Since the implementation of good manufacturing practices (GMPs) in the early 1970s, major improvements have been achieved in the control of microbial contamination in pharmaceutical environments. However, microbial contamination of pharmaceutical products is one of the major reasons for product recall and manufacturing problems. Knowledge of the distribution and survival of microorganisms in pharmaceutical environments is critical in the process control of nonsterile and sterile pharmaceutical products. This knowledge is somewhat limited by the ubiquitous distribution of microorganisms in manufacturing facilities, the diversity of microorganisms in environmental samples, and the flexibility of microorganisms in surviving under different environmental fluctuations. Optimization of pharmaceutical manufacturing has led to more efficient testing systems to monitor the analysts, environment, water, raw materials, and finished products that are the major sources of introduction of microorganisms into the processes. However, to avoid microbial contamination, adherence to GMP is the foundation for manufacturing safe and efficacious pharmaceutical products.

With the latest developments in computer science, automation, genomics, combinatorial chemistry, and process control, the manufacture and quality control analysis of pharmaceuticals will be changed significantly. Therefore, optimization of quality control analysis in pharmaceutical operations has become an interdisciplinary endeavor that requires communication and cooperation between microbiologists and other scientists. This book discusses major issues regarding testing and quality control in pharmaceutical manufacturing, which will ensure product and process integrity. Why is it important to control the presence of microorganisms in a manufacturing facility? What systems do we need to prevent this contamination? What tests do we perform to guarantee the safety and efficacy of the products manufactured under those conditions? What new technologies are available to optimize sample analysis and manufacturing? What regulations must be followed to provide quality products? We hope to provide answers to all these questions. This book is aimed at pharmacy students, chemists, engineers, pharmaceutical scientists, and microbiologists working in or associated with the pharmaceutical industry, with the intention of being a first step toward the understanding of microbial control in pharmaceutical environments.

Luis Jimenez

Contents

Preface *iii*
Contributors *vii*

1. Microorganisms in the Environment and Their Relevance to Pharmaceutical Processes 1
   Luis Jimenez

2. Microbial Limits 15
   Luis Jimenez

3. Microbial Monitoring of Potable Water and Water for Pharmaceutical Purposes 45
   Anthony M. Cundell

4. Sterility Test and Procedures 77
   Luis Jimenez

5. Environmental Monitoring 103
   Luis Jimenez

6. Biological Indicator Performance Standards and Control 133
Jeanne Moldenhauer

7. Rapid Methods for Pharmaceutical Analysis 147
Luis Jimenez

8. Endotoxin: Relevance and Control in Parenteral Manufacturing 183
Kevin L. Williams

9. Proper Use and Validation of Disinfectants 251
Laura Valdes-Mora

10. Antimicrobial Effectiveness Test and Preservatives in Pharmaceutical Products 283
Luis Jimenez

Index *301*

Contributors

Anthony M. Cundell, Ph.D. Wyeth Pharmaceuticals, Pearl River, New York, U.S.A.

Luis Jimenez, Ph.D. Genomic Profiling Systems, Inc., Bedford, Massachusetts, U.S.A.

Jeanne Moldenhauer, Ph.D. Vectech Pharmaceutical Consultants, Inc., Farmington Hills, Michigan, U.S.A.

Laura Valdes-Mora, M.S. Elite MicroSource Corporation, Panama City, Florida, U.S.A.

Kevin L. Williams, B.S. Eli Lilly and Company, Indianapolis, Indiana, U.S.A.

1

Microorganisms in the Environment and Their Relevance to Pharmaceutical Processes

Luis Jimenez
Genomic Profiling Systems, Inc., Bedford, Massachusetts, U.S.A.

1. INTRODUCTION

Microorganisms on Earth are widely distributed across different environmental habitats [1]. They are present in water, air, sediments, and soil. One of the reasons for the wide distribution of microorganisms in the environment is the great physiological diversity regarding the utilization of inorganic and organic compounds to sustain microbial viability, maintenance, reproduction, and growth [1]. Microbial cells degrade organic and inorganic compounds to sustain microbial metabolism. Some microbial species do not require high concentrations of organic or inorganic compounds to survive and grow. Microbial species such as *Pseudomonas* spp., *Acinetobacter* spp., *Burkholderia* spp., and *Stenotrophomonas* spp. exhibit a tremendous physiological versatility by using a wide variety of organic and inorganic compounds to support microbial metabolism.

Microorganisms carry the energy needed for metabolic processes in the phosphate energy-rich molecule called adenosine triphosphate (ATP) [2].

ATP is the most important energy compound in the microbial cell. Enzymatic reactions are an important part of the catabolic pathways used by microorganisms to generate ATP. For instance, organic compounds such as carbohydrates are converted to pyruvate through a process called glycolysis. Some microbes use glycolysis to generate ATP, in the absence of oxygen. The end products of that process (e.g., fermentation) are alcohols and acids. Other microorganisms utilize inorganic compounds such as sulfate and nitrate to generate ATP. In other cases, solar energy is utilized to generate ATP by bacterial photosynthesis. When oxygen is present in the environment, microorganisms develop metabolic reactions driven by inorganic or organic compounds to generate ATP. Furthermore, respiratory metabolism is also used. Respiratory metabolism is based upon the transfer of electrons from different types of electron donors and acceptors such as nicotinamide-adenine dinucleotide hydrogen (NADH), flavin-adenine dinucleotide hydrogen (FADH), and cytochromes. Some bacteria use oxygen as the ultimate electron acceptor (e.g., aerobic respiration) whereas others use different types of inorganic compounds (e.g., anaerobic respiration). However, other types of bacteria can live in the presence or absence of oxygen (facultative).

Although microbial populations are present in all types of habitats, there are several major limiting factors that affect microbial distribution, survival, and proliferation in the environment. These factors are:

- Temperature
- Available water
- Concentration of organic compounds
- Concentration of hydrogen ions (pH)
- Concentration of inorganic compounds
- Concentration of particulates in the air
- Redox potential (Eh)
- Pressure
- Light intensity

Because of the different environmental fluctuations encountered, natural microbial communities do not exist in a state of perpetual proliferation and growth. There are major seasonal fluctuations regarding temperature, light intensity, available water, and concentration of organic and inorganic compounds on the basis of the geographical location of a given microbial community. For instance, water habitats in tropical locations do not undergo the same temperature fluctuations observed in temperate habitats. Therefore, microorganisms in temperate habitats exhibit a higher tolerance to increased temperatures when compared to microorganisms in tropical climates. The environment is always changing and microorganisms respond to these changes by adapting and surviving. Some of these adaptations allow microbial cells to grow very slowly or remain dormant for long periods of time.

2. STRATEGIES FOR MICROBIAL SURVIVAL IN THE ENVIRONMENT

How do microorganisms respond to different environmental fluctuations in the environment? They respond to these fluctuations by adopting different survival strategies [3]. These strategies are based upon the minimal utilization of energy to support microbial metabolism and growth. Growth is defined as an increase in the number of cells over time. However, microbial populations do not grow continuously because of the fluctuations in the amount of available water, food, etc. When laboratory cultures are prepared, microbial cells are inoculated into rich growth media with high concentrations of carbon, nitrogen, and phosphate. After inoculation and a brief phase, where microbes do not grow (lag phase), microorganisms grow exponentially until they utilize all available food sources (log phase). At that time, the numbers of cells stabilize. This is called stationary phase. If the culture media is not replenished with fresh growth media, the number of cells decreases due to the lack of nutrients and cell death. For instance, laboratory cultures of *Escherichia coli* double every 20 min when grown in rich nutrient media. However, when cell suspensions of the same microorganism are introduced into growth chambers immersed in a low-nutrient environment, doubling time is significantly slower [4]. Evidently, different growth dynamics are found between a high-nutrient and low-nutrient environment.

Some of the survival strategies are based upon the formation of bacterial spores as a response to nutrient deficiency and high temperature. *Bacillus* spp. and *Clostridium* spp. are commonly known as spore formers. These bacterial species are widely distributed in air, water, and soil samples. Germination of the spores is triggered by environmental factors indicating the presence of optimal conditions for microbial growth.

Other microorganisms respond to environmental fluctuations by changes in the enzymatic and protein profiles [3]. These changes are generally found in a wide variety of microbial species such as *Acinetobacter* spp., *Arthrobacter* spp., *Agrobacterium* spp., *Pseudomonas* spp., and the family Enterobacteriaceae. When low-nutrient concentration environments are encountered, microbial cells produce new types of enzymes and proteins, which are essential for microbial survival and maintenance.

Another survival strategy is when microbial cells reduce their size and metabolism. Along with size reduction, there is a decrease in respiration and cell numbers. In some cases, an increased adhesion to surfaces has been reported. This results in the formation of biofilms. Biofilm formation concentrates the cells on a surface and creates a microenvironment where nutrient utilization is optimized.

To support microbial metabolism under low-nutrient concentration conditions, four classes of carbon, phosphate, and nitrogen compounds are used for potential storage of food sources. These compounds are:

- Carbohydrates
- Lipids (poly-β-hydroxybutyrate and polyalkanoates)
- Polyphosphates
- Cyanophycin/phycocyanin.

These compounds are degraded by microorganisms under stress-induced conditions to provide endogenous sources of energy to maintain microbial viability and growth.

Arthrobacter spp. are a good example of these types of bacterial populations. They are pleomorphic bacteria undergoing different cell morphologies under different nutritional conditions. A transition from rods to cocci is observed when cultures go from exponential growth phase to stationary growth phase. The cells are capable of long-term survival under hostile environmental conditions by utilizing endogenous sources of energy.

Gram-negative bacterial species undergo a viable but nonculturable stage [3]. When microbial cells enter this stage, several changes take place. It has been reported that cell size, enzymatic profile, membrane proteins, and microbial metabolism are dramatically reduced. New enzymes and proteins are produced as a response to the environmental fluctuations encountered. This response is commonly triggered by the lack of carbon, nitrogen, and phosphate sources. Furthermore, microorganisms undergoing this transitional stage do not grow on regular growth media (e.g., uncultured). However, they have been enumerated and proven to be physiologically viable by alternative methods with increasing sensitivity and resolution.

When microbial populations adopt some of these survival strategies, identification by standard methods is difficult and might lead to erroneous conclusions. This is because standard methods are based upon the phenotypical analysis of microorganisms. Macroscopical and microscopical analyses are based upon colony morphology, cell size, enzymatic profiles, and carbon utilization profiles.

Standard methods are used in clinical, environmental, pharmaceutical, and food microbiology to diagnose microbial pathogenesis and contamination [5]. However, the development of better analytical methods has provided an accurate and sensitive representation of the distribution and activity of microorganisms in the environment. This new information has supplemented the knowledge obtained using traditional culture and enrichment methods.

3. ISOLATION, ENUMERATION, AND IDENTIFICATION OF MICROORGANISMS

In the beginning of the field of microbiology, microbial isolation and identification were based upon the phenotypical analysis of microbial cells by microscopical analysis of water, fermentation products, and clinical samples by Leeuwenhoek [6], Koch [7], and Pasteur [8]. After several years, the plate count was invented in the laboratory of Koch [7]. Up to that point, most of the works were basically concentrated on infectious disease analysis for diagnosis and prognosis. It was not until the significant contributions of Winogradsky and Beijerinck that the enrichment culture technique was developed to isolate microorganisms from environmental samples. Optimizing the enrichment media to enhance the growth of microorganisms with specific metabolic activity leads to the isolation of specific microbes present in low numbers. The role of microorganisms in the cycling of materials in the environment and the common metabolic reactions between microorganisms and macroorganisms was demonstrated by the works of Kluyver, van Niel, and Stainer [2].

Further developments in microbial methodology lead to selective agar media for pathogen isolation from clinical samples. Membrane filtration analysis was introduced after the Second World War. The development of membrane filtration allowed the concentration of large volumes of liquid on a filter. Larger sample volumes were analyzed by optimizing assay sensitivity and resolution. In some cases, water samples contain low numbers of microorganisms, which would not be detected unless large volumes (e.g., 100 mL) are analyzed.

Up to that point, all analyses were based upon enumeration and detection of colonies based on morphology, color, differential staining, cell morphology, and biochemical reactions of isolated colonies. For instance, macroscopical and microscopical analyses of microbial communities from clinical and environmental samples relied on the above characteristics.

During the late 20th century, molecular biology techniques provided a clearer picture of the distribution and complexity of microbial communities in environmental and clinical samples [9–13]. Some of the techniques used are:

- Gene probes
- Polymerase chain reaction (PCR) technology
- DNA sequencing
- Nucleic acid extractions from environmental matrices.

Further studies also demonstrated the use of specific biochemical indicators for the presence of microorganisms. These analyses provide information on the microbial community, microbial population, and individual cells. For instance, microbial biomass can be determined by:

- Direct microbial counts
- ATP and total adenylate
- Cell wall components (lipids and muramic acid)
- Bacteriochlorophyll and other pigments
- DNA
- Proteins.

The application of these molecular biology techniques and biomass measurements to environmental and clinical analysis demonstrated that the majority of microorganisms in the environment are unculturable but viable.

Studies demonstrated that when individual cells are counted and analyzed by direct microscopy, different growth dynamics are observed. For example, direct microbial counts using epifluorescence microscopy yield higher counts than standard plating techniques [2]. However, overestimation of the numbers is a result of the inability to distinguish between living and dead microorganisms. Direct microscopy with fluorochromes (dyes) such as acridine orange (AODC), 4′,6-diamidino-2-phenyl-indole (DAPI), Hoechst 33258, and fluorescein isothiocyanate (FITC) provided an alternative to the plate count. However, it was difficult to determine cell viability. Are these cells viable? Are we just counting dead cells?

Several modifications of the direct count method allow the determination of the numbers of viable cells [e.g., combining the direct count method with INT (2-[*p*-iodophenyl]-3-[*p*-nitrophenyl]-5-phenyl tretrazolium chloride) staining]. Respiring microorganisms reduce INT to INT-formazan by accumulating intracellular dark red spots visible through a microscope. Other methods rely on the inhibition of cell division by nalidixic acid (DVC) where microscopical observations show elongated cells [3]. Another method counts the numbers of cells dividing actively [14].

Combining direct microscopy with radioactive substrates to analyze incubated microorganisms was also used [15]. Specific types of microorganisms can be also be detected by fluorescent antibody techniques [3]. All these studies consistently indicated that a high percentage of the microorganisms in a sample did not grow on standard plate media but were viable. Up to that point, viability was understood as the capacity of a microorganism to grow on plate media. Growth on plate media requires duplication of microbial cells to a stage where colonies are visually detected. The minimum numbers of cells required for a colony to be visible ranges from 1×10^6 to

5×10^6 cells. However, on the basis of these and other studies, viability was defined as an indication of bacterial activity, not growth [3].

Direct extraction of DNA and RNA from environmental and clinical samples further confirmed that the great majority of the microbial community in a given sample do not grow on standard plate media [16–18]. Furthermore, important microbial populations were detected and characterized using molecular biology techniques. It seems that because of the extensive physiology of microbial populations, no single medium or defined set of growth conditions can provide all the requirements for most of the organisms present in a given environmental sample. It seems that many of the microbial species dominating natural environments are not adapted to grow in media containing high concentrations of organic compounds. However, when low-nutrient media is used, a higher microbial recovery is found in some environmental samples [19,20]. Different types of low-nutrient media have recovered a previously unculturable segment of microorganisms from water, soil, and clinical samples. These populations do not grow on blood agar, soybean casein digest agar, soybean casein digest broth, nutrient broth, and nutrient agar, but have been shown to be metabolically active.

4. ANALYSIS AND CONTROL OF PHARMACEUTICAL ENVIRONMENTS TO MINIMIZE MICROBIAL SURVIVAL

One of the most important areas in pharmaceutical process control is the development of systems to control the numbers, survival, and proliferation of microorganisms during manufacturing of nonsterile and sterile pharmaceutical products. The facility where products are manufactured is basically a closed environment where people and materials will move in and out to carry out different processes.

Microorganisms, as previously mentioned, have a great catabolic capacity to derive energy from any type of organic or inorganic compounds. Therefore, having microorganisms in a product can cause spoilage of the formula by breaking down active ingredients and excipients. This might compromise the potency and efficacy of the drug. Furthermore, the presence of high numbers of microorganisms and pathogens represents a serious health threat to consumers because products will be ingested, injected, or applied to human skin. Pharmaceutical products are commonly used after a pathological condition (e.g., disease) is diagnosed. The disease can be based upon microbial infection or metabolic disorders.

Therefore, minimizing the numbers or preventing the introduction of significant numbers of microorganisms into pharmaceutical facilities and processes becomes the most important aspect of process control during pharmaceutical manufacturing [21]. What are the critical areas where microorganisms can be introduced?

First, some of the raw materials utilized for the development of pharmaceutical formulations are based upon natural products that contain a high microbial load. The production processes for these raw materials do not eliminate all microorganisms. Therefore, they are not sterile. Testing must be performed to determine the quality of these materials. The absence of *E. coli*, *Staphylococcus aureus*, *Pseudomonas aeruginosa*, and *Salmonella typhimurium* is required before raw materials can be used in pharmaceutical products. However, some of the manufacturing processes are designed to significantly reduce the number of microorganisms. Different types of bacteria commonly found in pharmaceutical raw materials are *Lactobacillus* spp., *Pseudomonas* spp., *Bacillus* spp., *Escherichia* spp., *Streptoccocus* spp., *Clostridium* spp., *Agrobacterium* spp., etc. and molds such as *Cladosporium* spp. and *Fusarium* spp.

A second critical area is the air in the facility. Air ventilation systems in manufacturing facilities are built to minimize the survival, distribution, reproduction, and growth of microbes. This facility is provided with humidity, ventilation, and air conditioning units (HVAC), which control these parameters. The air is filtered through a 0.5-μm filter to prevent the introduction into the facility of any particle higher than 0.5 μm. Microorganisms are commonly associated with particles in the air. Therefore, the exclusion of these particles in the facility minimizes the chances of microbial distribution and contamination by air. Air flow and pressure are controlled to exclude any nonviable and viable particle from entering critical areas. Humidity also controls the number of microorganisms in a room. The more humid is the room, the more chances there are for microorganisms to be carried by droplets of moisture. Therefore, a dry room provides a more hostile condition for microbes to grow than a humid room. A general practice in pharmaceutical environments is to apply ultraviolet light (UV) to reduce microbial contamination by air. Some of the microbial species commonly found in air samples in pharmaceutical environments are bacteria such as *Bacillus* spp., *Staphylococcus* spp., *Corynebacterium* spp. Common mold species are *Aspergillus* spp. and *Penicillium* spp.

A third critical area is the personnel in the plant and testing laboratories. Microorganisms are part of the normal flora of the human skin and body. Therefore, operators and laboratory analysts are the major sources of contamination during manufacturing and testing [22]. Some of the species living in the human skin are *Staphylococcus epidermidis*, *Staphylococcus capitis*, *Staphylococcus hominis*, *Propionibacterium* spp., *Propionibacterium acnes*, *Micrococcus* spp., etc. The normal flora for the human oral cavity is comprised of *Streptococcus salivarius*, *Streptococcus mutans*, etc. Molds can also be possible contaminants. Common molds from human flora are *Tricho-*

phyton spp., *Epidermophyton* spp., *Microsporon* spp., etc. To protect critical areas from human microbial flora, personnel wear gowns, hair covers, hoods, shoe covers, laboratory coats, face masks, gloves, boots, etc.

A fourth area of concern is water. Water is the most common raw material in pharmaceutical manufacturing. Drinking water is physically and chemically treated to reduce microbial numbers and pathogenic microorganisms. Water for pharmaceutical processes is further treated to minimize microbial numbers, endotoxin substances, and organic and inorganic compounds. The less organic compounds there are in the water, the fewer microorganisms will be found. Bacterial species such *Pseudomonas* spp., *Alcaligenes* spp., *Stenotrophomonas* spp., *Burkholderia cepacia*, *Burkholderia picketti*, *Serratia* spp., and *Flavobacterium* spp. are commonly found in water samples. Other types of bacteria can also be present but when found, they indicate fecal sources of contamination. These bacteria are *E. coli*, *Entero bacter* spp., *Klebsiella* spp., *Salmonella* spp., *Shigella* spp., *Clostridium perfringes*, and *Enterococcus* spp. Recent studies using 16S ribosomal analysis, PCR amplification, and denaturing gradient gel electrophoresis (DGGE) testing demonstrated the presence of the following culturable bacterial species: *Bradyrhizobium* spp., *Xanthomonas* spp., and *Stenotrophomonas* spp. However, the predominant bacterial type in the water system could not be detected on culture media.

A fifth area of concern is the equipment and building areas. Unless equipment is cleaned and sanitized, there is always the risk of microbial contamination. However, cleaning and sanitization of the equipment must provide a hostile environment for microorganisms to survive and grow. Bacteria such as *Pseudomonas* spp., *S. epidermidis*, *Bacillus* spp., etc. are commonly found in equipment. Molds are commonly found in walls and ceilings. Continuous sanitization and disinfection of floors, drains walls, and ceilings are advised to avoid the microbial colonization of these areas. Some of the mold species are *Aspergillus* spp., *Penicillium* spp., and *Aureobasidium* spp., etc. Using 16S ribosomal DNA analysis and sequencing, other microbial species found are *Taxeobacter* spp., *Flexibacter* spp., *Cytophaga* spp., *Ultramicrobacterium* spp., *Stenotrophomonas* spp., *Streptococcus* spp., *Sphingomonas* spp., and *Comamonas* spp.

Quality control analysis in the pharmaceutical industry relies on standard enrichment and/or plating of the different types of pharmaceutical raw materials and finished products [23–28]. Environmental monitoring of all critical areas also relies on standard microbiological assays [21]. When microorganisms contaminate pharmaceutical products, standard methods are performed to quantify, detect, and identify the numbers and types of microorganisms present in a given pharmaceutical batch. Standard, compendial methods are based upon the enrichment, incubation, and isolation of microorganisms from pharmaceutical samples. Because of the long incubation times, continuous manipulation, and time-consuming procedures, results are normally obtained within 6–8 days for nonsterile products and 14 days for sterile products. It has been recently reported that standard methods, as found in environmental samples, underestimate the numbers and diversity of microbial communities present in pharmaceutical environments [29–33]. This has been demonstrated in samples of water, contact plates, and air from different pharmaceutical manufacturing facilities and clean room environments. ATP bioluminescence, flow cytometry, direct viable counts, DNA, and PCR technology have demonstrated that a nonculturable portion of the microbial community in pharmaceutical environments is viable and not detected by standard methods. Therefore, these new technologies complement standard methods by providing higher resolution and discrimination between microbial species. Accurate information of the types and numbers of microorganisms in pharmaceutical environments will lead to the optimization of processes that minimize microbial distribution, viability, growth, and proliferation.

Furthermore, identification of several environmental isolates from pharmaceutical environments using standard identification procedures is proven to be incorrect [34]. When identification is performed by biochemical, lipids, and DNA analyses, DNA analysis provides the best reproducibility, sensitivity, accuracy, and resolution. To develop the proper corrective action when out-of-specification (OOS) results are obtained, accurate microbial identification is needed if the contamination source has to be determined and tracked. A corrective action is not effective if wrong information is used to develop a proper solution to a given problem.

On the basis of these studies, it is evident that in some cases, standard methods are not accurate and precise to optimize process control, leading to faster releasing time, sample analysis, and high-throughput screening of samples. Standard methods must be complemented by other technologies that can provide additional information on the processes and systems used in pharmaceutical manufacturing. Although standard methods are valuable and do provide information on the numbers, microbial genera, and species, they were developed as previously stated for the identification of microorganisms from clinical samples. Most clinical samples originate from human fluids or tissues, which are rich in nutrients and exhibit temperatures of 35–37°C. Environmental samples (e.g., raw materials, finished products, air, water, equipment swabs, and contact plates) taken from production facilities are not rich in nutrients (oligotrophic) and temperature fluctuates below and above ambient temperature. Low water activity and dramatical changes in pH also contribute to microbial stress. Furthermore, manufacturing of pharmaceutical products comprises physical processes such as blending, compression, filtration, heating, encapsulation, shearing, tableting, granulation, coating, and drying. These processes expose microbial cells to extensive environmental stresses. The facility where manufacturing takes place is designed to create an environment where microorganisms will not survive. Air flow, temperature, pressure, air particulates, etc. are optimized to reduce the numbers of microorganisms.

Microorganisms, as previously stated, survive under those conditions by adapting to the lack of nutrients and other environmental fluctuations by undertaking different survival strategies. Furthermore, bacterial cells that do not grow on plate media but retain their viability by going through the viable but culturable stage are still capable of causing severe infections to humans. Several studies have shown that microbial cells in pharmaceutical environments have changed the cell size and enzymatic and physiological profiles as a response to environmental fluctuations [35,36]. Similar responses have been reported by bacteria exposed to drug solutions where significant morphological and size changes are observed. Bacterial cells spiked into different types of injectable products have shown different changes in their metabolism, enzymatic profiles, and structural changes, which interfered with their identification using standard biochemical assays [35]. Furthermore, bacteria undergoing starvation survival periods are capable of penetrating 0.2/0/22 μm rated filters, which are supposed to retain all bacterial species [36].

Therefore, using enzymatic and carbon assimilation profiles (e.g., biochemical identification) to discriminate and identify microorganisms from pharmaceutical samples might, in some cases, yield unknown profiles that will not provide any significant information on the microbial genera and species. In pharmaceutical environments, information on the genera and species of a microbial contaminant will provide valuable information on the possible sources of the contamination, allowing the implementation of effective corrective actions.

It has been also shown that the recovery of microorganisms from environmental samples in pharmaceutical clean room environments is enhanced by using low-nutrient media [31,33]. The recovery of microorganisms from pharmaceutical water samples has been shown to be increased by the use of a low-nutrient media [30]. Similar results are observed for other environmental samples when low-nutrient media is used. The need for a stress recovery phase is demonstrated by longer incubation times and low-nutrient media.

Evidently, pharmaceutical environments are subjected to microorganisms originating from air, water, personnel, and materials introduced into the different facilities where products are manufactured and tested. New methods and additional information on the distribution, survival, and growth of microorganisms in pharmaceutical facilities provide additional information to enhance our understanding of the factors controlling the presence of microbial communities in pharmaceutical environments.

5. CONCLUSION

The progress and development of new analytical technologies to enumerate, isolate, and characterize microorganisms from the environment have provided a greater resolution and sensitivity to describe the composition, distribution, and biomass of microorganisms on Earth. The great majority of microbes in nature do not grow on plate media. Similar results have been observed in pharmaceutical environments. New information on the distribution, survival, growth, and reproduction of microorganisms in pharmaceutical environments will lead to the optimization of process control by optimizing the systems used for controlling microbial contamination.

REFERENCES

1. Dagely S. Chemical unity and diversity in bacterial catabolism. In: Poindexter JS, Leadbetter ER, eds. Bacteria in Nature. Vol. 3. Structure, Physiology, and Genetic Adaptability. New York: Plenum Press, 1989:259–291.
2. Karl DM. Determination of in situ microbial biomass, viability, metabolism, and growth. In: Poindexter JS, Leadbetter ER, eds. Bacteria in Nature. Vol. 2. Methods and Special Applications in Bacterial Ecology. New York: Plenum Press, 1986:85–176.
3. Roszak DB, Colwell RR. Survival strategies of bacteria in the natural environment. Microbiol Rev 1987; 51:365–379.
4. Muniz I, Jimenez L, Toranzos G, Hazen TC. Survival and activity of *Streptococcus faecalis* and *Escherichia coli* in tropical freshwater. Microb Ecol 1989; 18:125–134.
5. Jimenez L. Molecular diagnosis of microbial contamination in cosmetic and pharmaceutical products—a review. J AOAC Int 2001; 84:671–675.
6. Leeuwenhoek AV. Observations concerning little animals observed in rain, well, sea, and snow water. Philos Trans R Soc XI 1677; 821–831.
7. Koch R. Zur untersuchung von pathogenen organismen. Mitt Kaiser 1881; 1:1.
8. Cundell A. Historical perspective on methods development. In: Easter MC, ed. Rapid Microbiological Methods in the Pharmaceutical Industry. New York, NY, USA: Interpharm/CRC, 2003:9–17.
9. Stapleton RD, Rip S, Jimenez L, Koh S, Gregory I, Sayler GS. Use of nucleic acid analytical approaches in bioremediation: site assessment and characterization. J Microbiol Methods 1998; 32:165–178.
10. Drancourt M, Bollet C, Carlioz A, Martelin R, Gayral JP, Raoult D. 16S ribosomal DNA sequence analysis of a large collection of environmental and clinical unidentifiable bacterial isolates. J Clin Microbiol 2000; 38:3623–3630.
11. Small J, Call DR, Brockman FJ, Straub TM, Chandler DP. Direct detection of 16S rRNA in soil extracts by using oligonucleotide microarrays. Appl Environ Microbiol 2001; 67:4708–4716.

12. Pillai SD. Rapid molecular detection of microbial pathogens: breakthroughs and challenges. Arch Virol Suppl 1997; 13:67–82.
13. Sayler GS, Shields MS, Tedford ET, Breen A, Hooper SW, Sirotkin KM, Davis JW. Application of DNA–DNA colony hybridization to the detection of catabolic genotypes in environmental samples. Appl Environ Microbiol 1985; 49:1295–1303.
14. Hagstrom A, Larsson U, Horsted P, Normark S. Frequency of dividing cells, a new approach to the determination of bacterial growth rates in aquatic environments. Appl Environ Microbiol 1979; 37:805–812.
15. Tabor P, Neihof J. Improved microautoradiography method to determine individual microorganisms active in substrate uptake in natural waters. Appl Environ Microbiol 1982; 44:945–953.
16. Kroes I, Lepp PW, Relman DA. Bacterial diversity within the human subgingival crevice. Proc Natl Acad Sci U S A 1999; 96:14547–14552.
17. Torsvik V, Goksøyr J, Daae FL. High diversity in DNA of soil bacteria. Appl Environ Microbiol 1990; 56:782–787.
18. Torsvik V, Øvreås L, Thingstad TF. Prokaryotic diversity—magnitude, dynamics, and controlling factors. Science 2002; 296:1064–1066.
19. Connon SA, Giovannoni SJ. High-throughput methods for culturing microorganisms in very-low-nutrient media yield diverse new marine isolates. Appl Environ Microbiol 2002; 68:3878–3885.
20. Zengler K, Toledo G, Rappé M, Elkins J, Mathur EJ, Short JM, Keller M. Cultivating the uncultured. PNAS 2002; 99:15681–15686.
21. United States Pharmacopeial Convention. Microbiological evaluation of clean rooms and other controlled environments. In: US Pharmacopoeia. Rockville, MD: United States Pharmacopeial Convention, 2002:2206–2212.
22. Hyde W. Origin of bacteria in the clean room and their growth requirements. PDA J Sci Technol 1998; 52:154–164.
23. European Pharmacopoeial Convention. Microbiological examination of non-sterile products. In: European Pharmacopoeia. 3rd ed. Strasbourg, France: Council of Europe, 2001:70–78.
24. The Japanese Pharmacopoeia. Microbial Limit Test. 13th ed. Tokyo, Japan: The Society of Japanese Pharmacopoeia, 1996:49–54.
25. United States Pharmacopeial Convention. Microbial limit test. In: US Pharmacopoeia. Rockville, MD: United States Pharmacopeial Convention, 2002: 1873–1878.
26. European Pharmacopoeial Convention. Sterility. In: European Pharmacopoeia. 3rd ed. Strasbourg, France: Council of Europe, 2001:63–67.
27. United States Pharmacopeial Convention. Sterility tests. In: US Pharmacopoeia. Rockville, MD: United States Pharmacopeial Convention, 2002:1878–1883.
28. The Japanese Pharmacopoeia. Sterility Test. 13th ed. Tokyo, Japan: The Society of Japanese Pharmacopoeia, 1996:69–71.
29. Kawai M, Matsutera E, Kanda H, Yamaguchi N, Tani K, Nasu M. 16S ribosomal DNA-based analysis of bacterial diversity in purified water used in

pharmaceutical manufacturing processes by PCR and denaturing gradient gel electrophoresis. Appl Environ Microbiol 2002; 68:699–704.

30. Nagarkar P, Ravetkar SD, Watve MG. Oligophilic bacteria as tools to monitor aseptic pharmaceutical production units. Appl Environ Microbiol 2001; 67: 1371–1374.
31. Kawai M, Yamaguchi N, Nasu N. Rapid enumeration of physiologically active bacteria in purified water used in the pharmaceutical manufacturing process. J Appl Microbiol 1999; 86:496–504.
32. Venkateswaran K, Hattori N, La Duc MT, Kern R. ATP as a biomarker of viable microorganisms in clean room facilities. J Microbiol Methods 2003; 52:367–377.
33. Wallner G, Tillmann D, Haberer K. Evaluation of the ChemScan system for rapid microbiological analysis of pharmaceutical water. PDA J Pharm Sci Technol 1999; 53:70–74.
34. Montgomery S. A comparison of methods for identification of microorganisms in the pharmaceutical manufacturing environment. First Annual Rapid Micro Users Group, Validation Requirements for Rapid Microbiology, Chicago, IL, September 1–7, 2002.
35. Papapetropoulou M, Papageorgakopoulou N. Metabolic and structural changes in *Pseudomonas aeruginosa*, *Achromobacter* CDC, and *Agrobacterium radiobacter* cells injured in parenteral fluids. PDA J Pharm Sci Technol 1994; 48:299–303.
36. Sundram S, Mallick S, Eisenhuth J, Howard G, Brandwein H. Retention of water-borne bacteria by membrane filters: Part II. Scanning electron microscopy (SEM) and fatty acid methyl ester (FAME) characterization of bacterial species recovered downstream of 0.2/0.22 micron rated filters. PDA J Pharm Sci Technol 2001; 55:87–113.

2

Microbial Limits

Luis Jimenez
Genomic Profiling Systems, Inc., Bedford, Massachusetts, U.S.A.

1. INTRODUCTION

This chapter will discuss the microbiological analysis of nonsterile pharmaceutical products with emphasis in the microbiological test requirements and test methods. When a nonsterile pharmaceutical product is manufactured, quality control evaluation includes the microbiological testing of raw materials, excipients, active ingredients, bulk, and finished products. However, because of their nature, nonsterile samples contain high numbers of microbes and objectionable microorganisms that might represent a serious health threat to consumers. High number of microorganisms can also change the chemical composition of a given pharmaceutical formulation by spoilage, affecting the stability and integrity of the product and package. Furthermore, since these products are not sterile, a microbial bioburden is allowed based upon the product specifications. This means that although there are microorganisms present in the sample, their quantity and types will determine the safety of that particular pharmaceutical product and efficacy of the manufacturing process. Therefore the microbiological testing of nonsterile pharmaceuticals is defined as microbial limits [1]. How and when to define those limits is based upon:

- Chemical composition of product
- Production process

- Route of application
- Intended use of product
- Delivery system of product

Nonsterile pharmaceuticals are manufactured under aseptic conditions, but the processes used during production are not monitored on a regular basis. Furthermore, the criteria for manufacturing nonsterile pharmaceuticals are completely different when compared to sterile products. To date, there are no regulatory or compendial guidelines. However, according to the code of federal regulations (CFR) part 211.113, companies must have appropriate written procedures, designed to prevent the presence of objectionable organisms from drug products not required to be sterile [2]. This includes standard operating procedures (SOPs) for manufacturing and quality control analysis of each nonsterile product. Written procedures for manufacturing, packaging, and quality control analysis allow reproducibility, continuity, accuracy, and process control.

For instance, in sterile manufacturing, water, air, and environmental monitoring are performed on a routine basis preventing sterility failures and system breakdown. However, nonsterile manufacturing does not monitor these areas, if they monitor at all, as frequent as sterile processes. Therefore to control the presence, viability, and proliferation of microorganisms, effective environmental control, equipment and personnel sanitation, aseptic techniques, and good manufacturing practices (GMP) are needed [3]. However, pharmaceutical companies follow different strategies during the manufacturing of nonsterile products. For instance, some companies perform environmental monitoring of production facilities and equipment sporadically, while others perform it on a regular basis or none at all [4,5]. Microbial identification of environmental isolates from nonsterile manufacturing environments varies from company to company. In some cases, companies pursue microbial identification by only a gram stain reaction, e.g., gram negative or positive. Other companies take the identification one step further when the environmental isolate is completely identified by genera and species such as gram-negative rod, *Pseudomonas aeruginosa*. Because of the infrequent and inconsistent monitoring of equipment, personnel, and environment, microbial limits testing of raw material and finished product is a critical step for the quality control analysis of nonsterile pharmaceuticals.

The three major pharmacopoeias, U.S. (USP), European (EP), and Japanese (JP), have divided microbial limit testing into two different tests: the quantitative test and qualitative test [1,6,7]. The quantitative test ascertains the numbers of microorganisms, bacteria, yeast, and mold present in a given pharmaceutical sample. The qualitative test determines the presence of specific pathogen indicators, e.g., *Salmonella* spp., *Staphylococcus aureus*,

Escherichia coli, *P. aeruginosa*, and the Enterobacteriaceae family which might cause disease to consumers or indicate the presence of other pathogenic bacteria. These indicators are representative microbial species of different types of bacterial populations. For instance, *Salmonella* spp. and *E. coli* are gram-negative rods, capable of lactose fermentation, commonly found in fecal sources. *Salmonella* spp. are virulent pathogens associated to intestinal disorders, while *E. coli* in general is not a virulent pathogen. However, some strains of *E. coli* are known to be producers of toxins associated to gastrointestinal diseases. *P. aeruginosa* is a gram-negative nonfermentative rod, which is typically associated to opportunistic infections. *S. aureus* is a gram-positive cocci commonly associated to skin, gastrointestinal, and toxic shock syndrome conditions. The Enterobacteriaceae family comprises genera such as Escherichia, Salmonella, Shigella, Citrobacter, Enterobacter, Klebsiella, Proteus, etc. Most of the members of this family, other than *Salmonella* spp. and *Shigella* spp., are opportunistic pathogens. They are widely distributed in the environment.

The use of the four pathogen indicator bacteria does not mean that the presence of other bacteria might not be a problem during quality evaluations. However, as previously mentioned, route of application and intended use of a given product will determine if there is a risk involved when these other microorganisms are present.

2. MICROBIAL CONTAMINATION OF NONSTERILE PRODUCTS

Unfortunately, at the beginning of the 21st century, microbial contamination of nonsterile products is one of the major reasons for product recalls, production shutdowns, and losses in labor and manufacturing. Millions of dollars are lost due to the lack of quality control, process control, and proper testing. When a contaminant is found in a production batch, an investigation is rapidly started to determine the contamination source, the numbers, and the types of microorganisms. For nonsterile products, this is critical because as previously mentioned, the presence of microorganisms in a product is not a reason to invalidate the test. The accurate numbers and types of microorganisms must be determined to ascertain the risk of batch release and the efficiency of process control.

This investigation must be fast and accurate so rapid corrective actions can be taken to prevent further contamination of production samples, huge financial losses, and release of contaminated product that can cause disease to consumers. Shutdown of production facilities during microbial contamination leads to major disruptions in the distribution and marketing of an important drug that can save critical patients. Therefore strict adherence to GMP optimizes product manufacturing. If a pharmaceutical process is in control, all the environmental conditions necessary to minimize microbial viability, survival, and proliferation must be in place. These processes rely on the strict control of temperature, particulates, airflow, personnel, and humidity to develop a production environment that minimizes microbial insult. Instrumentation to determine the different parameters to control environmental systems is calibrated and certified on a regular basis.

What are the sources of microbial contamination during the production of nonsterile pharmaceuticals? The great majority of the microbial contamination for nonsterile products has been reported to be due to the presence of microorganisms in raw materials or water or from poor practices during product manufacturing [8]. For instance, manufacturing under nonsterile conditions requires operators to follow specific GMP practices such as raw material testing, equipment sanitization, and wearing of gloves, masks, hats, and laboratory uniforms. To provide continuity and reliability during the performance of all processes, written instructions and procedures are developed for personnel use. Training of manufacturing and laboratory personnel is an important aspect of GMP compliance. Proper documentation of all training is necessary.

Water, the most common raw material in pharmaceuticals, is also a major source of contamination. The water system used during production must be validated and monitored to minimize the microbial bioburden. The system must be sanitized to prevent the formation of biofilms. Bacteria are known to be capable of colonizing surfaces under flowing conditions. The sanitization of the water system by heat or chemical treatment prevents microbial colonization of water lines.

Air quality must also be ascertained to prevent aerosol contamination by bacterial spores and mold. Unfortunately, operators and companies tend to underestimate the risk of microbial contamination in nonsterile pharmaceutical manufacturing. In some cases, the facilities where products are tested and manufactured have been shown to be inadequate for GMP process control [9]. In other cases, the lack of properly trained personnel along with the lack of a functional microbial limit-testing program increased the risk of microbial contamination [4]. This underestimation results in the lack of adequate controls and monitoring programs on the part of the manufacturer, which allows objectionable microbes to contaminate products. Because of the nonsterile nature of the products, less stringent GMP compliance can result in systems failure to detect microbial contamination. Microorganisms are widely distributed across the environment. If a process does not prevent microbial insult nor controls environmental parameters to minimize it, microbial colonization of equipment, water, and ventilation systems might result in frequent microbial contamination of products and processes.

A summary of the frequency and wide distribution of microbial contamination of nonsterile pharmaceutical products in the Unites States is shown in Table 1. Some of the product categories recalled by FDA from 1995 to 2002 range from liquids, tablets, capsules, oils, drops, creams, emulsions, water-based, and anhydrous products [9–14]. The pH of the recalled formulations range from acidic to alkaline. Evidently, microorganisms are capable of contaminating a given pharmaceutical formulation regardless of water content, pH, or manufacturing process. Gram-negative rods are the most commonly found bacterial isolates in tablets, topical products, oral solutions, gel products, medicated shampoos, and soaps. Molds and yeasts are also common contaminants, although not generally speciated (or at least not reported by species). Of the 112 recalls by FDA, *Pseudomonas* spp., *Burkholderia cepacia*, and *Ralstonia pickettii* account for 50%. These types of bacterial species are widely isolated when water is compromised by the formation of biofilms inside the water lines. In some cases, incoming city water is improperly treated to eliminate all microorganisms. Holding products for several days inside water lines not sanitized regularly is another major factor for microbial contamination. Contamination by mold and yeast is found in 21% of the samples. Of the USP indicators, *P. aeruginosa* is detected in 14% and *Salmonella* spp. in 4%. None of the recalls reported have indicated the presence of *E. coli* or *S. aureus* (Table 1).

Microbial contamination of nonsterile products has also been reported from other geographical areas around the world [15–28]. For instance, gram-negative bacteria are also found in samples from Africa, Asia, and Europe (Table 2). Samples from Africa and Europe demonstrate a higher frequency of microbial contamination by gram-positive bacteria than samples from America and Asia. That indicates that human intervention might be one of the major reasons for product contamination, while presence of gram negative bacteria might indicate lack of process control in water systems and raw materials.

A comparison of published scientific studies showed that bacteria from the Enterobacteriaceae family, *Pseudomonas* spp., and *B. cepacia* are the most frequently found microorganisms in samples of pharmaceutical products from all over the world. Other nonopportunistic gram-positive bacteria also found are *Staphylococcus* spp., *Bacillus* spp., *Clostridium* spp., and *Streptococcus* spp. Of the four USP, EP, and JP bacterial indicators, *S. aureus*, *P. aeruginosa*, and *E. coli* were found in samples of toothpastes, topical products, shampoos, oral solutions, and disinfectants. On the basis of published scientific studies and government reports, gram-negative bacteria are found

Table 1 FDA Product Recalls in the United States (from 1995 to 2002)

Product	Reason
Acetaminophen	Aerobic microorganism
Aminocaproic syrup	Yeast
Benzyl peroxide solution	*Burkholderia cepacia*
Topical cream	*Pseudomonas putida*
Triclosan lotion	*Pseudomonas aeruginosa*
Acne cream	*Burkholderia cepacia*
Albuterol sulfate inhalation solution	*Burkholderia cepacia*
Albuterol sulfate syrup	*Burkholderia cepacia*
Barium sulfate	Mold
Ursodiol cap	Potential microbial contamination
Vera Gel	*Enterobacter gergoviae*
Nonalcoholic body spray	*Burkholderia cepacia*
Triple S gentle wash	*Pseudomonas aeruginosa*
Amicar syrup	*Candida parapsilosis*
Sodium chloride cleanser	*Pseudomonas aeruginosa*
Albumin human 5%	*Enterobacter cloacae*
Eye gel	*Pseudomonas aeruginosa*
Mouth rinse antiplaque alcohol-free	*Burkholderia cepacia*
Medical food nutrition supplement	*Pseudomonas aeruginosa*
Dialysate concentrate	Bacterial contamination
Tylenol gelcaps	Aerobic microorganisms
Brand baby oil	*Burkholderia cepacia*
Wet and wild liquid makeup	*Pseudomonas aeruginosa*
Topical product	*Pseudomonas aeruginosa*
Dial brand dialyte concentrate	Mold
F12 nutrient mixture	Bacterial contamination
Gelusil liquid anti gas antacid	*Bacillus* spp.
Hydrox alcohol-free mouthwash	*Burkholderia cepacia*
Electrolyte solution	*Aspergillus niger*
Dry skin creme	Mold
Neoloid emulsfied castor oil	Exceeds microbial limits
Mouth rinse alcohol-free	*Burkholderia cepacia*
Fresh breath plus mouthwash	*Pseudomonas aeruginosa*
Fresh moment alcohol-free mouthwash	*Burkholderia cepacia*
Children's cologne	*Pseudomonas aeruginosa*
Vinegar and water douche	Mold
Skin creme	Mold
Preparation H ointment	Mold
Penecare lotion	*Candida lipolytica*
Aidex spray cleaner	Mold
Mouth rinse antiplaque alcohol-free Oral B	*Burkholderia cepacia*

TABLE 1 Continued

Product	Reason
Aloe vera cream	*Burkholderia cepacia*
Antacid–antigas liquid suspension	Bacterial contamination
Sea therapy mineral gel	*Pseudomonas aeruginosa* *Pseudomonas fluorescens*
Shampoo exotic fruits	Bacterial contamination
Mouth wash alcohol-free	*Pseudomonas aeruginosa*
Medical food nutrition supplement	*Pseudomonas aeruginosa*
Panama jack tanning lotion	Bacterial contamination
Acne treatment cream	*Burkholderia cepacia*
Astringent pad	Mold
Oral suspension	Yeast
Clinical resource food supplement	*Pseudomonas aeruginosa*
Nystatin oral suspension	Possible microbial contamination
Kenwood brand emulsified castor oil	Exceeds microbial limits
Fluoride mouth rinse	*Burkholderia cepacia*
Benzoyl peroxide wash	Potential for microbial contamination
Shampoo (antidandruff)	*Burkholderia cepacia*
Misoprostal tablets	*Burkholderia cepacia*
Simethicone drops	*Burkholderia cepacia*
Vitamin E-lanolin lotion	Mold
Nutritional beverage powders	May contain *Salmonella* spp.
Formance	May contain *Salmonella* spp.
Hand and body lotion with lanolin	Mold
Cytotec tablets	*Pseudomonas* spp.
Propac protein supplement	May contain Salmonella
Sodium fluoride oral mouth	Mold
Soylac infant formula	May contain Salmonella
Ben-Agua wash	Potential for contamination
HEB cream base	Mold
Kayolin pectin suspension	Microbial contamination
Antacid oral liquid suspension	Bacterial contamination
Body wash and shampoo	*Klebsiella oxytoca*
Hygienic wipe pads	Molds
Eye shadow	*Pseudomonas stutzeri*
Soy protein infant formula	*Klebsiella pneumoniae* *Pseudomonas aeruginosa*
Cream base	Mold
Oral suspensions	Yeast
Antacid–antigas oral	Bacterial contamination
Aloe skin cream	*Burkholderia cepacia*
Food industry sanitizing soap	*Burkholderia cepacia*

Table 1 Continued

Product	Reason
Hand disinfectant and body lotion	*Burkholderia cepacia*
Shampoo	*Burkholderia cepacia*
Alcohol free mouthwash	*Pseudomonas aeruginosa*
Cough syrup	Exceeds microbial limits
Disinfectant first aid treatment	*Burkholderia cepacia*
Sunburn gel and spray	*Burkholderia cepacia*
Antiplaque alcohol free mouth rinse	*Burkholderia cepacia*
Infant formula	Nonpathogenic spoilage microorganisms
Boric acid solution	Exceeds microbial limits
Minocycline capsules	Microbial contamination
Myla-care antacid antigas liquid	Bacterial contamination
Sodium chloride	*Ralstonia pickettii*
Benzalkonium chloride towelette	*Burkholderia cepacia*
Calcitriol	*Bacillus cereus*
Syrup	*Staphylococcus warneri*
Haloperidol oral solution	Microbial contamination
Hydrocortisone polistirex suspension	Microbial contamination
Lidocaine HCl/epinephrine injection	Microbial contamination
Colostrum cream	*Pseudomonas putida*
Eye and ear drops	*Pseudomonas fluorescens*
Ophthalmic solution	*Burkholderia cepacia*
Antiseptic solution	*Pseudomonas aeruginosa*
Nystatin oral suspension	*Acinetobacter baumanii*
Povidone–iodine solution	*Pseudomonas putida*, *Salmonella* spp. *Aeromonas sobria*
Bactroban ointment	*Ralstonia pickettii* *Pseudomonas fluorescens*
Gel	Microbial contamination
Bicarbonate concentrate	Mold contamination
Simethicone solution	Microbial contamination
Ampicillin suspension	Mold contamination
Antacid liquid	*Bacillus licheniformis*
Eye and nasal drops	*Pseudomonas mendocina* *Klebsiella pneumoniae*
Progesterone cream	Mold contamination
Mouthwash	*Pseudomonas alcaligenes* *Pseudomonas baleurica*

Source: Refs. 10–14.

TABLE 2 Distribution of Microorganisms as Microbial Contaminants in Pharmaceutical Samples Around the World

Product	Microorganism	Country
Toothpaste	*Staphylococcus aureus*	Africa
	Pseudomonas aeruginosa	
	Escherichia coli	
	Pseudomonas spp.	
	Mold	
Mouthwash	*Staphylococcus* spp.	
	Mold	
Shaving creams	*Staphylococcus aureus*	
	Staphylococcus spp.	
	Mold	
Medicated shampoos	*Staphylococcus aureus*	
	Pseudomonas aeruginosa	
	Staphylococcus spp.	
	Mold	
Cream	*Pseudomonas aeruginosa*	Asia
	Acinetobacter spp.	
	Klebsiella spp.	
	Bacillus spp.	
	Enterobacter spp.	
	Mold	
Antiseptics for vaginal douching	*Burkholderia cepacia*	
Antiseptic cotton product	*Burkholderia cepacia*	
	Pseudomonas aeruginosa	
	Pseudomonas putida	
	Stenotrophomonas maltophilia	
Gastrointestinal gel	*Burkholderia cepacia*	Europe
Topical cream	*Streptococcus faecalis*	
	Pseudomonas aeruginosa	
Suppository	*Streptococcus faecalis*	
Solution	*Streptococcus faecalis*	
Tablets	*Clostridium* spp.	
Liquid soaps	*Enterobacteria*	
	Clostridium spp.	
	Streptococcus faecalis	
Oral solution	*Pseudomonas aeruginosa*	
	Pseudomonas spp.	
	Staphylococcus aureus	
Medicated shampoos	*Pseudomonas fluorescens*	
	Moraxella urethralis	

TABLE 2 Continued

Product	Microorganism	Country
Oral liquids	*Burkholderia cepacia* *Pseudomonas* spp. *Stenotrophomonas maltophilia* *Citrobacter freundii*	
Disinfectant solution	*Pseudomonas aeruginosa*	
Hydrocortisone cream	*Serratia liquefaciens* *Achromobacter xylosoxidans* *Klebsiella pneumoniae* *Pseudomonas fluorescens* *Enterobacter cloacae*	
Balsam shampoo	*Burkholderia cepacia*	
Eye lotion	*Pseudomonas* spp.	
Nose drops	*Stenotrophomonas maltophilia* *Pseudomonas* spp. *Pseudomonas putida* *Klebsiella pneumoniae*	
Lanolin cream	*Enterobacter agglomerans*	
Halciderm cream	*Enterobacter agglomerans*	
Skin cream	*Enterobacter agglomerans*	
Savlon cream	*Pseudomonas* spp.	
Belnovo cream	*Enterobacter* spp.	
Calamine cream	*Pseudomonas pseudoalcaligenes*	
Tyrotrace ointment	*Enterobacter cloacae*	
Lasonil ointment	*Pseudomonas fluorescens*	
The ointment	*Pseudomonas aeruginosa*	
Molivate ointment	*Pseudomonas stutzeri*	
Albucid eye ointment	*Enterobacter agglomerans*	
Oral liquid	*Staphylococcus aureus* *Enterobacter agglomerans*	
Medicated hand soap	*Escherichia coli* *Proteus* spp.	United States
Topical ointment	Gram-positive cocci Gram-negative rods	
Poloxamer iodine solution	*Pseudomonas aeruginosa*	
Pharmaceutical products	*Burkholderia cepacia* *Pseudomonas pickettii* *Pseudomonas acidovorans* *Pseudomonas aeruginosa* *Pseudomonas putida*	

TABLE 2 Continued

Product	Microorganism	Country
	Pseudomonas fluorescens	
	Pseudomonas stutzeri	
	Pseudomonas mendocina	
	Pseudomonas diminuta	
	Pseudomonas vesicularis	
	Comamonas testosteroni	
	Acinetobacter calcoaceticus	
	Agrobacterium radiobacter	
	Flavobacterium breve	
	Flavobacterium meningosepticum	
	Flavobacterium odoratum	
	Flavobacterium multivorum	
	Bordetella bronchiseptica	
	Stenotrophomonas maltophilia	
	Sphingomonas paucimobilis	
	Alteromonas putrefaciens	
	Pasteurella pneumotropica	
	Chryseomonas luteola	
	Flavimonas oryzihabitans	
	CDC group IV c-2	

Source: Refs. 15–28.

to be the most common microbial contaminant in nonsterile pharmaceutical samples regardless of geographical location or time. This indicates that the lack of process control in pharmaceutical environments represents the major factor for nonsterile product contamination.

What is the clinical significance of the presence of microorganisms in nonsterile pharmaceutical formulations? Of the four USP bacterial indicators, *Salmonella* spp. and some virulent strains of *E. coli* and *S. aureus* can cause disease when administered to healthy persons by a natural route. More generally, the USP bacterial indicators and other common pharmaceutical contaminants may cause disease in immunocompromised people or in other classes of susceptible persons [29–38]. These classes include patients with severe preexisting disease, immunocompromised people, and newborn infants. For products intended for immunocompromised patients or infants, the limits must be lower than for people with functional immune systems [39]. This is because the presence of any objectionable microorganism can be fatal for these patients. The USP does not list any other risk indicators, but in the absence of regulatory guidelines, nonsterile pharmaceuticals are manufactured using good manufacturing practices (GMP) as the primary regulatory requirement. Because these bacterial indicators do not include all the opportunistic bacteria present in the environment, microbiological guidelines have been established based upon the intended use of the product, route of administration, nature of the product, and potential risk to the consumer [40]. Whether infection occurs, and the form it takes, depends on the route of administration, the dose of organisms, and the class of person as mentioned above [41,42]. Almost all studies reported on illness attributed to contaminated pharmaceuticals products are from hospital practice, parenteral drugs, and ophthalmic solutions, although investigations carried out by the Swedish National Board in 1965 revealed that a wide range of products were routinely found to be contaminated with coliforms, yeasts, molds, and *Bacillus subtilis*. However, two nationwide outbreaks of infection were traced to the use of contaminated products; in one case, 200 patients were involved in an outbreak of salmonellosis caused by contaminated tablets.

3. RECOMMENDED MICROBIAL SPECIFICATIONS AND LIMITS

What are the threshold limits for the development of microbial specifications for objectionable microorganisms in pharmaceutical products? How many microorganisms are acceptable in a sample? What types of microorganisms are acceptable in a given pharmaceutical raw material and finished product? Are microorganisms, by the numbers and types, present in a sample dangerous to consumers and will they also affect the integrity of the product? There is no comprehensive list of microorganisms, which are called objectionable. Opportunistic pathogens cause disease in children with an infective dosage of 100 colony forming units (CFU), while for adults, 10^6 CFU are needed to colonize the gut [42]. However, the U.S., European, and Japanese pharmacopoeias recommend different guidelines for the development of microbiological attributes for nonsterile pharmaceutical products. For instance, the USP suggests that some product categories such as plant-, animal-, and mineral-based formulations must be tested for *Salmonella* species [43]. When products are designed to be administered orally, *E. coli* should also be tested. With topical pharmaceutical formulations, *S. aureus* and *P. aeruginosa* must also be part of the routine microbiological testing. Vaginal, rectal, and urethral formulations are to be tested for yeast and mold.

The EP recommends more detailed guidelines on the quality of nonsterile pharmaceutical preparations [44]. For the purpose of this chapter, category 2 includes all nonsterile formulations. For topical, transdermal patches, and respiratory tract drugs, a total viable count of not more than 100

CFU/g or mL is recommended. Absence of enterobacteria and other gram negatives, *P. aeruginosa*, and *S. aureus* is also recommended.

For category 3 formulations such as taken by oral and rectal route, recommendations specify a total viable count of not more than 1000 CFU/g or mL and not more than 100 CFU yeast and mold/g or mL. When these formulations are based upon raw materials of mineral, animal, or plant origin, the limits for total counts must be no more than 10,000 CFU/g or mL. Furthermore, not more than 100 enterobacteria and other gram-negative bacteria and absence of *Salmonella* spp., *S. aureus*, and *E. coli* are also recommended. For herbal remedies formulated on one or more vegetable drugs, total viable counts should range from 10^5 to 10^7 CFU/g or mL for bacteria and from 10^4 to 10^5 CFU/g or mL for mold and yeast. If the formulation is added to boiling water before use, not more than 10^2 CFU/g or mL of *E. coli* are recommended. However, if boiling water is not added, *E. coli* and *Salmonella* spp. must be absent.

4. TEST REQUIREMENTS

What are the tests required by the different pharmacopoeias for the analysis of nonsterile pharmaceuticals? What kind of criteria do we use to evaluate the efficacy of the methods for detecting microbial contamination in nonsterile products?

According to the European (EP), Japanese (JP), and U.S. (USP) pharmacopoeias, for a nonsterile pharmaceutical product, microbial limit testing is performed in a stepwise manner; first, the sample is tested to determine the numbers of microorganisms [1,6,7]. This will indicate how many bacteria, yeast, and molds are present in a sample. This is called microbial bioburden. Second, for qualitative analysis, the sample is incubated in broth for at least 24 hr to enhance the isolation of some pathogenic microorganisms. The reason for incubating the samples for at least 24 hr is due to the fact that pathogenic microorganisms are present in lower numbers than nonpathogenic microbes. An enrichment step and growth on selective media will enhance the isolation of pathogenic microorganisms such as *Salmonella* spp. and *E. coli* [45].

Before sample testing is performed, the methods must be shown to be capable of detecting and isolating bacteria, yeast, and mold. This part of the procedure is called the preparatory testing. The preparatory testing involves the inoculation of different types of microorganisms into the samples to demonstrate the accuracy, efficacy, reproducibility, and sensitivity of a given method for detecting microbial contamination (Table 3). Because of the nonsterile nature of the products, the developing criteria for testing can be completely different for products with different applications. Prior to pro-

TABLE 3 Microbial Limits Test Microbiological Indicators for Preparatory Testing and Standard Analysis (As Per United States, European, and Japanese Pharmacopoeia)

	USP	EP	JP
Quantitative microorganisms	*Staphylococcus aureus*	*Staphylococcus aureus*	*Staphylococcus aureus*
	Escherichia coli	*Escherichia coli*	*Escherichia coli*
	Pseudomonas aeruginosa	*Bacillus subtilis*	*Bacillus subtilis*
	Salmonella spp.	*Candida albicans*	*Candida albicans*
		Aspergillus niger	
Qualitative microorganisms	*Staphylococcus aureus*	*Staphylococcus aureus*	*Staphylococcus aureus*
	Escherichia coli	*Escherichia coli*	*Escherichia coli*
	Pseudomonas aeruginosa	*Pseudomonas aeruginosa*	*Pseudomonas aeruginosa*
	Salmonella spp.	*Salmonella* spp.	*Salmonella* sp.
		Enterobacteriaceae	
		Clostridium spp.	

duction, all raw materials are tested and qualified to be of a quality that will minimize the introduction of a significant number of microorganisms to the manufacturing process and finished product. For instance, an oral pharmaceutical product developed for transplant patients will have a completely different microbial limit approach than an oral dosage formulation for gas relief. Since the patients receiving the transplant drug may be immunocompromised, it might be safer to have zero counts of bacteria, yeast, and mold [39]. The pathogen indicator specification can be expanded to include absence of any gram-negative rods. However, for the gas relief formulation targeting a healthy population, a limit of less then 100 colony forming units and absence of four pathogen indicators and gram-negative rods might be a reasonable specification. Therefore to develop the microbiological specifications, we must account again for the intended use of the product, nature of product, target population, manufacturing process, and route of administration.

5. TEST METHOD VALIDATION

5.1. Quantitative Test

To determine the accuracy and sensitivity of the test methods used for microbial limit testing, according to the USP, 10 g or mL samples of the test material are inoculated with separate viable cultures of *S. aureus*, *Salmonella* spp., *E. coli*, and *P. aeruginosa* (Table 3). Some laboratories also use cultures of *Candida albicans* and *Aspergillus niger* to validate the quantitative recovery of yeast and mold. The EP recommends inoculating the samples with *S. aureus*, *E. coli*, *B. subtilis*, *C. albicans*, and *A. niger*. Same types of microorganisms are used in the JP with the exception of *A. niger*. Although compendial recommendations are not specific regarding the number of samples required for method validation, at least three different production batches are generally used [45]. That number will provide important information on the sensitivity, reproducibility, and accuracy of the validation data. When a validated formula has been modified or replaced, further validation work must be performed. Some companies also perform method validation on a yearly basis.

The procedure comprises the addition of no less than a 10^{-3} dilution of a 24-hr broth culture of the recommended microorganisms to different dilutions of the test material in diluents such as phosphate buffer, buffered sodium chloride peptone solution, Letheen broth (LB), soybean casein digest broth (SCDB), or lactose broth (LacB) (Table 4). The recommended sample size is 10 g or 10 mL of test material. However, when production batches do not have

TABLE 4 Microbial Limits Testing Growth Media as per USP, EP, and JP

		Incubation time (days)	Temperature (°C)
(A) Microbial counts			
Bacteria			
MCTA, SCDA, Letheen agar		2–5	35–37
Yeast and mold			
SDA, PDA, mycological agar		5–7	25
(B) Microbial enrichments			
SCD Broth		1–3	35–37
Lactose			
Lauryl tryptose broth			
(C) Pathogen isolation selective media and broth			
Pseudomonas aeruginosa	Cetrimide, Pseudomonas isolation	1–3	35–37
Escherichia coli	MaConkey	1–2	35–37
Staphylococcus aureus	Mannitol Salt, Bair Parker, Vogel Johnson	1–2	35–37
Salmonella typhimurium	Selenite, Tetrathionate, BSA, XLD, BGA	2–4	35–37

a significant amount of sample, volumes of less than 10 g or mL can also be used [1]. A positive control solution containing the microorganisms and the diluent without the test article is simultaneously analyzed. For instance, a 1:10 dilution of product suspension and control solution is inoculated with a given microbial culture, thoroughly mixed, and poured or spread plated on some of the most common bacterial growth media such as soybean casein digest agar (SCDA), microbial content test agar (MCTA), or Letheen agar (LA). Mold and yeast samples are plated on media such as Sabouraud dextrose agar (SDA), potato dextrose agar (PDA), or mycological agar (MA). Incubation times for bacterial plates range from 2 to 5 days at 32–35°C depending upon the company's specifications. Mold and yeast plates are normally incubated for 5–7 days at 22–25°C.

At least three replicas of the experiment must be performed and each should show that the average numbers of CFU recovered from the test article are not less than 70% of the inoculum control [45]. Table 5 demonstrates the validation of Letheen broth as the diluent and SCDA containing 1% lecithin as the media for quantitation of bacteria and yeast in a pharmaceutical product (tablet A). Ten grams of the product are diluted in Letheen broth and then plated on the agar media. As previously discussed, the minimum recovery for all microorganisms must be 70%. Unfortunately, the recoveries for *B. cepacia*, *P. aeruginosa*, and *Enterobacter gergoviae* are less than 70%. Therefore the media is not suitable for quantitation of all microorganisms. However, when a higher product dilution is used, recovery values for all microorganisms fall within 71–97% (Table 6). Evidently, a higher dilution of the product allows the recovery of all microorganisms. The testing conditions are then set for routine quality control analysis.

In the EP validation protocol, the sample is validated with a difference of no more than a factor of 5 between the sample with the test material and without it [6]. Failure to recover the recommended numbers of microorganisms suggests that a modification of the test method must be carried out. Diluting, filtrating, or inactivating the inhibitory substances by neutralization can recover the inoculated microorganisms. For instance, increasing the test article dilution to 1:100 and 1:1000 or adding different concentrations of inactivating agents such as polysorbate 20 and 80, lecithin, or sodium thiosulfate can overcome the inhibitory effect of the formulation. This might neutralize any antimicrobial activity of the test article. Due to the insoluble nature of some products, homogenization by heating at no more than 45°C might be necessary. Furthermore, the substance can be reduce to a fine powder by using a blender. When the antimicrobial nature of the product overcomes any of the strategies described, membrane filtration can be the last resort to remove any inhibitory substances. Membrane filters of about 50 mm in diameter and a pore size not greater than 0.45 μm are recommended [1].

TABLE 5 Validation of Microbial Limits Quantitative Test for a Pharmaceutical Tablet A

(A) Quantitative test: dilution 1:10
Diluent = Letheen broth
Plating media = soybean casein digest agar with 1% lecithin

	Colony forming units/g or mL			
Test organisms	Control	Average	Sample	Average
Burkholderia cepacia ATCC 25416	69, 59, 55	61	25, 24, 8	26
Escherichia coli ATCC 8739	48, 49, 44	47	40, 45, 39	41
Staphylococcus aureus ATCC 6538	69, 75, 63	69	70, 60, 61	64
Salmonella choleraesuis ATCC 10708	48, 41, 40	43	44, 39, 42	42
Pseudomonas aeruginosa ATCC 9027	42, 37, 41	40	19, 21, 22	21
Enterobacter gergoviae ATCC 33028	60, 63, 62	61	17, 23, 25	22
Candida albicans ATCC 10231	61, 59, 55	58	55, 60, 52	56

%recovery = (average sample count/average control count) × 100

	%Recovery
Burkholderia cepacia ATCC 25416	42
Escherichia coli ATCC 8739	87
Staphylococcus aureus ATCC 6538	93
Salmonella choleraesuis ATCC 10708	98
Pseudomonas aeruginosa ATCC 9027	53
Enterobacter gergoviae ATCC 33028	36
Candida albicans ATCC 10231	97

After filtration, the membrane is washed three or more times with a buffer solution to remove any residual antimicrobial substances. The membrane is then placed on agar media, which is incubated for a given period of time. When recovery values fall within the numbers mentioned above, the test article is considered to be validated by membrane filtration.

In cases when any of the above strategies are not capable of recovering the microorganisms from the test article, it can be assumed that the strong antimicrobial nature of the formulation will destroy any microorganism

TABLE 6 Validation of Microbial Limits Quantitative Test for a Pharmaceutical Tablet A

(A) Quantitative test: dilution 1:100
Diluent = Letheen broth
Plating media = soybean casein digest agar with 1% lecithin

	Colony forming units/g or mL			
Test organisms	Control	Average	Sample	Average
Burkholderia cepacia ATCC 25416	69, 59, 55	61	61, 56, 55	57
Escherichia coli ATCC 8739	48, 49, 44	47	40, 45, 39	41
Staphylococcus aureus ATCC 6538	69, 75, 63	69	70, 60, 61	64
Salmonella choleraesuis ATCC 10708	48, 41, 40	43	44, 39, 42	42
Pseudomonas aeruginosa ATCC 9027	42, 37, 41	40	31, 32, 37	33
Enterobacter gergoviae ATCC 33028	60, 63, 62	61	45, 43, 42	43
Candida albicans ATCC 10231	61, 59, 55	58	55, 60, 52	56

%recovery = (average sample count/average control count) × 100

	%Recovery
Burkholderia cepacia ATCC 25416	93
Escherichia coli ATCC 8739	87
Staphylococcus aureus ATCC 6538	93
Salmonella choleraesuis ATCC 10708	98
Pseudomonas aeruginosa ATCC 9027	83
Enterobacter gergoviae ATCC 33028	71
Candida albicans ATCC 10231	97

present. However, proper documentation of the validation work showing the inefficient neutralization of different methods must be maintained and filed.

As an alternative to the plate count and membrane filtration methods, the USP, JP, and EP recommend the most probable number method (MPN) when no other method is available. However, this method is rarely used by industry. The accuracy and precision of the MPN is less than the plate count and membrane filtration. This method consists in the inoculation of different dilutions of the product suspensions into a suitable medium for bacterial enumeration. The samples are then incubated for 5 days at 30–35°C. After incubation, each dilution tube is observed for the detection of microbial growth by turbidity. The MPN of bacteria per gram or milliliter is determined from specific tables [1]. However, the MPN method does not provide reliable results for the enumeration of yeast and mold.

The final interpretation of the quantitative results for the EP and JP is based upon the sum of the bacterial count and the fungal count. This sum of the two values is called the total viable aerobic count. For the USP, results are reported separately as total aerobic microbial count and total yeast/fungal counts.

5.2. Qualitative Test

Once the quantitative recovery of microorganisms has been validated, the next step is to inoculate the test articles with specific microbial species that might indicate the presence of objectionable microorganisms. These microbial species are called indicators. The USP and JP recommend using the following bacterial species for the validation of pathogen screening: *S. aureus*, *Salmonella* spp., *E. coli*, and *P. aeruginosa*, while the EP includes the same species along with *Enterobacter* spp. and *Clostridium* spp. Although these are the species recommended for validation purposes, as previously discussed, there are reports of microbial contamination and products recalls due to other types of pathogenic or opportunistic microorganisms. For instance, a survey of the scientific literature indicates that *B. cepacia* is one of the most frequently isolated bacterial contaminants in pharmaceutical samples around the world (Tables 1 and 2). However, *B. cepacia* is not listed by any of the pharmacopoeias. As previously discussed, other gram-negative bacteria can also possess a health threat to consumers if present in high numbers. Bacteria such as *Acinetobacter* spp., *Pseudomonas putida*, *Pseudomonas fluorescens*, *Enterobacter* spp., and *Klebsiella* spp. are frequently found in some samples. This indicates that the pathogen-screening test must not be limited to the recommended indicators but must include other pathogens that might generate serious health threats to consumers and compromise product integrity. The history of a given product or manufacturing facility regarding normal flora must be considered when pathogen screening testing is validated. It might be possible to use some of the frequently isolated microorganisms from a given production facility to expand the range of pathogen screening. The continuous presence of these microorganisms in the plant might indicate that a manufacturing process is out of control.

For the validation of the pathogen screening part of the USP, JP, and EP microbial limit test, a 10^{-3} dilution of a 24-hr culture of the indicators previously described or any other pathogenic species are inoculated into a dilution of the test article in SCDB, LB, and LacB with or without neutralizers. Again, sample dilution can range from 1:10 to 1:1000. If a 1:10 dilution does not recover the spiked microorganism, then further dilutions are tested, e.g., 1:100 and 1:1000, to determine the right dilution factor. Furthermore, as in the quantitative step, addition of neutralizers to the media might enhance the recovery of the microorganisms when antimicrobial ingredients are present.

After incubation, the samples are streaked onto different types of selective agar media. Incubation times range from 24 to 96 hr at 35–37°C. Different companies have different incubation times that must be properly validated and documented according to the company's procedures. Table 7A shows that when pharmaceutical tablet A is analyzed using a 1:10 dilution, three different types of microorganisms are not detected: *S. aureus*, *P. aeruginosa*, and *E. gergoviae*. However, when a 1:100 dilution is used, all microorganisms are detected (Table 7B).

For *Salmonella* spp., the USP and JP require a preenrichment step in lactose broth followed by transfer to fluid selenite–cysteine medium (FSCM) and fluid tetrathionate medium (FTM) (Table 4). However, the EP requires an enrichment step prior to the lactose enrichment by using buffered sodium

TABLE 7 Validation of Microbial Limits Qualitative Test for a Pharmaceutical Tablet A

Test organisms	MacConkey	Vogel-Johnson	PIA
(A) Qualitative test: dilution 1:10			
+ = growth			
− = no growth			
Burkholderia cepacia ATCC 25416	+	−	+
Escherichia coli ATCC 8739	+	−	−
Staphylococcus aureus ATCC 6538	−	−	−
Salmonella choleraesuis ATCC 10708	+	−	−
Pseudomonas aeruginosa ATCC 9027	−	−	−
Enterobacter gergoviae ATCC 33028	−	−	−
(B) Qualitative test: dilution 1:100			
+ = growth			
− = no growth			
Burkholderia cepacia ATCC 25416	+	−	+
Escherichia coli ATCC 8739	+	−	−
Staphylococcus aureus ATCC 6538	−	+	−
Salmonella choleraesuis ATCC 10708	+	−	−
Pseudomonas aeruginosa ATCC 9027	+	−	+
Enterobacter gergoviae ATCC 33028	+	−	−

chloride peptone solution. After enrichment in FSCM and FTM, all procedures recommend transferring an aliquot of the enrichments on brilliant green (BGA), bismuth sulfite (BSA), and xylose lysine deoxycholate (XLD) agar. Validation of the recovery of *Salmonella choleraesuis* is shown in Tables 8 and 9. All samples are validated for the detection of this microorganism using a 1:100 dilution of the products.

For *E. coli*, the USP and JP protocols require streaking the lactose broth enrichments onto MacConkey agar medium (Mac). After incubation, if brick-red colonies of gram-negative rods surrounded by a reddish precipitation zone are not found, the samples are negative for *E. coli* (Table 10). The EP procedure relies on the enrichment of the sample in SCDB followed by another enrichment in MacConkey broth and streaking on Mac agar.

The USP and the JP recommend Cetrimide and Pseudomonas isolation agar for isolating *P. aeruginosa* and other *Pseudomonas* spp. However, the EP recommends only Cetrimide agar. All procedures rely on the typical morphological characteristics of *P. aeruginosa*-type strains. These characteristics are shown in Table 10. When bacterial colonies are not phenotypically similar to the "normal" typical colony morphology, it is assumed that the bacterial isolate is not the targeted pathogen.

Up to this point, the practice varies according to the company's specifications and procedures. In some cases, all bacterial growth on selective media is identified to the genus and species level regardless of colony morphology or color. However, in other cases, atypical colony morphology is assumed to be sufficient for final discrimination of bacterial isolates and no further identification is performed. A survey of identification practices by industry shows that only 30% of the people asked identify all microorganisms from plate counts and enrichments regardless of the typical or atypical colony

TABLE 8 Validation of *Salmonella typhimurium* Recovery in Pharmaceutical Samples (Preenrichment Step)

Dilution: 1:100

Microorganism	Product	Inoculum	Preenrichment (lactose)
Salmonella choleraesuis ATCC 10708	Tablet	89, 90	+
Salmonella choleraesuis ATCC 10708	Liquid	79, 92	+
Salmonella choleraesuis ATCC 10708	Powder	86, 93	+
Salmonella choleraesuis ATCC 10708	Emulsion	67, 89	+
Salmonella choleraesuis ATCC 10708	Liquid	78, 88	+

TABLE 9 Validation of *Salmonella choleraesuis* Recovery in Pharmaceutical Samples (Selective Enrichment and Isolation)

Dilution: 1:100

Microorganism	Product	Selective enrichment		Selective agar media		
		FSCM	FTM	BSA	XLD	BGA
Salmonella choleraesuis	Tablet	+	+	+	+	+
Salmonella choleraesuis	Liquid	+	+	+	+	+
Salmonella choleraesuis	Powder	+	+	+	+	+
Salmonella choleraesuis	Emulsion	+	+	+	+	+
Salmonella choleraesuis	Liquid	+	+	+	+	+

morphology [5]. In some cases, identification of atypical colonies is required when the plate count exceeds 100 CFU/g or mL. However, when microorganisms are subjected to environmental stresses, colony morphology on plate media might be atypical which indicates that phenotypical identification might not be a reliable presumptive identification of the environmental isolates. An alternative and more accurate practice is to identify all microbial

TABLE 10 Microbiological Characteristics of Colony Forming Units on Selective Agar Media Used for Pathogen Screening of Nonsterile Pharmaceutical Samples

Selective agar	Colony morphology	Bacteria
Baird-Parker	Black shiny with clear zones	*Staphylococcus aureus*
Vogel-Johnson	Black colonies with yellow zones	*Staphylococcus aureus*
Mannitol Salt	Yellow colonies with yellow zones	*Staphylococcus aureus*
Cetrimide	Greenish colonies	*Pseudomonas aeruginosa*
Pseudomonas	Yellowish colonies	*Pseudomonas aeruginosa*
MacConkey	Brick-red colonies	*Escherichia coli*
Xylose–lysine–desoxycholate	Red colonies with or without black centers	*Salmonella* spp.
Bismuth sulfite	Black or green colonies	*Salmonella* spp.
Brilliant green	Small transparent or pink colonies with pink-red zones	*Salmonella* spp.
Violet red bile glucose	Red or reddish colonies	*Enterobacteriaceae*

growth obtained on selective media regardless whether or not colonies demonstrate atypical colony morphology.

For *S. aureus*, all protocols recommend enrichment in SCDB. However, the EP subcultures the enrichment on Baird Parker agar, while the USP and JP subculture on the same media along with Vogel-Johnson and mannitol-salt agar. Again, the protocols rely on morphological characteristics based upon colony morphology, color, and type (Table 10).

As previously mentioned, the EP goes one step further than the other pharmacopoeias by requiring a specific test for *Clostridium* spp. and Enterobacteriaceae. Product enrichment is made in Clostridium broth and then subcultured on Columbia agar with gentamicin. Enrichment is incubated under anaerobic conditions. A semiquantitative test for *Clostridium perfringens* requires the addition of samples to lactose monohydrate sulfite medium containing a Durham tube. After incubation for 48 hr at 45–46.5°C, all samples showing a blackening due to iron sulfide and gas formation are considered positive.

For Enterobacteriaceae, a preenrichment in lactose broth for 5 hr is the standard procedure. After preenrichment, subculturing in Entobacteriaceae enrichment broth (18–48 hr at 35–37°C) followed by streaking plates of violet red bile glucose agar (VRBG) (18–24 hr at 35–37°C) complete the procedure. Absence of growth indicates absence of gram-negative bacteria.

In conclusion, all regulatory agencies rely on standard microbiological assays. As described in this chapter, these assays are labor-intensive, require different types of media, time-consuming, and require continuous manipulation of samples and reagents.

6. HISTORY AND HARMONIZATION OF MICROBIAL LIMITS TESTING

An excellent article on the development of the USP Microbial Limit test chapter ⟨61⟩ has been published by Cundell [46]. Initial testing, during the 1940s, has consisted of a total count on tryptone glucose yeast extract (TGYE) and eosine methylene blue (EMB) or Endo agar for *E. coli* detection. Major changes are implemented in 1970 with the addition of tests for *Salmonella* species, *P. aeruginosa*, and *S. aureus*. Quantitation methods for yeasts and molds enumeration are added in 1985 with the use of SDA. Evidently, harmonization of the test with international guidelines seems to be the next evolutionary step.

Because of the globalization of the pharmaceutical sciences and industry, harmonization of compendial tests between different pharmacopoeias has been a priority for the last 5 years. This harmonization will prevent the duplication of microbiological testing and GMP compliance for products and raw materials analyzed in the United States, the European Community, and Japan. Several articles are published on the harmonization of microbial limits tests [47–49]. In order to expand these efforts in the United States, a proposal submitted in 1999 stated that the Microbial Limits chapter ⟨61⟩ must be broken down into three different chapters. The first chapter ⟨61⟩ comprises the enumeration test with a new coliform and Enterobacterial count test. These changes are necessary to harmonize the USP with the EP. Objectionable microorganisms are described in a new chapter, ⟨62⟩. Additional objectionable microorganisms are *Clostridium* spp., *B. cepacia*, and *C. albicans*. Several guidelines are described in another new chapter ⟨1111⟩ to further define the microbiological attributes of nonsterile products. With these new guidelines, microbiologists will have a common and better understanding of microbial testing procedures and specifications for nonsterile pharmaceuticals worldwide. However, the harmonization process is on going and may bring more changes to the different chapters and sections.

7. SAMPLING AND TESTING

7.1. Testing Conditions

A microbiology laboratory testing nonsterile products usually complies with class 10,000 or 100,000 room requirements [50]. However, in some cases, the laboratory does not fall into any of these categories. These classifications are based upon the numbers of particles retained on a 0.50-μm filter. To prevent microbial contamination by analysts and environment, aseptic techniques must be used during testing with all work performed inside a laminar flow cabinet. The laminar flow cabinet provides a class 100 testing environment. Sanitization and cleaning of working areas are performed during tests and are properly documented. Calibration of scales, incubators, and water bath is performed daily or weekly according to written procedures. All personnel performing testing must be properly trained. Training is documented and reviewed every 3 years if the analyst continues performing analysis on a regular basis.

During testing inside the laminar flow cabinets, environmental plates are placed on the right- and left-side corners of the cabinet. The plates can be SCDA or blood agar for bacteria and SDA for yeast and mold. Incubation times for bacteria range from 2 to 5 days at 35–37 °C, while mold and yeast are read after 5–7 days at 25 °C. The use of these plates during testing provides another level of security to ascertain that proper aseptic techniques have been practiced and that conditions inside the hood are driven to minimize microbial contamination.

7.2. Sampling

The distribution of microorganisms in a given pharmaceutical production batch is not homogenous [3]. Furthermore, it has been demonstrated that microbial distribution in pharmaceutical production batches is heterogeneous or patchy [3]. Therefore microorganisms are frequently lumped together following a negative binomial distribution [3]. Examination of only one sample might result in the overestimation or underestimation of microbial contamination. Microbial distribution in a production batch is affected by:

- The composition of the pharmaceutical formulation, e.g., raw materials and actives
- Delivery system of the pharmaceutical formulation, e.g., tablet, cream, liquid
- Manufacturing process, e.g., blending, compression, filtration, heating, encapsulation, shearing, tableting, granulation, coating, and drying
- pH of the pharmaceutical formulation
- Water activity of the pharmaceutical formulation
- Quality of the water system
- Aseptic techniques of the analysts

According to the USP, EP, and JP, the sample volume for a microbial limit test must be a composite sample of a production lot by sampling a number of containers and compositing the sample. After thorough mixing from the composite, 10 g or mL are sampled. However, they keep the door open for interpretation since it is also mentioned that other appropriate quantity can also be used. Furthermore, the USP recommends sampling 10 finished product containers for a sample size of 10 g or mL.

However, there are several interpretations on the numbers of finished product samples needed, the frequency of the sampling, how many individual samples, which samples are composited, etc. Regardless of the strategy, the importance of proper documentation and validation of the sampling schedule is highly recommended to justify a given practice.

The nature and the frequency of the testing will depend on the nature of the product, manufacturing process, facility size, and environmental conditions. It might be necessary to monitor different stages during the manufacturing process to minimize the chance of microbial contamination. For example, some companies schedule a 24-hr production day into three 8-hr shifts. Each shift is subdivided into beginning, middle, and end of the shift. Microbiological samples are taken at the beginning (beg), middle (mid), and end (e) of each shift. That will give an indication of the quality of the different batches during that shift cycle. In some cases, companies combined all three stages (beg, mid, and e) into a composite sample that is tested for microbial limits. However, if contamination is found in the composite sample, there is no way to determine whether the bad sample came from the beginning, middle, or end of the shift. In other cases, companies composite all three shifts (beg, mid, and e) into a large big composite sample for a total of 10 g.

A resample must be tested by analyzing all three production stages to track the sources of contamination. Regarding resampling, it has been established that the volume must be 2.5 times of the original [1]. For example, if you tested 10 g in 100 mL of broth, the retest must be 25 g. However, appropriate volume adjustments must be performed to account for the larger sample size. The volume of the broth for the retesting will be 250 mL. Resampling, however, cannot be used to eliminate a positive result since it must be properly justified and documented the reasons why resampling is necessary. For instance, the fact that a positive result for *P. aeruginosa* cannot be repeated during resampling is not a good reason to eliminate the first data point. That will indicate that the system does have a problem and unless it is demonstrated that the analyst introduced the microorganism during testing, it is not a safe practice to release the batch for general distribution.

8. MICROBIOLOGICAL TESTING OF HERBAL AND NUTRITIONAL SUPPLEMENTS

Nutritional supplements and herbal medicines are also tested to determine the microbiological quality of the raw materials and formulations [51]. Because of the continuous health-related claims of these products, regulatory agencies are currently recommending the application of GMP to their manufacturing and quality control. This is done to control the quality, efficacy, and safety of these products. A wide variety of nutritional supplements are based upon the use of natural ingredients such as botanicals and mineral oils. These materials contain large number of microorganisms.

The test methods are based upon the same requirements and methods described for nonsterile pharmaceutical products [1]. However, these tests are not mandatory since the chapter is part of the informational sections of the USP [51]. The only difference between the nonsterile test and the supplements is that yeast and mold are required to be part of the preparatory test (validation test), while nonsterile pharmaceuticals do not require these two microorganisms to be part of it.

9. CONCLUSION

Validation of microbiological testing for nonsterile pharmaceuticals provides a reliable way to determine the potential risk of high microbial bioburden and objectionable microorganisms in finished products and raw materials. Because a bioburden is allowed in nonsterile pharmaceutical products, their microbiological risk is based upon the nature of the product, intended use, and route of application. Monitoring of critical areas such as facilities, equipment, raw materials, air, and water must be part of a testing plan to determine the efficacy of process control to minimize microbial contamination and the presence of objectionable microorganisms. A good microbiological program for nonsterile pharmaceuticals relies on cGMP practices to provide safe, stable, and efficacious products.

REFERENCES

1. United States Pharmacopeial Convention. Microbial limit test. In U.S. Pharmacopoeia. Rockville, Maryland: United States Pharmacopeial Convention, 2002:1873–1878.
2. Federal Register of the United States of America. Code of Federal Regulations. Subpart F: Production and Process Control, 21 Part 211.113: Control of microbiological contamination. Rockville, Maryland, 1996.
3. Underwood E. Ecology of microorganisms as its affects the pharmaceutical industry. In: Hugo WB, Russell AB, eds. Pharmaceutical Microbiology. 6th ed. Oxford, England: Blackwell Science, 1998:339–354.
4. Reich RR, Miller MJ, Paterson H. Developing a viable environmental program for non-sterile pharmaceutical operations. Pharm Technol 2003; 27:92–100.
5. Mestrandrea LW. Microbiological monitoring of environmental conditions for non-sterile pharmaceutical manufacturing. Pharm Technol 1997; 21:59–74.
6. European Pharmacopoeial Convention. Microbiological examination of non-sterile products. European Pharmacopoeia. 3rd ed. Strasbourg, France: Council of Europe, 2001:70–78.
7. The Japanese Pharmacopoeia. Microbial Limit Test. 13th ed. Tokyo, Japan: The Society of Japanese Pharmacopoeia, 1996:49–54.
8. Baird R. Contamination of non-sterile pharmaceuticals in hospital and community environments. In: Hugo WB, Russell AD, eds. Pharmaceutical Microbiology. 6th ed. Oxford, England: Blackwell Science, 1998:374–384. Chapter 19.
9. FDC Reports. Quality control reports “The Gold Sheet.” 1997; 31, Number 1.
10. FDC Reports. Quality control reports “The Gold Sheet.” 1998; 32, Number 1.
11. FDC Reports. Quality control reports “The Gold Sheet.” 1999; 33, No. 8.
12. FDC Reports. Quality control reports “The Gold Sheet.” 2000; 34, No. 2.
13. FDC Reports. Quality control reports “The Gold Sheet.” 2001; 35, No. 3.
14. FDC Reports. Quality control reports “The Gold Sheet.” 2002; 36, No. 3.
15. Desvignes A, Sebastien F, Benard J, Campion G. Etude de la contaminacion microbienne de diverses preparations pharmaceutiques. Ann Pharm Fr 1973; 31:775–785.
16. Coppi G, Genova R. Controllo della contaminazione microbica in preparati

farmaceutici orally e topici col metodo delle membrane filtranti. Farmaco 1970; 26:224–229.
17. Zani F, Minutello A, Maggi L, Santi P, Mazza P. Evaluation of preservative effectiveness in pharmaceutical products: The use of a wild strain of *Pseudomonas cepacia*. J Appl Microbiol 1997; 43:208–212.
18. Palmieri MJ, Carito SL, Meyer J. Comparison of rapid NFT and API 20E with conventional methods for identification of gram-negative nonfermentative bacilli from pharmaceutical and cosmetics. Appl Environ Microbiol 1988; 54:2838–3241.
19. De La Rosa MC, Medina MR, Vivar C. Microbiological quality of pharmaceutical raw materials. Pharm Acta Helv 1995; 70:227–232.
20. Abdelaziz AA, Ashour MSE, Hefnai H, El-Tayeb OM. Microbial contamination of cosmetics and personal care items in Egypt-shaving creams and shampoos. J Clin Pharmacol Ther 1989; 14:29–34.
21. Archibald LK, Corl A, Shah B, Schulte M, Arduino MJ, Aguero S, Fisher DJ, Stechenberg BW, Banerjee SN, Jarvis WR. *Serratia marcescens* outbreak associated with extrinsic contamination of 1% chlorxylenol soap. Infect Control Hosp Epidemiol 1997; 18:704–709.
22. Baird R, Shooter RA. *Pseudomonas aeruginosa* infections associated with use of contaminated medicaments. Br Med J 1976; 2:349–350.
23. Ferguson A, Patel A, Bair RM. A comparison of two incubation temperatures for the isolation of gram negative contaminants from raw materials and nonsterile pharmaceuticals. J Clin Pharmacol Ther 1987; 12:249–254.
24. Lambin MS, Desvignes A, Kiger JL, Azaria M. Etude de la contamination microbienne des sirops pharmaceutiques. Ann Pharm Fr 1972; 30:161–168.
25. Na'was T, Alkofahi A. Microbial contamination and preservative efficacy of topical creams. J Clin Pharmacol Ther 1994; 19:41–46.
26. Oie S, Kamiya A. Microbial contamination of antiseptics and disinfectants. Am J Infect Control 1996; 24:389–395.
27. Wargo EJ. Microbial contamination of topical ointments. Am J Hosp Pharm 1973; 30:332–335.
28. Zembrzuska-Sadkowska E. The danger of infections of the hospitalized patients with the microorganisms present in preparations and in the hospital environment. Acta Pol Pharm 1995; 52:173–178.
29. Bergogne-Berezin E. The increasing significance of outbreaks of *Acinetobacter* spp.: The need for control and new agents. J Hosp Infect 1995; (suppl):441–452.
30. Parrot PL, Terry PM, Whitworth EN, Frawley LW, Ceble RD, Wachsmuth IK, McGowan JE Jr. *Pseudomonas aeruginosa* peritonitis associated with intrinsic contamination of poloxamer–iodine solution. Lancet 1982; ii:683–685.
31. Anderson RL, Bland LE, Favero MS, McNeil BJ, Davis DC, Mackel DC, Gravelle CR. Factors associated with *Pseudomonas pickettii* intrinsic contamination of commercial respiratory therapy solutions marketed as sterile. Appl Environ Microbiol 1985; 50:1343–1348.
32. Coyle-Gilchrist MM, Crewe P, Roberts G. *Flavobacterium meningosepticum* in the hospital environment. J Clin Pathol 1976; 29:824–826.

33. Drelichman V, Band JD. Bacteremias due to *Citrobacter diversus* and *Citrobacter freundii*. Incidence, risk factors, and clinical outcome. Arch Intern Med 1985; 145:1808–1810.
34. Hamill RJ, Houston ED, Georghiou PR, Wright CE, Koza MA, Cadle RM, Goepfert PA, Lewis DA, Zenon GJ, Clarridge JE. An outbreak of *Burkholderia* (formerly *Pseudomonas*) *cepacia* respiratory tract colonization and infection associated with nebulized albuterol therapy. Ann Intern Med 1995; 122:762–766.
35. Hawkins RE, Moriarty RA, Lewis DE, Oldfield EC. Serious infections involving the CDC group Ve bacteria *Chryseomonas luteola* and *Flavimonas oryzihabitans*. Rev Infect Dis 1991; 13:257–260.
36. Hsueh PR, Teng LJ, Yang PC, Chen YC, Pan HJ, Ho SW, Luh KT. Nosocomial infections caused by *Sphingomonas paucimobilis*: Clinical features and microbiological characteristics. Clin Infect Dis 1998; 26:676–681.
37. Okharavi N, Ficker L, Matheson MM, Lightman S. *Enterobacter cloacae* endophthalmitis: report of four cases. J Clin Microbiol 1998; 36:48–51.
38. VanCouwenberghe CJ, Farver TB, Cohen SH. Risk factors associated with isolation of *Stenotrophomonas* (*Xanthomonas*) *malthophilia* in clinical specimens. Infect Control Hosp Epidemiol 1997; 18:316–321.
39. Manu-Tawiah W, Brescia BA, Montgomery ER. Setting threshold limits for the significance of objectionable microorganisms in oral pharmaceutical products. PDA J Pharm Sci Technol 2001; 55:171–175.
40. Dabbah R, Knapp J, Sutton S. The role of USP in the assessment of microbiological quality of pharmaceuticals. Pharm Technol 2001; 25:54–60.
41. Parker MT. The clinical significance of the presence of microorganisms in pharmaceutical and cosmetic preparations. J Soc Cosmet Chem 1972; 23:415–426.
42. Rusin PA, Rose JB, Haas CN, Gerba CP. Risk assessment of opportunistic bacterial pathogens in drinking water. Rev Environ Contam Toxicol 1997; 152:57–83.
43. United States Pharmacopeial Convention. Microbiological attributes of non-sterile pharmaceutical products. In: U.S. Pharmacopoeia. Vol. 25. Rockville, Maryland: United States Pharmacopeial Convention, 2002:2205–2206.
44. European Pharmacopoeial Convention. Microbiological quality of pharmaceutical preparations. In: European Pharmacopoeia. 3rd ed. Strasbourg, France: Council of Europe, 2001:294–295.
45. United States Pharmacopeial Convention. Validation of microbial recovery from pharmacopeial articles. U.S. Pharmacopoeia, United States Pharmacopeial Convention. Rockville, Maryland, 2000; 25:2259–22261.
46. Cundell AM. Review of the media selection and incubation conditions for the compendial sterility and microbial limit tests. Pharmacop Forum 2002; 28:2034–2041.
47. Anonymous. Pharmacopeial Reviews. Microbial enumeration tests. Pharmacop Forum 1999; 25:7761–7773.
48. Anonymous. Pharmacopeial Reviews. Microbiological procedures for the absence of objectionable microorganisms. Pharmacop Forum 1999; 25:7774–7784.

49. Anonymous. Pharmacopeial Reviews. Microbiological attributes of non-sterile pharmacopeial articles. Pharmacop Forum 1999; 25:7785–7791.
50. United States Pharmacopeial Convention. Microbiological evaluation of clean rooms and other controlled environments. In: U.S. Pharmacopoeia. Vol. 25. Rockville, Maryland: United States Pharmacopeial Convention, 2002:2206–2212.
51. United States Pharmacopeial Convention. Microbial limits tests-nutritional supplements. In U.S. Pharmacopoeia. Rockville, Maryland: United States Pharmacopeial Convention, 2003:2659–2663.

3

Microbial Monitoring of Potable Water and Water for Pharmaceutical Purposes

Anthony M. Cundell
Wyeth Pharmaceuticals, Pearl River, New York, U.S.A.

1. INTRODUCTION

Water is a major pharmaceutical ingredient and has been identified as a significant potential source of microbial contamination. This view is supported by the prevalence of pseudomonads in nonsterile pharmaceutical drug product recalls. For example, for the 10-year period of 1991–2001, the average number of recalls per annum for microbial contamination of nonsterile pharmaceutical and over-the-counter drug products was six recalls. The emphasis on water-borne gram-negative bacteria of the species *Burkholderia* (*Pseudomonas*) *cepacia* (10 recalls), *Pseudomonas putida* (5 recalls), *Pseudomonas aeruginosa* (4 recalls), *Pseudomonas* spp. (2 recalls), *Ralstonia* (*Pseudomonas*) *pickettii* (1 recall), *Pseudomonas alcaligenes* (1 recall), and *Pseudomonas baleurica* (1 recall) is notable. Pseudomonads represent 40% of the recalls and this reflects the Food and Drug Administration (FDA) concern for bacteria derived from water, which are capable of growth in liquid oral dosage forms and of overwhelming the preservative system.

Three grades of water are employed by the pharmaceutical industry. They are potable water, which may be used for equipment and facility cleaning, as raw material for the preparation of water for pharmaceutical purposes [purified water, United States Pharmacopoeia (USP), and water for injection, USP], and for personnel lavatories and food preparation; purified water, USP, for the final rinsing of equipment and ingredient water in non-sterile pharmaceutical products; and water for injection, USP, for equipment cleaning and manufacture of parenteral products. A comprehensive discussion of water used in the pharmaceutical and biotechnology industries can be found in USP General Informational Chapter 1231, *Water for Pharmaceutical Purposes* [36].

The scope of this chapter covers the application of bacterial monitoring of water to the pharmaceutical industry. The article will discuss the historical background of water monitoring, potable water monitoring methods, water for pharmaceutical purposes monitoring, effects of media selection, incubation temperature and incubation time on microbial recoveries, setting alert and action levels for pharmaceutical-grade water monitoring, a suitable quality control (QC) program for water testing, application of new microbial testing methods to water monitoring, potable water testing regulations, and a brief discussion of the relationship of water monitoring to microbial ecology.

It should be emphasized that water monitoring is a continuous process of measuring, recording, and detecting adverse changes in the microbial population of water supply or distribution system that would impact the intended use of the water, whether it is potable water or pharmaceutical ingredient water. Changes in the number and composition of the water using a standard method, relative to past monitoring and established quality requirements, will be more important than detecting the absolute number and full biodiversity of the microbial population within the water. Emphasis on the use of total coliforms as an indicator of fecal pollution in water monitoring is discussed, whereas endotoxin and chemical monitoring is outside the scope of the chapter.

2. A BRIEF HISTORICAL REVIEW OF WATER MONITORING

> "The water is collected in sterilized vessels. Then 1 mL is drawn into a sterilized pipette and thoroughly mixed in a test tube with about 10 mL of a 10% Nutrient Gelatin liquefied at 30°C. This mixture is poured out upon a sterilized glass plate, which, after the solidification of the gelatin, is placed in a moist jar. If the water is rich in germs, it must be diluted in a definite manner with 10, 50, or 100 to 1000 mL of sterilized distilled water. The number of germs capable of development in 1 mL of the original water is computed." [Hueppe, F. The Methods of Bacteriological Investigation (translated by Herman M Briggs). New York: D. Appleton and Co., 1886]

In 1883, Koch lectured to a medical group on the application of plating methods to the systematic study of microorganisms in air, water, and soil. His coworker, Hesse—by drawing air through tubes lined by gelatin and incubating them so the colonies could be counted—achieved the counting of airborne bacteria. The organisms isolated from Berlin air (e.g., micrococci, bacilli, aspergilli, etc.) are those familiar to microbiologists monitoring the air inside and out of buildings. Counts of bacteria in water were achieved by adding 1 mL of water to the plate and pouring in the nutrient gelatin, mixing the inocula and media, incubating the solidified plate, and counting the colonies that grew on the plate. This is the now-familiar pour plate (PP) method that continues to be the bedrock of microbiological techniques. Similar approaches to counting bacteria in soil revealed that the number of bacteria decreased as they went from the organic-rich upper soil layers to the organic-poor lower soil layers [2].

In 1885, Frankland reported the first routine examination of water in London using gelatin plate counts and recognized that organisms from sewage were evidence of water pollution. Soon after the classical 1885 work by Escherich on the microflora of the human intestine, in which he discovered the facultative anaerobe *Bacillus* (*Escherichia*) *coli*, it was suggested in 1892 by Schardinger that coliform bacteria be used as indicators of recent fecal contamination of water. In 1904, Eijkman developed the fermentation of glucose or lactose at the elevated temperature of 45°C as a highly selective detection method for the detection and enumeration of *Escherichia coli* [14]. These discoveries are the basis of total and fecal coliform counts.

Historically, water samples were analyzed for coliforms using the most probable number (MPN) multiple-tube fermentation test that is based on the ability of coliforms to grow in a selective broth at 35°C, producing acid or gas from lactose within 24–48 hr. The number of coliforms and the their 95% confidence limit can be determined using MPN tables for the test volumes and number of fermentation tubes used [33].

The most significant post-Second World War advancement in microbiological enumeration methods was the introduction of membrane filtration techniques to count microorganisms. The German filter manufacturer Sartorius-Werke AG developed membrane filters commercially. Prior to the war, membrane filters were primarily used for sterile filtration of air and liquids. In a response to the need to determine water quality after wartime bombing, the German Hygiene Institutes used membrane filtration for culturing coliforms. In 1947, the German microbiologist Muller [24] published a method of counting coliforms on membrane filters using a lactose–fushsin broth. The membrane filtration technique was first introduced into the United States in 1951 by Clark et al. [5]. Today, membrane filtration is widely in use for counting microorganisms in beverages, water and wastewater, food, and pharmaceutical products. It is especially useful in enumerating low numbers of microorganisms, as the sample is concentrated as it is filtered.

The technique is based on passing the sample through a 47-mm-diameter membrane filter and entrapping the microorganisms with a 0.45-μm-sized pores. The filter is placed on the appropriate microbiological culture medium and incubated. The colonies are counted after 48–72 hr. The method is more precise than the MPN multiple-tube technique, with the ability to vary the sample size with the density of the bacterial population and also isolate discrete colonies, but has the disadvantages of reduced surface area of the filters reducing the countable number of colonies, lack of contrast of the colonies and the filter surface, potential stress to the bacteria due to drying of the filter, and, as with all microbiological culture methods, the unintended selectivity of microbiological culture media [30,31].

This limitation of microbiological culture media has been recognized by the pioneering American microbiologists, father and son H. W. Conn and H. J. Conn:

> Another common application of Koch's technique is the counting of bacteria. If in the material that is mixed with gelatin or agar every microorganisms [sic] is separate from every other one and grows into a colony, it is obvious that the number of colonies represents the number of microorganisms in the material plated. This method is commonly used in estimating the number of bacteria in water, milk, soil, or other materials. Although to get a small enough number of colonies on the plates to count, it is often necessary to dilute the materials. This method is so convenient that it is widely used, but unfortunately many bacteria do not grow on the culture media commonly used, while those grow often occur in large clumps that do not break up when plated—both of which fact cause the plate count to be considerably below the actual number of bacteria present [in] the material investigated. [Conn HW, Conn HJ. Bacteriology: A Study of Microorganisms and Their Relation to Human Welfare. Baltimore, MD: Williams and Wilkins, 1922]

In 1987, Colwell et al. introduced the concept of viable but not culturable bacteria to describe latent human pathogen *Vibrio cholerae* in estuarine waters. These organisms could not be enumerated using standard methods but were detected by direct viable counting methods. Viable but nonculturable bacteria may be present in potable water and they have the potential to cause human infection. This belief may be supported by: (1) the well-known observation that cultural techniques used to isolate bacteria from water samples may underestimate the number of viable bacteria determined by vital staining techniques within an environmental sample by up to two to three magnitudes; (2) the size of these bacteria, which tends to approach the 0.45-μm retention dimension used to rate membrane filters used in water monitoring; and (3) reports in the scientific literature that that nonculturable pathogens from the environment or laboratory studies may cause infection in humans [6].

In relationship to the issue of viable but nonculturable bacteria, the following questions need to be explored:

- To what extent do standard plate, spread plate, and membrane filtration counts with nutrient-rich media underestimate the numbers of bacteria in potable water, purified water, and water for injection? For routine monitoring of pharmaceutical-grade waters, is this underestimation of any practical significance?
- What is the effect of enumeration method, media selection, incubation temperature, and incubation time on bacterial recoveries? Can the cultural conditions be standardized?
- Because bacteria in water (with low levels of nutrients and/or stressed by low temperatures) tend to cease division or divide forming cells of significantly reduced dimensions that approach the smallest size that may be retained on a 0.22-μm membrane filter, will a 0.45-μm membrane filter, when challenged by high numbers of these bacteria, have a lower probability of retaining the bacteria that may be present in the water sample?

3. POTABLE WATER MONITORING

A fundamental requirement for public health is access to clean water. Table 1 describes the test parameters used for monitoring potable water in the United States [7,39].

3.1. Heterotrophic Plate Count (HPC)

Although the HPC, formally termed standard plate count, has generally no direct relationship to health effects and there is no current HPC requirement in the National Primary Drinking Water Regulations [7], it may be used to measure the number and variety of bacteria that are common to the water supply. In general, the lower is the plate count, the better maintained the water system is in terms of distribution system flushing, absence of dead legs, and residual chlorine levels. In addition, the HPC may provide supporting data on the significance of the coliform test results by determining locations in a distribution system where residual chlorine levels are not maintained and/or bacteria persist. The HPC may be determined using plate, spread, or membrane filtration methods and provides an approximation of the total viable

TABLE 1 Biological Test Parameters Used to Monitor Potable Water

Test parameter	Method	Regulatory requirement
Heterotrophic plate count	Pour plate	None
Total coliforms	MPN multiple tube and membrane filtration methods	Zero per 100 mL; not more than 5.0% of samples are total coliform-positive per month; every sample that contains total coliforms must be analyzed for either fecal coliforms or *E. coli*; if there are two consecutive TC-positive samples, and if one or more is positive, the system is in violation and would be reportable
Fecal coliforms (e.g., *E. coli*)	MPN multiple-tube and membrane filtration methods Presence–absence tests	Future requirement
Cryptosporidium spp., *Giardia lambdia*	Immunofluorescence	99% Removal 99.9% Removal/inactivation
Legionella spp.	Selective culture	No limit
Viruses (enteric)	Cytotoxicity	99.9% Removal/inactivation

bacterial population. In a survey of U.S. drinking water quality found in 969 public water supplies, an HPC equal or less than 10 colony-forming units (cfu)/mL occurred with 60% of the distribution systems that had detectable chlorine residual [12].

The literature summarized below suggests that there is an indirect relationship between the heterotrophic plate count and the presence or absence of total and fecal coliforms. In contrast, there appears to be a direct relationship between total and fecal coliform isolation.

3.2. Total Coliform Count

Coliforms are bacteria present in the intestinal tract and feces of warm-blooded animals, including humans, that are capable of producing gas and acid from lactose broth at 35 ± 0.5°C. The MPN multiple-tube method for coliforms uses lauryl tryptose broth with 0.5% lactose incubated at 35 ±

0.5°C for 24 ± 2 hr when the tubes are examined for growth, gas, and acidic reaction. The tubes are reincubated for 48 ± 3 hr and reexamined (Standard Method 9221). The results are confirmed using brilliant green lactose bile broth with complete identification using endo and/or MacConkey agar [33]. The National Primary Drinking Water Regulations [7] requirement is zero in a 100-mL sample, with no more than 5% of monthly samples being positive, as indicative of the potential presence of pathogens in the water supply. If one of two consecutive total coliform-positive samples contains fecal coliforms/*E. coli*, it is a maximum contaminant level (MCL) violation and would be reportable.

The relationship between HPC (Plate count agar (PCA), 48-hr incubation at 35°C) and frequency of occurrence of total and fecal coliforms in 969 public water supplies has been reported [9]. As the HPC increases, the percentage of samples positive for total coliforms increases up to an HPC of 500 cfu/mL. Of the total coliform occurrences, between 15% and 50% were positive for fecal coliforms. The lower frequency of occurrences when the HPC exceeds 500 cfu/mL is probably caused by false-negative results due to interference caused by the high noncoliform count (Table 2).

Coliform bacteria are a diverse group of organisms capable of fermenting lactose with acid and gas production, composed of members of the genera *Citrobacter*, *Escherichia*, *Enterobacter*, and *Klebsiella* (Table 3). It should be noted that many water-borne pathogens (e.g., *Salmonella*, *Shigella*, and *Vibrio*) are not detected during coliform monitoring but coliform-positive results are considered to be indicative of recent fecal contamination.

TABLE 2 The Relationship Between Heterotrophic Plate Count, Total Coliforms, and Fecal Coliforms in Potable Water with Increasing Heterotrophic Plate Counts

HPC (cfu/mL)	Number of samples (cumulative %)	Total coliforms occurrences (%)	Fecal coliforms occurrences (%)
<1–10	1013 (41.4)	47 (4.6)	22 (2.2)
11–30	317 (54.4)	28 (7.5)	12 (3.2)
31–100	396 (70.6)	72 (18.2)	28 (7.1)
101–300	272 (81.7)	48 (17.6)	20 (7.4)
301–500	120 (86.6)	30 (25.0)	11 (9.2)
501–1000	110 (91.1)	21 (19.1)	9 (8.2)
>1000	164 (100)	31 (18.9)	5 (3.0)
Total	2446	277 (11.3)	107 (4.4)

Source: Ref. 9.

TABLE 3 Identity of the Members of Different Genera Composing the Classification of Coliform Bacteria

Citrobacter	Escherichia	Enterobacter	Klebsiella
C. freindii	*E. coli*	*E. aerogenes*	*K. pneumoniae*
C. diversus		*E. agglomerans*	*K. rhinoscleromatis*
		E. cloacae	*K. oxytoca*
			K. ozaenae

Source: Ref. 8.

Infectious agents other than bacteria are found in potable water. They include protozoa, viruses, and helminthes (Table 4). Contemporary water treatment methods have virtually eliminated the infectious agents for typhoid fever, cholera, and dysentery. However, emerging pathogens that are not eliminated or occur with a malfunctioning treatment system have given rise to some recent large outbreaks of water-borne disease. The 1993 contamination of the Milwaukee municipal water supply with the protozoan *Cryptosporidium* resulted in an estimated 50 deaths, 4000 hospitalizations, and 400,000 outbreaks of intestinal illness. Another notable case was the 2000 *E. coli* O157:H7 outbreak in Walkerton, Ontario, Canada, where seven people died and more than 1000 others became ill when heavy rains washed cow manure into the town wells at a time when the chlorination system was not operating due to mechanical failure [38].

TABLE 4 Major Infectious Agents Found in Contaminated Water Supplies

Bacteria	Protozoa	Viruses	Helminthes
Campylobacter jejuni	*Balantidium coli*	Adenovirus	*Ancylostoma duodenale*
Enteropathogenic E. coli	*Entamoeba histolytica*	Enterovirus	*Ascaris lumbricoides*
Salmonella	*Giardia lambdia*	Hepatitis A	*Echinococcus granulosis*
Shigella	*Cryptosporidium* spp.	Norwalk agent	*Necator americanus*
Vibrio cholerae		Reovirus	*Strongyloides stercoralis*
Yersinia enterocolitica		Rotavirus	*Taenia* spp.
		Coxsackie virus	*Trichuris trichiura*

A recent review article on the microbiological safety of drinking water [34] stated that the Environmental Protection Agency (EPA) acceptable risk for infectious disease from potable water is an annual risk of 10^{-4} (one infection in 10,000 per year). The article reports that the 1% infectious dose for *V. cholera*, *Salmonella typhi*, and *C. jejuni* is 1425, 263, and 1.4 cells, respectively. The tolerable concentration for *V. cholera*, *S. typhi*, and *C. jejuni* would be 71, 13, and 0.1 cfu per 100 mL of drinking water if the daily water consumption was 2 L.

3.3. Fecal Coliform Count

The fecal coliform procedure (Standard Method 9221 E) uses EC medium incubated in a water bath at 44.5 ± 0.2°C for 24 ± 2 hr with acid and gas production as presumptive positive result [33].

3.4. *E. coli* Count

E. coli is a dominant member of the fecal coliform group of bacteria that is found in potable water and indicative of fecal contamination. Methods for the detection of *E. coli* are based on their possession of the enzyme β-glucuronidase, which is capable of hydrolyzing the fluorogenic substrate 4-methyl umbelliferyl-β-D-glucuronide (MUG) with the release of the fluorogen 4-methyl umbelliferone when grown in EC-MUG medium at 44.5°C within 24 ± 2 hr [33].

3.5. Presence/Absence of *E. coli*

Chromogenic substrate coliform tests that utilize hydrolyzable substrates may be used as either an MPN multiple-tube count or a presence–absence test (Standard Method 9223). Because all coliforms possess the hydrolytic enzyme β-D-galactosidase, chromogenic substrates such as *ortho*-nitrophenyl-β-D-galactopyranoside (ONPG) are used to detect total coliforms, whereas the fluorogenic substrate MUG is used to detect the enzyme β-glucosidase found in *E. coli* [33].

Commercially available Presence–Absence (P–A) Coliform kits containing the two substrates are inoculated with a 100-mL water sample and incubated at 35 ± 0.5°C for 24 hr. A yellow color change is indicative of the ONPG hydrolysis liberating *ortho*-nitrophenol due to the presence of coliforms. Positive coliform tubes are examined with a ultraviolet lamp and tubes positive for *E. coli* will fluoresce due to liberated 4-methyl umbelliferone.

These products would be suitable for period monitoring of incoming potable water used in the pharmaceutical industry for coliforms.

3.6. Fungal Counts

Although fungi would not be expected to proliferate in pharmaceutical-grade water due to their nutritional requirements, the ability of media used for water monitoring to support fungal growth may be a consideration. The Standard Methods for the Examination of Dairy Products [32] recommends a standard methods agar with antibiotics chlortetracycline HCl and chloramphenicol for the enumeration of yeast and molds in dairy products. The media is identical to the plate count agar described in the Standard Methods for the Examination of Water and Wastewater, The role of antibiotics is to suppress the growth of bacteria when enumerating fungi in the presence of a high background of bacteria, and a requirement is not needed when monitoring pharmaceutical-grade water. Other reports in the literature support the use of plate count agar for fungal enumeration [1]. Because a review of the water monitoring literature suggests that both plate count agar and R2A medium give equivalent results, if there is a concern about the ability to count fungi, then plate count agar may be preferable to R2A.

4. EFFECT OF MEDIA SELECTION, INCUBATION TEMPERATURE, AND INCUBATION TIME ON MICROBIAL RECOVERIES

As emphasized earlier, it is widely recognized by microbiologists that media selection, incubation temperature, and incubation time will profoundly affect bacterial recovery from water. In general, a less rich media, lower incubation temperatures, and longer incubation times will result in higher recoveries of bacteria from water. The literature was reviewed to support this generalization. Commonly used microbiological culture media varies greatly in the amounts of utilizable organic material. It is expected that bacteria found in water physiologically suited to growth at the expense of low-nutrient concentrations will grow more favorably in less rich media at below ambient temperature. The composition of soybean–casein digest (SCD) agar, plate count agar, R2A agar ,and m-HPC agar is listed in Table 5.

The utilizable organic content in soybean–casein digest agar, plate count agar, R2A agar, and m-HPC agar is 2.3%, 0.85%, 0.28%, and 5.6 % by weight, respectively. It is widely established in the technical literature that soybean–casein digest agar, because of its high nutrient content, is not appropriate for water monitoring.

Recent examples of reports that the microbial count of water is generally higher when less rich media is used include reports on hemodialysis and semiconductor water microbial monitoring.

TABLE 5 Composition of Common Media Used for Water Monitoring

Ingredient	Soybean–casein digest agar	Plate count agar	R2A agar	m-HPC agar
Peptone	—	5 g	—	20.5 g
Gelatin	—	—	—	25.0 g
Tryptose	15 g	—	—	—
Soytone	5 g			
Proteose peptone no. 3	—	—	0.5 g	—
Beef extract	3 g	—	—	—
Casein digest	—	—	0.5 g	—
Yeast extract	—	2.5 g	0.5 g	—
Glucose	—	1 g	0.5 g	—
Glycerol	—	—	—	10.0 mL
Soluble starch	—	—	0.5 g	—
NaCl	5 g	—	—	—
K_2HPO_4	—	—	0.3 g	—
$MgSO_4$	—	—	0.05 g	—
Sodium pyruvate	—	—	0.3 g	—
Agar	15 g	15 g	15 g	15 g
Purified water	1000 mL	1000 mL	1000 mL	1000 mL

Although the Association for the Advancement of Medical Instrumentation (AAMI) recommended tryptic soy agar (TSA) as the standard agar, several studies have resulted in a general preference for R2A agar, as it appeared to be more sensitive in demonstrating the contamination of typical hemodialysis-associated bacteria. In the Netherlands, TSA is still used for culturing dialysate, whereas dialysis water is cultured on R2A [37]. Van Der Linde et al. [37] evaluated the bacterial yields of dialysis fluids on both media, and qualified their use in routine microbiological monitoring within their hemodialysis center. Between April 1995 and March 1996, 229 samples of pretreated and final purified dialysis water, and samples of dialysates were collected. The specimens were aseptically taken from the tap, various points of the reverse osmosis water treatment system, and the effluent tubes of 32 bicarbonate hemodialysis machines. Samples of 0.1 mL were inoculated in duplicate on spread plates with TSA and R2A agars. After 10 days of incubation at 25 ± 2°C, the numbers of colonies were counted. The ranges of spread were taken as 0–100 and 0–200 cfu/mL. The R2A medium had significantly higher colony counts than TSA medium for both dialysis water and dialysates.

As stated above, the recommended culture methods for monitoring the bacterial contamination of water, dialysate, and bicarbonate concentrate in dialysis centers in the United States involve culturing these fluids for 48 hr at 37°C. A variety of media and commercial culture methods are accepted for monitoring these fluids. Over a 3-month period, a comparison was made by Pass et al. [25] between an acceptable culture method, TSA employing the PP technique at 37°C for 48 hr, and PP cultures on PCA and R2A agar, incubated at ambient temperature (23°C) for 48, 72, and 168 hr. Increases in the colony counts over time occurred for all three fluids. However, counts were greater on PCA and R2A than on TSA. The increases over the standard 48-hr TSA cultures ranged as high as 10^4 times for 23°C cultures at 7 days of incubation. Bacterial colonies that appeared at 48, 72, and 168 hr were isolated and identified. *Pseudomonas*, *Moraxella*, *Acinetobacter*, and CDC group VI C-2 were among some of the common bacteria isolated. This study indicates that the media utilized, the time of incubation, and the temperature of incubation may result in a significant underestimation of the bacterial population of water and dialysis fluids, thus potentially placing the patient at a higher risk.

TSA medium was used in the semiconductor industry to determine the concentration of viable oligotrophic bacteria in ultra-pure water systems. Deionized water from an ultra-pure water pilot plant was evaluated for bacterial growth at specific locations, using a nonselective medium (R2A) designed to detect injured heterotrophic as well as oligotrophic bacteria. Governal et al. [10] compared the results obtained with R2A to those obtained using TSA medium. Statistically greater numbers of bacteria were observed when R2A was used as the growth medium. Total viable bacterial numbers were compared both before and after each treatment step of the recirculating loop to determine their effectiveness in removing bacteria. The reduction in bacterial numbers for the reverse osmosis unit, the ion exchange bed, and the ultraviolet sterilizer was 97.4%, 31.3%, and 72.85, respectively, using TSA medium, and 98.4%, 78.4%, and 35.8% using R2A medium. The number of viable bacteria increased by 60.7% based on TSA medium and 15.7% based on R2A medium after passage of the water through an in-line 0.2-μm pore size nylon filter, probably because of the growth of bacteria on the filter. Their results suggest that R2A medium may give a better representation of the microbial water quality in ultra-pure water systems and therefore a better idea of the effectiveness of the various treatment processes in the control of bacteria.

Less recent studies support the use of 28°C as an optimum incubation temperature when using R2A agar and the spread plate technique [13,18,19].

Although the plate count agar continues to be the recommended medium for the standard bacterial plate count (35°C, 48-hr incubation) of water and wastewater, Reasoner and Geldreich [26] reported that plate count agar does not permit the growth of many bacteria that may be present in treated potable water supplies. They developed a new medium for use in the heterotrophic plate count and for subculture of bacteria isolated from potable water samples. Their new medium, designated R2A, contains 0.5 g of yeast extract, 0.5 g of Difco Proteose Peptone no. 3 (Difco Laboratories), 0.5 g of casamino acids (Difco Laboratories), 0.5 g of glucose, 0.5 g of soluble starch, 0.3 g of K_2HPO_4, 0.05 g of $MgSO_4 \cdot 7H_2O$, 0.3 g of sodium pyruvate, and 15 g of agar per liter of laboratory quality water. The pH is adjusted to 7.2 with crystalline K_2HPO_4 or KH_2PO_4 and sterilized at 121 °C for 15 min.

Results from parallel studies with spread, membrane filter, and pour plate procedures showed that R2A medium yielded significantly higher bacterial counts than did plate count agar. Studies of the effect of incubation temperature showed that the magnitude of the count was inversely proportional to the incubation temperature. Longer incubation time, up to 14 days, yielded higher counts and increased detection of pigmented bacteria. Maximal bacterial counts were obtained after incubation at 20 °C for 14 days. As a tool to monitor heterotrophic bacterial populations in water treatment processes and in treated distribution water, R2A spread or membrane filter plates incubated at 28 °C for 5–7 days was recommended by Reasoner and Geldreich [26]. However, extended incubation times are not compatible with routine water monitoring in the pharmaceutical industry especially when we need to react to adverse trends in a timely fashion.

The pour plate method, although a well-accepted and simple technique, is limited to 1 mL by the inoculum size of the sample or diluted sample; colonies grow embedded in the agar and may be difficult to retrieve for subculture whereas the temperature of the molten agar (i.e., 40–50 °C) may reduce the recovery due to heat stock. Thc data presented in Table 6 clearly support this view.

A number of other reports support the view that R2A agar yields higher counts than PCA for a range of untreated and treated drinking water and source water [3,13,22,26,33]. For example, Brozel and Cloete [3] found that at 25 °C, the minimum incubation time to achieve the maximum colony count was 72 hr, whereas at 30 °C, it was 48 hr.

In general, spread plates consistently give higher counts than both pour plates (because of the adverse effect of the above ambient temperatures of molten agar) and the membrane filtration method. The bacterial counts increase with increasing incubation time irrespective of the media, method, or incubation temperature. However, studies do suggest that 28 °C is a better incubation temperature than either 20 °C or 35 °C.

The use of spread plates was supported by higher recoveries with this method compared to the pour plate method. Furthermore, the higher is the

TABLE 6 Effect of Incubation Temperature and Time on Colony Counts

		Incubation time (days)			
Temperature (°C)	Media/method	2	4	6	7
20	PCA/PP	22	130	570	900
	R2A/SP	90	1100	4700	6100
	R2A/MF	75	650	3000	4900
	M-HPC/MF	48	400	1600	2000
28	PCA/PP	90	640	950	1000
	R2A/SP	**360**	**2800**	**6700**	**7200**
	R2A/MF	160	2200	3500	4000
	M-HPC/MF	140	1000	1700	1900
35	PCA/PP	22	100	110	115
	R2A/SP	200	340	500	510
	R2A/MF	41	200	270	280
	M-HPC/MF	32	140	150	150

PCA = plate count agar; R2A = R2A agar; M-HPC = m-HPC agar; PP = pour plate; SP = spread plate; MF = membrane filter. Highest counts are in bold.
Source: Ref. 27.

holding temperature of the molten agar prior to pouring the plates, the lower is the bacterial recovery with incubation times of up to 21 days (Table 7).

Reasoner and Geldreich [26] compared the mean microbial counts from 10 water distribution system samples obtained with four different culture methods incubated at 35°C (Table 8).

This suggests that the spread plate method using R2A agar incubated at 28°C for 5–7 days is the preferred culture condition for water monitoring. However, several disadvantages are apparent with this scheme. Spread plates require specialized preparation in that they need to be air-dried to limit surface moisture that may promote convergent growth instead of discrete colony formation, and the inoculum must be spread over the agar surface using a sterile glass hockey stick. The inoculum size is limited to 1 mL, which is a serious disadvantage when monitoring purified water, USP, or water for injection, USP, that may have low bacterial counts. In addition, many microbiologists have experienced occasional difficulties in recovering colonies through subculture from R2A media, presumably because the isolates are acclimatized to the low-nutrient medium whereas additional incubators maintained at 25–30°C would be required in the testing laboratory.

At first sight, one would expect that the use of membrane filters with a porosity tighter than the 0.45-μm pore size, recommended in the Standard Methods for water monitoring, would result in the retention of greater

TABLE 7 Effect of Media Temperature and Incubation Time on Recoveries

Standard method	Incubation time at 35°C (days)				
	2	4	7	14	21
Spread plate	110 ± 31[a]	130 ± 38	200 ± 40	250 ± 75	300 ± 61
Pour plate 42°C agar	**33 ±11**	**59 ± 13**	**76 ± 16**	**82 ± 14**	**84 ± 15**
Pour plate 45°C agar	23 ± 6	50 ± 9	62 ± 15	78 ± 17	79 ± 17
Pour plate 50°C agar	12 ± 2	32 ± 4	43 ± 7	55 ± 6	58 ± 9

All cfu × 10^{-2}/mL.
Highest PP results are in bold.
[a] Standard deviation.
Source: Ref. 13.

numbers of bacteria, especially those with reduced dimensions that are found in water with low-nutrient levels and reduced temperature, and higher counts. There are a number of reports in the literature that suggest that this is not always the case. For example, the pore size may actually affect cell retention, microbial recovery, and colony size. The cell retention will depend on the bacterial population with the probability of passage of smaller bacterial cells through the membrane increasing with the increasing pore size and bacterial count. In the range of the requirements for potable water, purified water, and

TABLE 8 Composition of the Mean Microbial Counts from 10 Distribution Systems Using Four Different Counting Methods

Medium/method	Counts (cfu/mL; incubation at 35°C)		
	48 hr	72 hr	168 hr
PCA/PP	210	320	860
PCA/MF	50	380	1200
R2A/MF	250	750	1500
R2A/SP	**1200**	**2300**	**4300**

PCA = plate count agar; PP = pour plate; MF = membrane filter; .SP = spread plate. Highest counts are in bold.

water for injection (i.e., 500, 100, and 10 cfu/100 mL, respectively), the passage of bacterial cells through a membrane filter with a retention rating of 10^7 *Brevundimonas diminuta* per square centimeter may not be an issue. The recovery for bacteria capable of growth on microbiological media will depend on the media selection, incubation temperature, and membrane properties. There are some evidence that decreasing the pore size may limit the diffusion of nutrients through the membrane and limit both colony development and growth.

A study was conducted by the Millipore Corporation (Bedford, MA) to confirm the validity of enumerating affluents from 0.2-μm membrane filters in bacterial challenge studies with a 0.45-μm membrane [4]. Membrane filters with pore size ratings of 0.22 and 0.45 μm were tested for their ability to recover *Brevundimonas diminuta* ATCC 19146, the organism typically used in bacterial retention testing of sterilizing-grade membrane filters. For each of the two pore size ratings, filters of two membrane filter polymer materials, hydrophilic PVDF (Millipore Durapore), and mixed esters of cellulose were tested, resulting in an evaluation of four potential recovery filters. The 0.45-μm mixed esters of cellulose filter are the currently accepted membrane for this purpose. The data show no difference in the ability of the four filters to recover freshly cultured *P. diminuta.* Moreover, the membrane filter method was shown to provide high bacterial recovery efficiency, equivalent to that of the spread plate method. The author concluded that 0.22-μm filters, despite their ability to retain higher levels of bacteria, proved not to have an advantage over 0.45-μm membranes in terms of bacterial recovery.

A recent publication by Lillis and Bissonnette [17] addressed the ability of water-borne bacteria to pass through 0.45-μm membrane filters. Individual groundwater supplies in rural West Virginia were examined by a double-membrane filtration procedure to determine the presence of HPC bacteria capable of escaping detection on conventional pore size (0.45 μm) membrane filters but are retained on 0.22-μm pore size filters (i.e., filterable bacteria). Because the authors believed that the optimum cultural conditions for recovery of filterable bacteria are not well defined, their initial efforts focused on evaluation of various media (R2A, m-HPC, and HPCA) and incubation temperatures (15°C, 20°C, 28°C, and 35°C) for specific recovery of filterable bacteria. As reported in the earlier literature, maximum recovery of small-sized HPC bacteria occurred on low-nutrient concentration R2A agar incubated for 7 days at 28°C. Similarly, identical cultural conditions gave enhanced detection of the general HPC population on 0.45-μm pore size filters. A 17-month survey of 10 well water supplies conducted using R2A agar incubated for 7 days at 28°C resulted in detection of filterable bacteria (ranging in density from 9 to 175 cfu/mL in six of the groundwater sources). The proportion of filterable bacteria in any single sample never exceeded 10% of the total HPC population. This difference would appear to be below the variability of the membrane filtration method (i.e., 15–35% RSD) [20]. A majority of the colonies appearing on the 0.22-µm membrane filters were pigmented (50–90%), whereas the proportion of colonies demonstrating pigmentation on the larger porosity filters failed to exceed 50% for any of the samples (19–49%). Identification of randomly selected isolates obtained on the 0.22-µm filters indicated that some of these filterable bacteria have been implicated as opportunistic pathogens.

5. PURIFIED WATER AND WATER FOR INJECTION MONITORING

5.1. Total Aerobic Microbial Count

The types of water for pharmaceutical use and their recommended methods and specifications are listed in U.S. Pharmacopeial Informational Chapter <1231> *Water for Pharmaceutical Purposes* [36]. The USP references the Standard Methods for the Examination of Water and Waste Water (APHA), 20th edition, 1998, for information on specific test methods (Table 9). In supplement 2002 of the Pharmacopoeia Europa (Ph. Eur.), there are two water monographs. One is water for injection, subdivided into sections on water for injection in bulk and a section for sterilized water for injections. The second monograph is on purified water. This monograph is also divided into two sections: purified water in bulk and purified water in containers. For the microbiological examination (if relevant), R2A medium incubated at 30–35°C for 5 days and not plate count agar is prescribed (Table 10). This difference in methods is problematic for routine water monitoring by pharmaceutical companies.

With pharmaceutical companies manufacturing drug products for the international market, a water monitoring strategy that accommodates both the USP and Ph. Eur. requirements must be developed. The author recommends the USP-recommended methods because of their 48- to 72-hr incubation time for use in routine monitoring whereas the Ph. Eur.-recommended methods encourage that a 5-day incubation time be run periodically (i.e., monthly). This would enable the company to certify that, if monitored, the water system meets Ph. Eur. requirements.

During the validation of a new pharmaceutical water system, extensive microbial monitoring would be conducted to demonstrate a state of control of the system. During routine operation, the monitoring program may involve sampling the storage tank and each loop daily, and each point of use over a 1- or 2-week period. The sampling method should reflect the usage of the water system using manufacturing. The tap should be flushed and a sample collected

TABLE 9 USP-Recommended Methods

Method	Minimum sample size (mL)	Recommended membrane filter pore size	Incubation conditions	Recommended microbial limit
Potable water—pour plate using plate count agar	1	NA	48–72 hr at 30 to 35°C	NMT 500 cfu/mL
Purified water—pour plate or membrane filtration using plate count, R2A, or m-CPC agar	1	0.45 μm gridded membrane filters	48–72 hr at 30 to 35°C	NMT 100 cfu/mL
WFI—membrane filtration using plate count agar	100	0.45 μm gridded membrane filters	48–72 hr at 30 to 35°C	NMT 10 cfu/100 mL

Source: Standard Methods for the Examination of Water and Wastewater, APHA, 20th ed., 1998.

into a sterile container from the water stream. Sanitization of the taps with alcohol, flaming the taps, etc. are not necessary to aseptically collect a sample and may compromise the water sample and/or the point of use. Samplers need to be trained in aseptic technique and suitably clothed to take a sample. If the sample cannot be processed within 1 hr after collection, transport and/or storage of the sample at refrigeration temperature (2–8°C) for up to 6 hr is recommended [33].

Questions arise about the necessity for validating the standard methods used for water monitoring. A procedure contained in the Standard Methods for the Examination of Water and Wastewater qualifies as a standard method by either undergoing development, validation, and collaborative testing that meets the requirements as set out in Sections 1040B and C of the Standard Methods, or being accepted as widely used by members of the Standards Committee and has appeared in two previous editions of Standard Methods [33]. There is no validation requirement for the method, but a laboratory should be qualified to run the method by demonstrating that the

TABLE 10 Ph. Eur. Required Methods (2000:0169 Water for Injections and 2000:0008 Water, Purified)

Method	Minimum sample size (mL)	Recommended membrane filter pore size	Incubation conditions	Microbial limit
Potable water—pour plate using plate count agar	1	NA	48–72 hr at 30–35°C	NMT 500 cfu/mL
Purified water in bulk—total viable aerobic count (2.6.12) membrane filtration using agar medium S (R2A agar)	1 (size of the sample is to be chosen in relation to the expected result)	Nominal pore size not greater than 0.45 μm	5 days at 30–35°C	NMT 100 cfu/mL
Water for injection in bulk—total viable aerobic count (2.6.12) membrane filtration using agar medium S (R2A agar)	At least 200	Nominal pore size not greater than 0.45 μm	5 days at 30–35°C	NMT 10 cfu/ 100 mL

Note that coliform monitoring of purified water and water for injection is not a compendial requirement and is not indicated for routine monitoring.

equipment is qualified, proving that written procedures are available, and proving that the microbiologists are trained to run the method.

Because the water for pharmaceutical use must meet USP requirements, it is unlikely that different water samples, unlike a pharmaceutical ingredient, will affect the recovery of bacteria within the sample. In contrast, the microbial limit testing of pharmaceutical ingredients and pharmaceutical drug products is validated by using USP preparatory testing to demonstrate that the test material does not inhibit the recovery of bacteria when diluted or treated with neutralizers, then enumerated. The author is aware of misguided attempts to demonstrate the equivalency of various media and incubation conditions. For example, USP-recommended quality control microorganisms and environmental isolates grown overnight in soybean–casein digest broth, harvested by centrifugation, washed, and resuspended in diluent were inoculated into potable water, purified water, and water for injection. The inocula were confirmed by plating on soybean–casein digest agar. After holding the inoculated water samples for up to 4 hr, the microbial content was enumerated using R2A pour plates incubated at 30–35°C for 48 hr (USP-recommended incubation time) and 5 days (Ph. Eur.-recommended incubation time). Predictably, the recoveries were 5–10% lower than the inocula controls and there was no difference between the recoveries from the same plates read at 48 hr and 5 days. Clearly, inocula prepared in soybean–casein digest broth will have a higher recovery on soybean–casein digest than R2A agar. In addition, the only difference between the colonies recovered after a 48-hr incubation and a 5-day incubation will be in the size of the colonies as they continue to grow during the longer incubation time.

A reasonable approach is to accept the standard method and demonstrate the suitability of the testing laboratory to conduct water monitoring. This qualification would include validating the temperature distribution in incubators and water baths, growth promotion testing the media, and training the microbiologists running the tests.

The identity of the water isolates that exceed the alert and action levels should be routinely determined. In practice, a limited range of bacterial species is routinely found in purified water systems. Bacteria commonly found in water for pharmaceutical purposes include *Sphingomonas paucimobilis*, *Comamonas acidovorans*, *Xanthomonas maltophilia*, *R. pickettii*, *B. cepacia*, and *Pseudomonas vesicularis*.

6. SETTING ALERT AND ACTION LEVELS FOR WATER MONITORING

Action levels are typically set at 500, 100, and 10 cfu/100 mL, respectively, for potable water, purified water, USP, and water for injection, USP. Exceeding these levels would trigger, first, a laboratory investigation to confirm that the sampling was satisfactory, enumeration was run correctly, and counts were considered valid; second, a manufacturing investigation to determine if the water system was performing within its operating parameters would be conducted. Typically, a point-of-use exceeding the action level would be taken out of production until the manufacturing investigation and repeat testing are completed. In addition, the potential impact of the water monitoring out-of-level result on the product made with the water would be assessed. However, the monitoring out-of-level result would not represent an isolated sample. Other samples from the circulation tank, distribution loop, and other points-of-use on the loop would be tested on the same day, so it can be readily determined if the result reflects the entire system, single loop, or individual tap. If the other samples and repeat samples are satisfactory, then the result may be attributed to sampling error. With circulating water systems that are routinely disinfected by ozone or the maintenance of a hot water system, it is unlikely that bacteria will persist within the system. In these cases, it is more likely that bacterial isolation will be due to a sampling error than a system contamination.

Alert or warning levels are typically set from the statistical evaluation of the historical water monitoring data. Because microbial counts are not normally distributed but show a positively skewed distribution with many zero or low counts and fewer high counts (i.e., a Poisson distribution), the use of the mean plus 1 or 2 SD is not appropriate in setting alert and action levels. The use of a nonparametrical tolerance limit with a 95% probability at a 95% confidence level for alert levels and a 95% probability at the 99% confidence level for action levels is recommended.

Because the purpose of routinely monitoring a validated pharmaceutical water system is to determine when the microbial counts are out of trend, the rule to identify excursion from a state of control may be used based on control charting using Western Electric trend rules [23]. However, with modern water systems, typical results will be less than 1 cfu/mL purified water and less than 1 cfu/100 mL for water for injection/ so control charting may not be a fruitful activity. Another approach may be determining the time between isolation and/or alert and action level excursions and constructing a CUMSUM (cumulative sum) control chart of the time intervals [23]. This type of chart is particularly sensitive in detecting small but significant changes. A water system under control will have longer and longer time intervals between excursions/ whereas a water system moving out of a state of control will have shorter and shorter intervals between excursions.

7. QUALITY CONTROL PROGRAMS

The Standard Methods [33] outlines the details of a laboratory QC program for the microbiological examination of water. The recommended QC program for laboratory equipment is summarized in Table 11.

8. WATER TESTING REGULATION

The establishment of water monitoring standards has been instrumental in the promotion of public health.

Table 12 shows the sequence of introduction of standard water monitoring methods in the United States [27].

TABLE 11 A Recommended QC Program for Water Monitoring Equipment

Laboratory equipment	Control procedures
Temperature-recording devices	Check semiannually against an NIST- traceable thermometer. Whenever possible, use a chart recorder to provide a continuous temperature record.
Balances	Check monthly with certified weights. For those weighing 2 g or less, use an analytical balance with a sensitivity of 1 mg at a 10-g load. For larger quantities, use a pan balance with a sensitivity of 0.1 g at a 150-g load.
pH meter	Standardize with a least two standard buffers (pH 4.0, 7.0, or 10.0) and compensate for temperature before each series of measurements.
Water system	Use a deionization unit for reagent grade water. Monitor conductivity daily.
Media-dispensing equipment	Check the accuracy of the volume dispensed at the start of each run and periodically during an extended run.
Hot air oven	Monitor temperature in the 160–80°C range and run biological indicators quarterly.
Autoclave	Record hard copy of the temperature, time, and items sterilized for each load. Run biological indicators monthly.
Laminar flow hoods	Monitor pressure across the filters. Check airflow and for leaks semiannually. Expose air settling plates for up to 1 hr monthly.
Water baths	Monitor and record temperature daily. Clean bath as required.
Incubators	Use a temperature-recording device and alarm. Maintain in a general laboratory area at 16–27°C.
Microscope	Clean optics with lens tissue after each use. Monitor the life of the lamp and check lamp alignment as needed.

The 1974 Safe Drinking Water Act created the first ever mandatory national monitoring program to protect public health through drinking water safety administered by the U.S. Federal EPA. This act was amended in 1986 and 1996 to strengthen the regulations [39].

8.1. Highlights of the Safe Drinking Water Act of 1974

- Established a national structure for drinking water protection activities
- Authorized EPA to establish national enforceable health standards for contaminants in drinking water

TABLE 12 The Sequence of Introduction of Standard Water Monitoring Methods in the United States

Standard methods	Technology advance
First Standard Methods (1905)	Pour plate method with nutrient gelatin incubated at 20°C
Second Standard Methods (1912)	Introduction of agar as a solidifying agent. First U.S. Public Health Service (USPHA) Bacteriological Standards for Coliform Monitoring cited the 1912 edition of the Standard Methods for media and methods
Third USPHA Drinking Water Standards (1942)	Required that bacteriological samples be collected at representative points throughout the distribution system and the number of samples taken reflects the population served
Sixteenth Standard Methods (1985)	Added spread plate and membrane filter to the pour plate method and use of R2A agar as a low-nutrient alternative to plate count agar

Source: Ref. 27.

- Provided for public water system compliance through a federal–state partnership
- Established public notification to alert customers to water system violations
- Set up procedures to protect underground sources of drinking water.

8.2. The 1986 Amendments

- Required disinfection for all water systems
- Expanded the number of regulated contaminants and increased the pace of contaminant regulation
- Required filtration of all surface water supplies, unless strict criteria are met
- Established a monitoring program for unregulated contaminants.

In December 1998, new drinking water standards for *Cryptosporidium*, other disease-causing microbes, and potentially harmful by-products of the water treatment process were the first standards set under the 1996 Amendments. These new standards will prevent up to 460,000 cases of water-borne illness a year and reduce exposure to disinfection by-products by 25%.

9. APPLICATION OF NEW MICROBIAL TESTING METHODS TO WATER MONITORING

Three technologies that have been applied successfully to water monitoring include: (1) ATP bioluminescence, (2) polymerase chain reaction (PCR), and (3) solid-phase fluorescence laser scanning microscopy. Representative publications are discussed.

The enumeration of microorganisms in Water for Pharmaceutical Purposes using the MicroCount™ Digital System (Millipore Corporation) was compared to the USP-recommended Pour Plate and Membrane Filtration Count Methods [20]. A study, using a pure culture of *Burkholderia cepacia*, ATCC 25416, showed that the accuracy, precision, reproducibility, and linearity of the MicroCount™ ATP Bioluminescence System were equivalent to, or better than, the traditional methods. When the MicroCount™ System was used to monitor purified water and water for injection taps in a pharmaceutical plant over a month, comparable counts to the traditional methods were obtained within 24 hr compared to 48–72 hr with the other methods. The effectiveness of the memory device used for the isolation of colonies for characterization was demonstrated by comparing the number and pattern of the positive wells in the MicroCount™ plates with the isolation of colonies on the microbial count agar plates. The recovery on agar plates, although slightly higher, was not statistical different from the MicroCount™ plates. The predominated microorganisms isolated using all three methods were *R. pickettii*, *Bacillus sphaericus*, *Stenotrophomonas maltophilia*, and a *Staphylococcus* species.

During the study, information on the precision of the PCA/MF, PCA/PP, R2A/MF, R2A/PP, and MicroCount System for bacterial populations at a 20 cfu/mL level was obtained (Table 13).

It is widely accepted by microbiologists that the heterotrophic plate count method may not support the growth of all viable bacteria, which may be

TABLE 13 The Precision of Different Bacterial Counting Methods Used in Water Monitoring

Method	Incubation conditions	Relative SD %
MicroCount digital system	48 hr at 30–35°C	21.4
Plate count agar/membrane filtration	48 hr at 30–35°C	32.9
Plate count agar/pour plate	48 hr at 30–35°C	36.8
R2A agar/membrane filtration	48 hr at 30–35°C	25.4
R2A agar/pour plate	48 hr at 30–35°C	14.4

Source: Ref. 20.

present within a water sample and, as a result, will underestimate the bacterial population in a water sample. The use of alternative procedures using "viability markers" may yield additional information. In a study from an English water testing laboratory [28], bacteria were retained on a membrane filter and a fluorogenic substrate ChemChrome B (CB) was transported into the bacterial cells, converting the substrate to a fluorescent product by esterase activity, used to stain viable bacteria from potable water samples. The labeled bacteria from each sample were subsequently enumerated by using a novel laser scanning instrument marketed as Scan RDI (Chemunex, Inc.) in the United States. Furthermore, 107 potable water samples analyzed using the Scan RDI System gave a significantly greater number of bacteria than were detected by culture. The mean number of bacteria isolated on R2A agar incubated at 22°C for 7 days was around 25% of the total number of viable bacteria detected using the CB/Scan RDI enumeration. Additional analyses of 81 water samples using a 5-cyano-2,3,4-tolyl-tetrazolium chloride (CTC) viability assay also demonstrated the presence of many viable bacteria that were not capable of growth under the standard culture conditions. Moreover, the results with 75 of 81 samples indicated that CB had the ability to stain a significantly greater number of bacteria than the redox reagent CTC. Information on the precision of the method is available within the publication. For example, sample 13 ($n = 5$) had a mean of 1794 cfu/mL, range 1558–1932 bacteria/ml, SD = 145, and RSD 8.1%, whereas sample 4 ($n = 4$) had a mean of 336 bacteria/mL, range 48–336, SD = 48, and RSD = 14.3%. The authors concluded that the Scan RDI System was successfully used for rapid and accurate enumeration of labeled microorganisms, allowing information on the total viable microbial load of a water sample to be determined within 1 hr. The use of a scanning laser system in the routine microbiological quality control analysis of pharmaceutical grade water is described. In contrast, it was shown that the Scan RDI method provided the speed (less than 4 hr in all cases) and sensitivity (down to a single cell) required for routine real-time analysis, with microbial counts that correlated well with the plate count method [11].

Wallner et al. [35] evaluated the Scan RDI System for the testing of pharmaceutical water by comparing it to the standard plate count method. The Scan RDI system appeared to be at least as sensitive as the standard method. In some cases, the results were equivalent for both methods, but for most water samples, the Scan RDI results were higher than the standard plate count and sometimes exceeded the latter by an order of magnitude or more.

Japanese microbiologists have recently used nucleic acid-based techniques to analyze the bacterial population in water for pharmaceutical purposes [15]. The bacterial community in deionized water used in pharmaceutical manufacturing processes was analyzed by denaturing gradient gel electrophoresis (DGGE). 16S ribosomal DNA fragments, including V6, V7, and V8 regions, were amplified with universal primers and analyzed by DGGE. The bacterial diversity in purified water determined by PCR-DGGE banding patterns was significantly lower than that of other aquatic environments, confirming the selectivity of the purification processes. The bacterial populations with esterase activity sorted by flow cytometry and isolated on SCD and R2A medium were also analyzed by DGGE. The dominant bacterium in purified water possessed esterase activity but could not be detected on the SCD or R2A media. DNA sequence analysis of the main bands on the DGGE gel revealed that culturable bacteria on these media were *Bradyrhizobium* sp., *Xanthomonas* sp., and *Stenotrophomonas* sp., whereas the dominant bacterium was not closely related to previously characterized bacteria. The authors concluded that these data suggest the importance of culture-independent methods of quality control for pharmaceutical water.

10. WATER MONITORING IN RELATIONSHIP TO MICROBIAL ECOLOGY

It is a truism in ecology that as the complexity of ecosystems is reduced, the diversity of the fauna and flora is markedly reduced because of reduction in niches in the ecosystem. Because water for pharmaceutical purposes differs from potable water in seasonal variations in dissolved organic matter, temperature, and bacterial content, it is not unexpected that the microbial diversity in purified water and water for injection is considerably less than potable water and water for injection (Table 14). As stated above, the bacterial diversity in purified water determined by PCR-DGGE banding patterns was significantly lower than that of other aquatic environments [15].

A major concern of water companies is the persistence of coliform bacteria in biofilms formed on the interior of pipes used to distribute the water that may be shed into the water distribution system. Occasional failures of coliform testing during the summer months (i.e., one total coliform in a 100-mL sample) have been attributed to seasonal biofilm formation and may have public health implications. Similarly, there is a potential for biofilm development in the distribution system for purified water. The temperature, low dissolved organic matter, aeration, absence of chlorine, uneven levels of demand, and recirculation all favor biofilm formation. Strategies to reduce biofilm formation include high turbulence, absence of doglegs, flushing of taps when drawing off water, and periodic sanitization by hot water, steam, or ozone treatment. Biofilms were the subject of a recent review article [21].

With the amount of dissolved organic matter in potable water up to three magnitudes higher than water for pharmaceutical purposes, the ability of these waters to support bacterial growth is probably limited. For example,

TABLE 14 Parameters of Potable Water, Purified Water, and Water for Injection

Parameter	Potable water	Purified water	Water for injection
Temperature	Ambient temperature: 5–20°C	Room temperature: 20–25°C	>60°C
Total organic carbon	NMT 500 mg/L (total dissolved solids); typical ranges: 1–20 mg/L (surface water) and 0.1–2 mg/L (ground water)	NMT 0.5 mg/L	NMT 0.5 mg/L
Recirculation	Demand-driven	Recirculated	Recirculated
Residual chlorine	Greater than 0.2 mg/L	None	None
Microbial content	NMT 500 cfu/mL	NMT 100 cfu/mL	NMT 1 cfu/100 mL
Total coliforms	Zero in 100 mL	NA	NA
pH	6.5–8.5	5.0–7.0	5.0–7.0

if the dry weight of water-borne bacterium is 10^{-13} g, of which 50% is carbon, then water for injection containing 100 ppm of TOC (i.e., 10^{-9} g/L) should be able to support up to 2×10^4 bacteria. In general, water-borne bacteria adopt three different strategies to the nutrient level. When utilizable substrate is growth-limiting, then slow-growing bacteria with a high substrate affinity are favored, whereas when substrate is in excess, fast-growing bacteria are favored [29]. The third strategy used is pharmaceutical water systems in biofilm formation on surfaces. Microbial habitats as diverse as oceanic waters and water for injection share the characteristic of extremely low utilizable substrate where bacteria with high substrate affinity may predominate.

A perennial question asked in pharmaceutical discussion groups is whether water for injection that is maintained at 80°C needs to be monitored for thermophilic bacteria. Thermophiles [16] are a diverse group of Archaea and bacteria that include photosynthetic bacteria, chemolithoautotrophic and heterotrophic aerobic Archaea and bacteria, and anaerobic Archaea and bacteria (Tables 15 and 16). In general, the nutritional requirements of thermophiles (e.g., rich nutrients, vitamins, light, electron acceptors, anaer-

TABLE 15 Representative Thermophilic Archaea

Class of thermophiles	Representative organisms	Temperature maximum/ temperature optimum	pH optimum
Methanogenic anaerobes	*Methanobacterium thermoautotrophicum*	70–110°C/ 55–98°C	5.7–7.7
Aerobic thermoacidophiles	*Thermoplasma acidophilum* *Sulfolobus acidocaldarius* *Acidianus inferus*	65–96°C/ 60–90°C	1.5–3.0
Anaerobic thermoacidophiles	*Thermococcus celer* *Pyrococcus woesei* *Thermoproteus neutrophiles*	90–110°C	5.5–7.0

TABLE 16 Representative Thermophilic Bacteria

Class of thermophiles	Representative organisms	Temperature maximum/ temperature optimum	pH optimum
Aerobes	*Bacillus stearothermophilus* *Thermus aquaticus* *Thermoleophilum album*	65–85°C/ 55–75°C	2.0–8.0
Anaerobes	*Clostridium stercorarium* *Desulfovibrio thermophilus* *Thermotoga neapolitana*	65–90° C/ 60–75°C	5.7–8.0

obic conditions, and elevated temperatures) make it highly unlikely that thermophiles will exist in hot water of injection. If they did persist in water for injection, they would not grow in the human body, which has a temperature around 37°C. Given this situation, monitoring pharmaceutical-grade waters for thermophiles is not recommended.

11. CONCLUSIONS

As stated earlier for pharmaceutical companies manufacturing drug products for the international market, a water monitoring strategy that accommodates both the USP and Ph. Eur. requirements must be developed. The author recommends that the USP-recommended methods (because of their 48- to 72-hr incubation time, ease of subculture of isolates, and ability to readily isolate fungi) be used for routine monitoring, whereas the Ph. Eur.-recommended methods with a 5-day incubation time be run periodically (i.e., monthly) so that a testing history is available to certify that, if tested, the water system will meet the Ph. Eur. requirements.

The methods are for purified water–pour plate or membrane filtration using plate count, R2A, or m-CPC agar, with a minimum sample size of 1 mL, incubated at 30–35°C for up to 48 hr, and for water for injection membrane filtration using plate count or R2A agar with a minimum sample size of 100 mL incubated at 30–35°C for up to 48 hr. The recommended membrane filters are 0.45-μm gridded membrane filters.

REFERENCES

1. Beuchat LR, Frandberg E, Deak T, Alzamora SM, Chen J, Guerrero S, Lopez-Malo A, Ohlsson I, Olsen M, Peinado JM, Schnurer J, de Siloniz MI, Tornai-Lehoczki J. Performance of mycological media in enumerating desiccated food spoilage yeasts: an interlaboratory study. Int J Food Microbiol 2001; 70(1–2):89–96.
2. Brock TD, Koch R. A Life in Medicine and Bacteriology. Madison, WI: Science Tech Publishers, 1988.
3. Brozel VS, Cloete TE. Evaluation of nutrient agars for the enumeration of viable aerobic heterotrophs in cooling water. Water Res 1992; 28(8):111–1117.
4. Carter J. Evaluation of recovery filters for use in bacterial retention testing of sterilizing-grade filters. PDA J Pharm Sci Technol 1996; 50(3):147–153.
5. Clark HF, Geldreich EE, Jeter HL, Karbler PW. The membrane filter in sanitary bacteriology. Public Health Rep 1951; 66:951.
6. Colwell RR. Bacterial death revisited. In: Colwell RR, Grimes DJ, eds. Non-Culturable Microorganisms in the Environment. Washington, DC: ASM Press, 2000.

7. 40 CFR Part 141, Monitoring Requirements for Public Drinking Water Supplies.
8. Geldreich EE. Microbial Quality of Water Supply in Distribution Systems. CRC Press, Inc., 1996:504.
9. Geldreich EE, Allen MJ, Taylor RH. Interferences to coliform detection in potable water supplies. In: Hendricks CW, ed. Evaluation of the Microbiological Standards for Drinking Water. Washington, DC: U.S. Environmental Protection Agency, 13–30.
10. Governal RA, Yahya MT, Gerba CP, Shadman F. Oligotrophic bacteria in ultra-pure water systems media selection and process component evaluations. J Ind Microbiol 1991; 8(4):223–228.
11. Guyomard S. Validation of a scanning laser system for microbiological quality control (QC) analysis. Pharm Technol Int 1997; 9(Sept):50, 52, 54.
12. Haas CN, Meyer MA, Paller MS. Analytical note: evaluation of the m-SPC method as a substitute for the standard plate count in water microbiology. J AWWA 1982; 74, 322.
13. Klein DA, Wu SA. Factors to be considered in heterotrophic microorganism enumeration from aquatic environments. Appl Microbiol 1974; 27:429–431.
14. Leclerc H, Mossel DA, Edberg SC, Stuijk CB. Advances in the bacteriology of the coliform group: their suitability as markers of microbial water safety. Annu Rev Microbiol 2001; 55:201–234.
15. Kawai M, Matsutera E, Kanda H, Tani K, Yamaguchi N, Nasu M. Dominant bacteria in a viable but non-culturable state in pharmaceutical water. 2002: Abstracts of the General Meeting of the American Society for Microbiology. 2001; 101:639–640.
16. Kristjansson JK, Stetter KO. Thermophilic Bacteria. In: Kristjansson JK, ed. Thermophilic Bacteria. Boca Raton, FL: CRC Press, Inc., pp. 2–13.
17. Lillis, Bissonnette. Detection and characterization of filterable heterotrophic bacteria from rural groundwater supplies. Lett Appl Microbiol 2001; 32(4): 268–272.
18. Lombardo LR, West PR, Holbrook JL. A comparison of various media and incubation temperatures used in the Heterotrophic Plate Count analysis. Water Quality Technology Conference, AWWA, Denver, CO, 1985:251–270.
19. Maki JS, LaCroix SJ, Hopkins BH, Staley JT. Recovery and diversity of heterotrophic bacteria from chlorinated drinking waters. Appl Environ Microbiol 1986; 51:1047–1055.
20. Marino G, Maier C, Cundell AM. A comparison of the MicroCount Digital System to plate count and membrane filtration methods for enumeration of microorganisms in water for pharmaceutical purposes. PDA J Pharm Sci Technol, 1999.
21. Marshall KC. Starved and non-culturable microorganisms in biofilms. In: Colwell RR, Grimes DJ, eds. Non-Culturable Microorganisms in the Environment. Washington, DC: ASM Press, 2000.
22. Means EG, Hanami L, Ridway GF, Olson BH. Evaluating media and plating techniques for enumerating bacteria in water distribution systems. J AWWA 1981; 73:585.

23. Montgomery DC. Introduction to Statistical Quality Control. 4th ed. New York: John Wiley and Sons.
24. Muller G. Lactose–fushsin plate for detection of *E. coli* in drinking water. Z Hyg Infektionskr 1947; 127:187–190.
25. Pass T, Wright R, Sharp B, Harding GB. Culture of dialysis fluids on nutrient-rich media for short periods at elevated temperatures underestimate microbial contamination. Blood Purif 1996; 14(2):36–145.
26. Reasoner DJ, Geldreich EE. A new medium for the enumeration and subculture of bacteria from potable water. Appl Environ Microbiol 1985; 39(10):1–7.
27. Reasoner DJ. Monitoring heterotrophic bacteria in potable water. In: McFeters GA, ed. Drinking Water Microbiology: Progress and Recent Developments. Berlin: Springer-Verlag, 1990:452–477.
28. Reynolds DT, Fricker CR. Application of LASER scanning for the rapid and automated detection of bacteria in water samples. J Appl Microbiol 1999; 86(5):785–795.
29. Schlegel HG, Jannasch HW. Prokaryotes and their habitats. In: Starr MP, Stolp H, Balows A, Truper HG, Dworkin M, Schlegel HG, eds. Prokaryotes. Berlin: Springer-Verlag, 1981:43–82.
30. Sharpe AN. Development and evaluation of membrane filtration techniques in microbial analysis. In: Patel PH, ed. Rapid Analysis Techniques in Food Microbiology. Glasgow, Scotland: Blackie Academic and Professional, 29–60.
31. Shirley JJ, Bissonette GK. Detection and identification of groundwater bacteria capable of escaping entrapment on 0.45-micron-pore-size membrane filters. Appl Environ Microbiol 1991; 57(8):2251–2254.
32. Standard Methods for the Examination of Dairy Products. 16th ed. Washington, DC: American Public Health Association, 1992.
33. Standard Methods for the Examination of Water and Waste Water. 20th ed. Washington, DC: American Public Health Association, 1998.
34. Szewzyk U, Szewzyk W, Schleifer K-H. Microbiological safety of drinking water. Annu Rev Microbiol 2000; 54:81–127.
35. Wallner G, Tillman D, Haberer K. Evaluation of the ChemScan System for rapid microbiological analysis of pharmaceutical water. PDA J Pharm Sci Technol 1999; 53(2):70–74.
36. USP General Informational Chapter <1231> Water for Pharmaceutical Purposes.
37. Van der Linde BT, Lim JMM, Rondeel LPMT, Antonissen, De Jong GMT. Improved bacteriological surveillance of hemodialysis fluids: a comparison between Tryptic Soy Agar and Reasoner's 2A media. Nephrol Dial Transplant 1999; 14(10):2433–2437.
38. www.microbeworld.org/cissues/wqual
39. www.epa.gov/safewater

4

Sterility Test and Procedures

Luis Jimenez
Genomic Profiling Systems, Inc., Bedford, Massachusetts, U.S.A.

1. INTRODUCTION

Introducing microorganisms by a contaminated pharmaceutical product parenterally or through broken skin into the body cavities can result into disease and mortality. Pharmaceutical products such as injections, ophthalmic preparations, irrigation fluids, dialysis solutions, and medical devices implanted in the body must be and remain sterile. Therefore, sterilization is an essential stage in the manufacturing of any product that might be injected, or targeting mucosal surfaces, broken skin, and internal organs.

Sterilization can be defined as a process that removes and kills all microorganisms through a chemical agent or physical process [1,2]. However, when pharmaceutical products are manufactured, there is no absolute certainty that all the units will be sterile. This is because not all units are tested for sterility. To provide that kind of degree of assurance, all units must be shown to be sterile. This cannot be accomplished unless all units are destroyed. Therefore, the sterility of a pharmaceutical lot is described as a probability where the likelihood of a contaminated unit or article is acceptably remote. Such a state of sterility assurance level (SAL) can only be established through the use of adequate validated sterilization cycles and aseptic processing under appropriate good manufacturing process (GMP) practices. Furthermore, environmental monitoring of facilities, personnel, and processes is also a major component during process control of sterile manufacturing and testing. Sterility assurance means that there are no surviving microorganisms present in a product. Sterility, therefore, is not a subjective matter. A product is either sterile or not sterile. The likelihood of a product to be sterile is best illustrated in terms of the probability of microorganisms to survive the treatment process. For a parenteral pharmaceutical product, the standard probability is less than one in 1 million units processed ($<10^{-6}$). For instance, for a product containing 10^3 spores, an inactivation factor of 10^{-9} will be needed to give a sterility assurance level of 10^{-6}. This indicates that there is a probability of less than one in a million of microbial survivors to be present in a given sterile batch. Therefore, the sterilization process will need to produce a lethality level that will kill all microorganisms. Some of the most common procedures recommended to sterilize a product are as follows:

- Filtration
- Steam sterilization
- Dry heat sterilization
- Ionizing irradiation
- Ethylene oxide.

The choice depends on the capacity of the formulation and the package to resist the treatment applied by the sterility procedure selected. For instance, a liquid formulation can be sterilized by using autoclaving or filtration by aseptic processing, whereas medical devices are treated by ionizing irradiation. In some cases, the liquid is a biological product such as interferon or a small peptide that cannot be terminally sterilized. Therefore, filtration by aseptic processing is the only alternative. The efficacy of the treatment will depend upon the time the process was applied to the sample and the initial contamination level. Therefore, the fewer are the microorganisms present in a sample, the lesser is the time to make that sample "sterile." Validation and documentation of the treatment to develop consistent and reproducible sterilization results are two of the major requirements. However, the majority of sterile drugs are manufactured by aseptic processing because terminal sterilization degrades the chemical stability of a given formulation and damages the container/closure [3].

The increasing trend of product recalls due to lack of sterility assurance reflects the misunderstanding by different sectors in industry of the importance of sterilization processes, validation and aseptic processing. Improper validation and documentation of SAL is the number one reason for drug recalls. Furthermore, almost all recalls are from products manufactured using aseptic processing [3]. For instance, recalls due to the manufacturer's lack of support that the product was sterile are classified as class II recalls. Some other tested products have been recalled because they were found to be nonsterile. These are classified as class I recalls.

2. MICROBIAL CONTAMINATION OF STERILE PRODUCTS

Several cases of microbial contamination detected by sterility test have been reported. However, as shown in Table 1, the lack of sterility assurance appears to be the number one reason for product recalls from 1998 to 2002 [4–10]. Over the last 4 years, more than 135 drugs were recalled for lack of sterility assurance. The number of recalls increased from near 10 to 55 in 1999, 50 in 2000, and 52 in 2001 [11]. In other cases, contamination has been documented to be the result of filter penetration by stressed environmental microorganisms during aseptic processing [12]. Microbial infections in humans have been tracked to aseptically manufactured products that were supposed to be sterile [13,14]. Investigations of the contaminated products indicated that the source of contamination was a biofilm located inside the water system pipelines [12]. There are different types of microorganisms found in contaminated products. Gram-negative microorganisms such as *Serratia* species, *Methylobacterium* spp., *Stenotrophomonas maltophilia*, *Burkholderia cepacia*, and *Ralstonia pickettii* might indicate problems in the water system (Table 1). Molds such as *Penicillium* indicate improper sanitization of surfaces and lack of controls for air circulation. Products subjected to recall range from injectable solutions to medical devices (Table 1).

The fact that more cases of microbial contamination have been reported indicates that companies are not adhering to procedures that are fundamental to the validation and calibration of aseptic processing, or that there is some misunderstanding between regulatory agencies and industry on the proper requirements for aseptic processing validation. The absence of cycle validation and absence of cleaning procedures are two major reasons for noncompliance.

In some cases, even though validation studies have been performed, improper documentation has been a major reason for noncompliance with GMP [11]. The absence of standard operating procedures (SOPs) has also contributed to the problems in trying to develop a consistent and reliable approach in sterilization technologies, cycle validation, and aseptic processing. As a result of the increase in product recalls, the Food and Drug Administration (FDA) has developed an upgrade for a technical monograph on aseptic processing of sterile products [15]. This monograph further describes the critical control points during aseptic processing of pharmaceutical products. Furthermore, the document provides guidance in many areas where problems are persistent and redundant. Industry, however, is requesting

TABLE 1 Examples of Sterile Products Recalls from 1998 to 2002

Product	Reason for recall
Albuterol inhalation solution	*Serratia* species contamination
Baclofen injection	*Penicillium* mold *Methylobacterium* *Mycobacterium chelonae*
Methylprednisolone injection	*Penicillium* mold *Methylobacterium* *Mycobacterium chelonae*
Ceftazidime injection	Lack of sterility assurance
Cistracurium injection	Lack of sterility assurance
Mivacurium injection	Lack of sterility assurance
Doxorubicin injection	Lack of sterility assurance
Epirubicin injection	Lack of sterility assurance
Fluconazole injection	Lack of sterility assurance
Homeopathic eye drop	*Stenotrophomonas maltophilia*
Medroxyprogesterone injection	Lack of sterility assurance
Multivitamin injection	Lack of sterility assurance
Various antibiotic solutions	Lack of sterility assurance
Sodium chloride eye wash	Lack of sterility assurance
Succinylcholine injection	Lack of sterility assurance
Zidovudine injection	Lack of sterility assurance
Various injectable products	Lack of sterility assurance
Parenteral product	Mold, *Methylobacterium* *Mycobacterium chelonae*
Various injectable products	Lack of sterility assurance
Fluconazole injection	Lack of sterility assurance
Midazolam injection	Lack of sterility assurance
Technetium Tc99m albumin injection	Lack of sterility assurance
Vercuronium injection	Lack of sterility assurance
Various injectables	Lack of sterility assurance
Ophthalmic gel	Lack of sterility assurance
Inhalation solution	Lack of sterility assurance
Alcohol pads	Lack of sterility assurance
Aprotinin injection	Lack of sterility assurance
Cefuroxime injection	Lack of sterility assurance
Meperidine injection	Lack of sterility assurance
Methylprednisolone injection	Lack of sterility assurance
Polyvinyl alcohol ophthalmic solution	Lack of sterility assurance
Sodium bicarbonate injection	Lack of sterility assurance
Quinupristin/dalfopristin injection	Lack of sterility assurance
Saline ophthalmic solution	*B. cepacia* contamination
Heparin injection	Lack of sterility assurance
Living skin construct	*B. cepacia* contamination

TABLE 1 Continued

Product	Reason for recall
Serum	Bacterial contamination
Medical device	Microbial contamination
Medical device	Mold contamination
Medical device	Lack of sterility assurance
Medical device	Mold contamination
Medical device	Mold contamination
Ceftazidine injection	Lack of sterility assurance
Ceftazidine injection/cefazolin injection	Lack of sterility assurance
Lidocaine HCl/epinephrine injection	Lack of sterility assurance
Lidocaine HCl/epinephrine injection	Microbial contamination
Oxfloxacin otic solution	Lack of sterility assurance
Ticacillin disodium/clavulanate Potassium injection	Lack of sterility assurance
Various injectables	Microbial contamination
Glycyrrhizinic acid injection	Mold contamination
Sodium chloride respiratory therapy	*Ralstonia pickettii*

further guidance in several areas where problems are common and interpretation is vague (e.g., media fills and environmental monitoring of areas described as critical).

To develop a GMP process, validation of the different sterilization parameters must be performed [16]. Because there are different types of sterilization treatments, validation must show that the treatment used for a given pharmaceutical product will destroy any microorganism present in the samples on a regular basis using validated parameters without changing its chemical composition and effectiveness. As previously mentioned, documentation of these processes must be also a priority to comply with GMP. Furthermore, the validation of the system must provide evidence that the system is in control and that all potential routes of contamination are monitored and trended.

3. METHODS OF STERILIZATION

The major objective of a sterilization process is to destroy all microorganisms present in a given sample. Microbial metabolism is based upon the utilization of inorganic and organic compounds to drive cell growth, division, and maintenance [17]. Enzymatic reactions are essential to microbial growth, reproduction, survival, and distribution in the environment. All sterilization processes inactivate or interfere with these enzymatic reactions that support microbial metabolism. When exposing a microbial population to a sterilizing agent, the microbial inactivation follows an exponential death curve [16]. The probability of a population of microorganisms to survive a sterilization process is determined by their number, types, and resistance to the sterilization process. Furthermore, other factors such as moisture content, thermal energy, and time of exposure also affect microorganisms' survival. After the completion of a given sterilization cycle, for a pharmaceutical product, sterility means that the product has been sterilized where individual units have a probability of being nonsterile or have a SAL equal to 1×10^{-6} or more (terminally sterilized injectables). This indicates that there is a probability of one in a million that a microorganism can survive the sterilization process.

3.1. Steam Sterilization

When a sample is placed in an autoclave that employs saturated steam and pressure, that sample is sterilized using the most common method of sterilization. This method is called steam sterilization. The basic principle of operation is that the air in the chamber is displaced by the saturated steam, achieved by employing vents and traps. To displace the air more effectively from the chamber and from within articles, the sterilization cycle may include air and steam evacuation stages. The cycles for different products are based upon the heat penetration, distribution, and resistance of test articles. Temperatures of 121 °C and pressures of 15–21 psi are always used. However, the time required for complete sterilization must be determined during the validation process of different load configurations. These configurations are based upon the different types and numbers of materials treated by any particular sterilization process. During the validation, two parameters are measured. The first one is the mapping of the heat distribution inside the chamber to determine the "cold" spots. This will determine the uniformity and variability of the temperature inside the chamber. The second parameter is the heat penetration with real load configurations. These loads represent the types of material sterilized on a daily basis such as growth media, laboratory instrumentation, glassware, plastic containers, and biological waste. The placement of biological indicators (BIs) inside the autoclave near or inside the loads will allow the determination of the amount of temperature and pressure reaching into the different loads. It is important that the right temperature and pressure reach all materials inside the chamber for complete microbial kill. After incubating the different BIs, the absence of growth indicates the complete sterilization of all articles.

3.2. Dry Heat Sterilization

Dry heat sterilization utilizes a drying oven with heated filtered air. The air is distributed throughout the chamber by convection or radiation, and by employment of a blower system with devices for sensing, monitoring, and controlling physical parameters. Acceptable range for temperature in the empty chamber is + 15°C when the unit is operating at not less than 250°C.

A continuous process is employed for the sterilization and depyrogenation of glassware. Because dry heat is frequently used to eliminate pyrogenic substances from glassware and containers, a challenge with a given concentration of pyrogen must be part of the validation system. Standard methods require the inoculation of 1000 or more Unites States Pharmacopoeia (USP) units of bacterial endotoxin. The bacterial endotoxin test (BET) is used to demonstrate a 3-log cycle reduction [18]. Pyrogenic substances are bacterial components that cause fever and other pathogenic conditions in humans. Therefore, it is important to eliminate any of these substances from materials and equipment.

3.3. Gas Sterilization by Ethylene Oxide

The common agent used in gas sterilization is ethylene oxide. This kind of sterilization process is carried out when a sample cannot withstand the temperatures used in steam and dry heat sterilization procedures. Ethylene oxide is highly flammable, mutagenic, and levels the possibility of toxic residues in treated materials. The process is carried in a pressurized chamber similar to steam sterilization but with modifications unique to gas sterilizers. After sterilization is completed, the chamber must be degassed to enable microbial monitoring. Parameters such as gas diffusion, concentration, moisture content, holding time, and temperature are very important factors during the validation of gas sterilization processes. Moisture and gas concentration are also critical factors. Package design and chamber loading patterns must enhance gas diffusion to optimize gas penetration and microbial death.

3.4. Ionizing Radiation Sterilization

This kind of sterilization process is widely used with medical devices. Furthermore, several drugs have also been treated using this procedure. The advantages of ionizing radiation are as follows:

- Low chemical reactivity
- Low measurable residues
- Fewer variables to control during the sterilization process.

The process is controlled by adsorbed radiation dose. Irradiation increases temperature minimally but can affect plastic and glass materials. The two types of irradiation used are radioisotopic decay (gamma) and electron beam radiation. The dose to yield the sterility assurance level required must be determined during process validation. For gamma irradiation, validation procedures include material compatibility, loading patterns, identification of minimum and maximum doses, and timer setting. An effective sterilization dose tolerated without damaging the article must be selected. Specific product loading patterns must be determined with the minimum and maximum dosage distribution. Absorbed dose is determined by employing inoculated products with *Bacillus pumilus*. Other dosages are based upon the radiation resistance of the natural microbial population contained in the article to be sterilized.

3.5. Filtration

Filtration through microbial retentive materials is frequently used for the sterilization of heat-labile solutions by physical removal of the contained microorganisms [19]. A filter assembly generally consists of a porous membrane sealed or clamped into an impermeable housing. The effectiveness of a filter medium or substrate depends upon the filter's pore size and may depend upon adsorption of bacteria on or in the filter.

Rating the pore size of the filter membranes is based upon using microorganisms of the size represented by ascertaining the capability to retain the microbes. For instance, sterilizing filter membranes are capable of retaining 100% of a culture of 10^7 *Brevundimonas diminuta* ATCC 19146 per square centimeter of membrane surface under a pressure of not less than 30 psi (2 bar). These membranes are rated 0.22 or 0.2 μm, depending on the manufacturer's practice. This rating also applies to reagents and media. However, studies have demonstrated that 0.22-μm filters do not remove all microorganisms under all conditions [20–22]. Environmental bacterial isolates have been able to penetrate these filters more effectively than *B. diminuta*. These studies recommend the use of 0.1-μm filters. However, regulatory agencies and industrial practices are still based upon using 0.22-μm filters.

Filter membranes that are capable of retaining only larger microorganisms are labeled with a nominal rating of 0.45 um. There is no rating for these kinds of filters. However, they are able to retain *B. diminuta* and *Serratia marcescens* ATCC 14756. Test pressures vary from 5 psi, 0.33 bar for *S. marcescens* to 0.5 psi, 0.34 bar for *B. diminuta* to high 50 psi, 3.4 bar. Filter membranes rated 0.1 μm are tested using *Mycoplasma* strains at a pressure of 7 psi, 0.7 bar.

Other important parameters in the validation of a filtration process are as follows:

- Product compatibility
- Sorption
- Preservatives and other additives
- Effluent endotoxin content.

Microbial bioburden (BB) of the solution to be processed by filtration is a very important parameter to evaluate the effectiveness of a filtration process [23]. Determining the numbers of microorganisms in the sample prior and after the filtration step will indicate the efficiency of a given process. Furthermore, pressure, flow rate, and filter characteristics are also important. Membrane filters are based upon materials such as:

- Cellulose acetate
- Cellulose esters
- Cellulose nitrate
- Fluorocarbonate
- Acrylic polymers
- Polycarbonate
- Polyester
- Polyvinyl chloride
- Vinyl
- Nylon
- Polytef
- Metal.

A filter assemble must be tested for integrity prior to use and also after the filtration process is completed to demonstrate the integrity of the system. Some of the tests are as follows:

- Bubble point test
- Diffusive test
- Airflow test
- Pressure hold test
- Forward flow test.

There should be a correlation between these tests and microorganism retention for the process to be validated.

4. VALIDATION OF STERILIZATION PROCESSES

The goal of a sterilization process is the complete destruction of all microorganisms present in a test article. To perform the process in a reproducible, consistent, and reliable way, the sterilization process must be validated. Validation of a sterility process comprises the demonstration of the absence of microbial growth and the different parameters to achieve microbial death [24].

To determine the efficacy of the sterilization process, BIs are used [25]. BIs provide direct evidence that sterilization lethal conditions have been achieved during the treatment. Other process indicators such as temperature, gas concentration, pressure, humidity, etc. can be recorded by instruments and are critical parameters during the validation studies [26].

BIs are used during the validation process to determine the lowest probability to detect a nonsterile unit in a sterile load. BIs are standard preparations of bacterial spores specific to different types of sterilization processes. Table 2 shows the different types of BIs used for validating different sterilization treatments. For instance, if a sample is sterilized using irradiation processes, *B. pumilus* is the BI used, whereas for ethylene oxide treatments, *Bacillus subtilis* var. *niger* is the choice. Different types of BIs are used for wet (steam) sterilization validation studies. A chapter in this book describes the use and validation of BIs.

BIs are used to show a reproducible logarithmic inactivation of microorganisms due to their resistance to some of the sterilization processes. Bacterial spores are most resistant to these processes than vegetative bacteria. Therefore, if spores are inactivated, so are other types of vegetative bacteria. In sterilization science, the *D* value is used to measure the rate of microbial death. The *D* value is the time in minutes required at the specified conditions to reduce the numbers of viable microorganisms by 90%. The *D* values are obtained when the numbers of colony-forming units (CFU) (on a logarithmic scale) is plotted against the exposure sterilization time. A slope of the line will be the *D* value. The *D* value is used to predict the lethal effect of the sterilization process on the microorganism. If the conditions where the *D* values are

TABLE 2 Bacterial Spores Used as Biological Indicators for Different Sterilization Treatments

Wet heat	*Bacillus stearothermophilus*
	Bacillus subtilis
	Bacillus coagulans
	Clostridium sporogenes
Dry heat	*Bacillus subtilis*
	Bacillus subtilis var. *niger*
	Bacillus stearothermophilus
Ethylene oxide	*Bacillus subtilis* var. *niger*
Radiation	*Bacillus pumilus*

changed (e.g., temperature change from 121 °C to 105 °C), then the *D* values will also change. For instance, the *D* value for *Bacillus stearothermophilus* is approximately 2 min at 121 °C whereas at 105 °C, it will be closer to 35 min. When other sterilization processes such as gas sterilization are used, then other factors (e.g., relative humidity and gas concentration) affect the *D* values. For irradiation processes, the *D* value is sensitive to time of exposure and radiation dosage.

The *Z* value is the numbers of degrees of temperature required to produce a 10-fold change in the *D* value. The *Z* value is only important for thermal sterilization processes. The reason is that temperature is the main factor for the sterilization process to be effective. Using the *Z* value, we can predict the lethality of the treatment at different temperatures from which the *D* value was determined. Another indicator in the evaluation of moist and dry heat sterilization processes is the F_0 value, which can be used to estimate process lethality. The F_0 value indicates the integration of the instantaneous lethality over the duration of the sterilization process. More detailed information on *D*, *Z*, and F_0 values and their importance to sterilization processes is discussed elsewhere [16,24].

An example of a sterilization cycle is the overkill method. The overkill method provides a cycle with a minimum of a 12-log reduction of a resistant BI with a known *D* value of not less than 1 min. However, overkill ensures a greater log reduction than that. The assumption is that the natural bioburden in the product has less resistance to the sterilization process than the BI, and that the destruction of large numbers of resistant indicator organisms results in an even greater destruction of the biological bioburden. Cycle times are established by considering the time required to inactivate the indicators to achieve the 12-log reduction. Validation of an overkill cycle is based upon the use of BIs in a load adjacent to items at different locations inside the chamber.

The BB approach is a process commonly used for medical devices sterilization. It provides a probability of survival of less than one in a million for the most resistant microorganisms (BB) expected in the load. It requires information on the number and heat resistance of the BB and ongoing monitoring and control over the BB. BB sterilization requires knowledge of the quantity and resistance of any BB present in or on the items to be sterilized. Initial screening of the BB is performed to identify the most resistant microorganisms. The process involves the suspension and washing of the medical devices in a buffer. The buffer removes the bioburden from the devices. The buffer is then pooled and filtered through a 0.45-μm membrane. The membrane is then placed on growth media plates such as soybean–casein digest agar (SCDA) and Sabouraud dextrose agar (SDA). Incubation times range from 2 to 5 days. Colony-forming units on the plates are recorded and the final CFU per device is averaged. Once enumerated and identified, then these microorganisms are used as the BB. The BB approach is mostly used for medical devices sterilization. Continuous monitoring of the BB of medical devices prior to sterilization provides valuable information to determine the sterilization parameters that will deliver a reproducible and reliable sterilization process.

5. ASEPTIC PROCESSING

When terminal sterilization is not possible because of the heat-labile nature of the product, sterile filtration and aseptic processing are the choices to produce a sterile drug. For instance, a wide variety of products such as biologics (e.g., interferon) and vaccines are produced by aseptic manufacturing.

Manufacturing of pharmaceutical products by using aseptic processing comprises the individual sterilization of the components of a product with the final product assembled in an aseptic manner. This means that the final product is manufactured by a series of aseptic steps. These steps are designed to prevent the introduction of microorganisms into the processes. Because of the presence of these many steps, there are more chances for microbial contamination or/and human error to occur than in terminal sterilization. Because of this, validation studies for aseptic processing are more complex than terminal sterilization validations. Furthermore, process control of aseptic processing must involve constant monitoring of the environment and personnel to minimize the chances of microbial insult [1,27]. Basically, the process must be shown to be in control all the time to provide effective, reliable, reproducible, and continuous aseptic processing. Environmental monitoring programs comprised sampling of facilities, equipment, raw materials, air, water, and personnel. Furthermore, final product testing is also performed prior to release.

Microbial contamination for sterile products manufactured by aseptic processing is mainly caused by human interventions [15,28,29]. For instance, these include careless feeding of rubber closures, removing broken empty containers, and empty containers falling down. As mentioned above, all drug recalls during the last 10 years were produced by aseptic processing. The consistent noncompliance by similar recalls indicates the lack of monitoring and control of the sources of variability in a process. Lack of process control results in safety and efficacy failures. In some cases, contaminated products result in fatal infections and death [13,14].

For example, the bulk of a product is sterilized by filtration. The final containers have been sterilized by heat. The whole process involves different sterilization steps for components and products, which are combined in a highly controlled process within a controlled environment resulting in a sterile product. The areas of critical concern are the immediate microbial environment where these presterilized components are exposed during assembly to produce the final product. The critical areas for aseptic processing of a pharmaceutical product are air environment and equipment free of microorganisms, trained personnel who are adequately equipped and gowned, and validated processes.

To validate and certify an aseptic process, personnel, and facility, the efficiency of the system is ascertained by employing environmental monitoring procedures, and by processing sterile culture medium as a simulated product. The most common media used is soybean–casein digest broth (SCDB). Prior to use, the broth must be best shown to support microbial growth. This is performed by inoculating different types of microorganisms into the media and obtaining positive microbial growth.

After the filling process is completed, the filled containers are incubated at 25°C or 32°C for a minimum of 14 days. If microbial growth is present, identification of microbial contamination is needed to determine the source of the microorganisms. This simulated product media filling process is called media fills [27,30,31].

A successful media fill run campaign demonstrates that the facilities, personnel, process, and environment are capable of manufacturing the product in an aseptic way on the manufacturing line at that point in time.

During manufacturing, the environment and personnel are monitored by an environmental monitoring program. It is common practice to run three successful media fills. Several parameters to be considered are as follows:

- Procedures
- Fill volume
- Incubation time
- Temperature
- Inspection of filled units
- Documentation
- Interpretation of results
- Corrective actions required.

Process simulation runs are usually performed twice a year during different work shifts. In addition, media fills failures are run to determine the response to the possibility of microbial contamination and the corrective actions implemented to overcome that contamination.

New media fills are run when the following parameters are changing:

- New container closure
- New product

- New filling line
- New product facility
- New process
- New personnel.

The combination of videocamera monitoring and media fill test is considered to be one of the best methods for monitoring, evaluating the process, and investigating the results. In official compendia, the acceptance criteria for media fill test are specified to reject defectives of 0.1% with a 95% confidence level in media fills of more than 3000 units [27]. International guidelines, however, require zero positives out of 3000 media filled units at the 95% confidence level [32]. Environmental monitoring and media fills together are capable of detecting all events in aseptic processing that might compromise the microbiological quality of the environment.

6. TEST REQUIREMENTS FOR STERILITY TESTING

How many samples of a given batch do we need to test for sterility testing? The USP indicates the numbers of samples tested according to how many samples are in a lot [33]. However, these numbers are statistically low when compared to the total numbers of samples per lot. Therefore, if a small percentage of product containers in a given lot is contaminated, sterility testing might not detect it. For instance, because out of 3000 units in a given lot only 40 have been sampled to be tested, this imposes a tremendous statistical limitation to the test. Nevertheless, nondestructive alternatives are not currently available to ascertain the microbiological quality of every single unit in a lot. Therefore, the most important factors to demonstrate the sterility of all units in a lot must be documenting that the actual production and sterilization process inactivate the product bioburden and that any process supporting the sterilization process prevents microbial contamination. Sterility test is performed after the product is manufactured as a final quality control test [2,33,34]. As previously discussed, the number of samples selected for sterility testing depends upon the size of the lot. Table 3 shows the recommended numbers of test samples per batch of finished products as per USP. The recommended sample number varies depending upon the type of products. For instance, there are recommendations for medical devices, injections, antibiotic solids, etc. When a batch of injections contains not more than 100 samples, 10% or four samples of that batch must be tested for sterility. However, if there are more than 500 samples, 2% or 20 samples are tested. When products not intended for injection with a batch of not more than 200 samples are produced, only 5% or two samples are tested.

TABLE 3 Minimum Number of Samples Tested in Relation to the Number of Samples in the Batch

Numbers of sample in batch	Number of samples tested
Injections	
Not more than 100	10% or 4 samples, whichever is greater
More than 100 but not more than 500	10 samples
More than 500	2% or 20 samples, whichever is less
Antibiotic solids	
Pharmacy bulk packages (<5 g)	20 containers
Pharmacy bulk packages (>5 g)	6 containers
Bulks and blends	As solid bulk products
Products not intended for injection	
Not more than 200	5% or 2 samples, whichever is greater
More than 200	10 samples
Devices	
Not more than 100	10% or 4 samples, whichever is greater
More than 100 but not more than 500	10 samples
More than 500	2% or 20 samples, whichever is less
Solid bulk products	
Up to 4 containers	Each container
More than 4 but not more than 50	20% or 4 containers, whichever is greater
More than 50	2% or 10 containers, whichever is greater

Source: Reference 33.

7. TEST METHOD VALIDATION (BACTERIOSTASIS AND FUNGISTASIS)

In the USP, European Pharmacopoeia (EP), and Japanese Pharmacopoeia (JP) methods, to verify that the media and conditions used during sterility testing neutralize any antimicrobial activity and recover all microorganisms from the test articles, a small number of microorganisms (e.g., 10–100 CFU) from (Table 4) are inoculated into SCDB, for detecting aerobic microorganisms, and fluid thioglycollate media (FTM), for detecting anaerobic microorganisms [2,33,34] The microorganisms used during testing represent

Table 4 Microbiological Indicators for Use in Growth Promotion, Bacteriostasis, and Fungistasis Tests

USP	EP	JP
Fluid thioglycollate media		
Staphylococcus aureus	*S. aureus*	*C. albicans*
Pseudomonas aeruginosa	*P. aeruginosa*	
Clostridium sporogenes	*C. sporogenes*	*C. sporogenes*
Bacillus subtilis	*B. subtilis*	*B. subtilis*
Micrococcus luteus		*M. luteus*
Bacteroides vulgatus		*B. vulgatus*
Soybean–casein digest media		
	P. aeruginosa	
	S. aureus	
Bacillus subtilis	*B. subtilis*	*B. subtilis*
Aspergillus niger	*A. niger*	*M. luteus*
Candida albicans	*C. albicans*	*C. albicans*

different types of microorganisms commonly found in pharmaceutical environments. There are gram-positive bacteria, gram-negative bacteria, yeast, and molds (Table 4). The media is analyzed to demonstrate the capability to support microbial growth. The media is satisfactory if visual evidence of microbial growth is observed within 5 days.

The sample is then transferred into culture media. The recommended culture media volumes (dilutions) for specific product dosages are shown in Table 5 [33]. For instance, for liquid product samples with a range of 10–50 mL per container, 40-mL aliquots of the samples are used for the direct transfer method and 100 mL for the membrane filtration method. Similar recommendations are specified for solid products [33]. After samples are added to media with microorganisms, incubation times are 3 days for bacteria and 5 days for fungi. The samples in SCDB are incubated at 22.5 ± 2.5°C, whereas FTM is at 32.5 ± 2.5°C [35,36]. Usually direct transfer is the first method used during validation studies.

A control sample without the test article is run simultaneously. The control sample consists of the media inoculated with the specific microorganism. If the microbial growth is visually comparable between experimental and control, then the product does not have antimicrobial activity under the test conditions analyzed. If there is no comparable growth between both samples, the test conditions must be modified. One modification is to further dilute the tested sample in growth media. Further dilution of the sample usually results in acceptable microbial growth.

TABLE 5 Quantities of Samples for Liquid Products

Container content (mL)	Minimum volume sampled from each container	Minimum volume (mL)	
		Direct transfer	Membrane filtration
<10	1 mL, or entire contents if <1 mL	15	100
10 to <50	5 mL	40	100
50 to <100	10 mL	80	100
50 to <100 intended for intravenous administration	1/2 content	200	100
100–500	1/2 contents	NA	100
Over 500	500 mL	NA	100
Antibiotics (liquid)	1 mL	NA	100

As per the different Pharmacopoeias.
NA = not applicable.
Source: Reference 33.

Some other modifications currently used in laboratories are the addition of neutralizers such as polysorbate 20 and 80, sodium thiosulfate, and lecithin. In some cases, modification of the test included the use of other media for detecting aerobic bacteria. Some of the media available are Letheen and D/E broth. Alternative media can always be used as long as validation studies demonstrate good microbial recovery and effective neutralization of any antimicrobial activity.

When large volumes of liquid samples need to be analyzed, membrane filtration is the alternative method [37]. For membrane filtration analysis, the inoculum is added after transferring the sample through a 0.45-μm membrane filter. The filter has been previously washed with sterile diluent or diluting fluid three times [38]. A filter not treated with the test sample and inoculated with microorganisms represents the positive control. Again if the sample and positive control microbial growth are not visually comparable, further rinses and modification can be performed. Other modifications are increasing the number of washes during membrane filtration or adding enzymes such as penicillinases to inactivate antibiotic activity.

The validation must be performed again when there is a new product, or there is a change in the experimental conditions of the test. Documentation of all validation work must ensure proper practices and assay reproducibility.

8. STERILITY TESTING

Once the conditions of the sterility test (e.g., media, dilutions, etc.) have been shown to neutralize any antimicrobial activity against microorganisms, and positive microbial recovery and growth have been documented, the next step is to perform the test. A sterility test is basically a test that determines the complete absence of microorganisms from a pharmaceutical product. This is achieved by incubating parts of the whole product in a nutrient medium (Table 5) [33]. However, failure to detect microorganisms from the sample can be a result of the use of unsuitable media or inappropriate cultural conditions. Nevertheless, this kind of situation will not arise if validation studies are performed.

Sterility testing has been part of the USP since 1936 when the test consisted of a single enrichment broth incubated for 7 days at 37°C [36]. It was not until 1965 that the test was revised by the addition of FTM incubated at 30–35°C for 7 days and Sabouraud dextrose broth (SDB) at 20–25°C for 10 days.

In 1965, a validation requirement was added to the test by the addition of the bacteriostasis and fungistasis methods [36]. Thioglycollate and SDB were introduced to enhance the detection of anaerobes, yeasts, and molds. By 1970, SCDB has replaced SDB with incubation time extended to 14 days. At this point in time, membrane filtration was also added as an alternative to direct transfer (for a more thorough discussion of this topic, see Cundell [36]). Current procedures require incubation of SCDB for 14 days at 20–25°C for detecting aerobic microorganisms. Anaerobic microorganisms are detected by incubating the FTM sample at 30–35°C for 14 days.

Despite the lack of accuracy, sterility testing provides useful information for filter-sterilized, aseptically filled, and terminally sterilized products. Water, reagents, test solutions, equipment, and materials must be presterilized prior to testing to eliminate all possible false positives and cross-contamination incidents. All the operations must be conducted by taking rigid aseptic precautions in a clean room or a class 100 safety cabinet. The fundamental limitation of sterility testing is that the SAL for the test is lower than the sterilization process that is used to monitor. The SAL for sterility testing of liquid samples is on the order of 10^{-3} [33]. This means that for every 1000 samples tested, one false positive will arise. However, the actual SAL for a specific product depends upon the difficulty of the testing procedure. For instance, medical devices require extensive sample manipulation and large media volumes. Therefore, the SAL for testing a medical device might be less than the 10^{-3} level for other pharmaceutical products. Extensive sample manipulations increase the chances of microbial contamination and analyst errors. This can be overcome by a comprehensive training program for all personnel performing testing.

Because sterility test is performed by people and people are a source of microorganisms, proper gowning and aseptic techniques are thoroughly enforced. Complete gowning of personnel comprises the use of hair nets, gloves, boots, face masks, shoe covers, laboratory coats, safety glasses, etc.

However, people are still the major source of microbial contamination during clean room operations [28]. Furthermore, test sample packaging, media containers, and testing supplies can also be major sources of microbial contamination. Proper GMP practices and sanitization of these materials reduce the probability of microbial contamination. For instance, the sample exteriors of the test samples and media containers are disinfected before the materials are transferred to the laminar flow hood for sterility testing. In some cases, companies have double-door autoclaves connected to a sterility test room (class 1000) or an isolator. The hood's surfaces and the entire clean room including cabinets and laboratory benches are also disinfected. Disinfection and cleaning of the testing area must be performed on a weekly basis even if no testing is performed. Environmental monitoring of surfaces, air, and personnel during testing must also be part of the process control procedures for all testing facilities.

As previously discussed, the analyst is gowned to contain the microorganisms on their skin and clothing. As previously stated, proper training of the analysts in aseptic techniques and gowning reduce the probability of analyst error. The major reasons for sterility testing failures are as follows:

- Inadequate sterilization cycles
- Inadequate delivery of the sterilization process to the sample
- Underestimation of product bioburden
- Bioburden spikes
- Analyst error
- Resistant microorganisms to the product
- Compromised packaging.

Published scientific studies have reported that a large numbers of positive results are detected between days 7 and 14 days of incubation [31]. Typically, once samples are incubated, they are monitored every 3 days or on a daily basis to record the absence or presence of microbial growth. Microbial growth is defined as an increase in the turbidity of the media.

9. CLEAN ROOM ENVIRONMENTS

The facility where sterility testing is performed and where aseptic processing is conducted should not introduce microorganisms to the product. Furthermore, it should provide proper aseptic conditions to minimize and eliminate

TABLE 6 Classification of Clean Rooms Commonly Used in Pharmaceutical Facilities in the United States

Classification	Particles ≥0.5 μm	
	(m^3)	(ft^3)
Class 100	3530	100
Class 1000	35,300	1000
Class 10,000	353,000	10,000
Class 100,000	3,530,000	100,000

Source: Reference 27.

any microbial challenge. A clean room can be defined as an area in which the concentration of airborne particles is controlled to meet a specific criteria and where the concentration of microorganisms in the environment is also monitored [27]. As per USP, sterility test clean rooms are classified as class 1000 rooms (Table 6). Testing is performed in laminar flow cabinets classified as class 100. Table 6 shows the different classifications for controlled rooms used in pharmaceutical operations as per USP. The airborne cleanliness is defined by the concentration of airborne particles. This will include viable and nonviable particulates retained on a 0.5-μm high-efficiency particulate air (HEPA) filter. The less nonviable particulates are present in a room, the less are the microorganisms present in a clean room as long as the air flow, temperature, and humidity are the same. This is because microorganisms in air are associated with particles of different sizes. Therefore, they are attached to particles. The less are the particles, the less are the microorganisms present.

A clean room is also defined by the certification of filter integrity, air velocity, air patterns, air changes, and pressure differentials. These parame-

TABLE 7 At Rest EP Classification Requirements for Clean Room Environments Commonly Used in Pharmaceutical Facilities in the European Union

Classification	Particles			
	≥0.5 μm		≥5.0 μm	
	(m^3)	(ft^3)	(m^3)	(ft^3)
Class A	3530	100	0	0
Class B	35,300	1000	0	0
Class C	353,000	10,000	2000	57
Class D	3,530,000	100,000	20,000	570

TABLE 8 In Operation EP Classification Requirements for Clean Room Environments Commonly Used in Pharmaceutical Facilities in the European Union

	Particles			
	≥0.5 μm		≥5.0 μm	
Classification	(m^3)	(ft^3)	(m^3)	(ft^3)
Class A	3530	100	0	0
Class B	350,000	10,000	2000	57
Class C	3,500,000	100,000	20,000	570
Class D	Not defined		Not defined	

ters can affect the microbiological quality of a clean room operation. Testing is performed by individuals with proper training and documentation on aseptic techniques. Test performance records are documented and monitored. The extensive manipulation required for sterility testing may result in a probability of operator error of 10^{-3}. This means that after 1000 samples have been tested, analyst error might be a possibility.

The EP requirements for clean room environments are more dynamic because it divides the operation areas based upon at rest and in operation (Tables 7 and 8). Therefore, EP regulations want to monitor when the systems are in place but not in use. The types of particles monitored are not only those of 0.5 μm as the USP, but also those adding an additional requirement for particles retained on a 5.0-μm filter.

10. DIRECT TRANSFER METHOD

Once the pharmaceutical sample is obtained for sterility testing, different procedures are used to analyze that sample. Direct transfer is when the entire pharmaceutical test sample or an aliquot of it is transferred directly into the container with culture media. Before opening sample containers, suitable disinfectants are used to clean the exterior surfaces of packages. If solid samples such as creams or gels are analyzed, dissolution of the sample by heating or stirring, prior to transfer to the culture media, is performed. When samples are not easily dissolved, then agents such as polysorbate 20 and 80 and other organic solvents are used. However, it is important to document that the agents do not affect the growth of microorganisms. This is documented during the validation studies (bacteriostasis and fungistasis). The different regulatory agencies recommend the sample size to be transferred into

SCDB and FTM. For example, for a test article that contains less than 1 mL in a given container, the whole sample must be tested. Furthermore, this volume must be added to 15 mL of culture medium. When the test article contains between 2 and 10 mL, 1 mL of this is added to 15 mL of culture medium. After inoculation, samples are mixed and incubated for not less than 14 days for SCDB (20–25°C) and FTM (30–35°C). The samples are observed during different time intervals to determine the presence or absence of microbial growth as indicated by turbidity. However, because of the chemical composition of several pharmaceutical products, turbidity is developed after sample addition; in those cases, samples are streaked onto plate media or aliquots are transferred to fresh liquid media for not less than 7 days to confirm microbial growth.

11. MEMBRANE FILTRATION METHOD

An alternative method to direct transfer is membrane filtration. Filterable liquids, alcohols, oils, and solvents can be analyzed using this method. In some cases, biopharmaceuticals are also tested using membrane filtration. A sample of the pharmaceutical product is filtered through a filter. After filtration, the filter is rinsed using different rinsing solutions to remove product residue. As previously discussed, rinsing is performed with three 100-mL portions of fluid. Higher volumes can also be used but validation studies must be performed. The filter is then transferred to media containers. When only one membrane filter is used, it is divided into two parts. One half is added to the SCDB, whereas the other half goes into FTM. The diameter of the filters used during membrane filtration is 20–50 mm, with a pore size of 0.45 μm or smaller. Incubation conditions are similar to the ones used for the direct transfer method.

12. STERITEST METHOD

In 1974, a closed membrane filtration system to perform sterility test was introduced. The Millipore Steritest™ system (Millipore Corporation, Bedford, MA) has reduced the number of positive results by providing a closed system for sterility testing [39]. Sample filtration, media addition, and incubation are self-contained. However, sampling of the articles to be tested is still susceptible to human and environmental contamination.

13. ISOLATOR TECHNOLOGY

Sterility testing and manufacturing can also be performed inside isolators [40]. An isolator is a device that creates a controlled environment in which to conduct sterility test or aseptic manufacturing. Isolator systems have different sizes. They can be the size of a glove box or an entire room. They are sealed or supply air through a microbial retentive filter and are able to be reproducibly sterilized. Isolators do not exchange air with the surrounding environment. They are completely enclosed HEPA-filtered chambers interfacing with a vapor-phase hydrogen peroxide (VHP) sterilizer and/or steam sterilizer. When closed, it uses only sterilized interfaces or a specialized rapid-transfer port (RTP) for material transfer. When open, it allows the transfer of materials through a defined opening that has been validated and designed to preclude the entry of contamination. Isolators are constructed of flexible plastics, rigid plastics, glass, or stainless steel. They protect the test article by limiting direct contact between the analyst and the samples. All transfers are performed in an aseptic fashion while maintaining complete environmental separation. Aseptic manipulations are performed in half suits, which are flexible components of the isolator wall. The suits allow the operator a full range of motion within the isolator, or by gloves and sleeves. Operators are not required to wear special clean room clothing for conducting testing within isolators. The interior of the isolator is treated with sporicidal chemicals that result in the elimination of viable microorganisms. The air system in the isolator is processed microbial retentive filters (HEPA). The isolator meets the particulate air quality requirements of class 100 area but no requirements are needed for air velocity or exchange. Although the system is air leak-proof, it is not impermeable to gas exchange with the surrounding environment. Isolators are attached to sterilizers to enable direct transfer of solutions, sterile media, supplies, etc. RTPs or doors enable isolators to be connected to one another so that supplies can move aseptically. A compressed gasket assembly provides an airtight seal preventing microbial contamination. To switch to a sterility test using isolators, product validation must be performed. Testing will be more time-consuming when compared to the regular sterility test.

14. CONCLUSION

When validation studies are conducted, there are more variables to control in aseptic processing than in terminal sterilization. Process control allows the continuity, reproducibility, and optimization of a sterile procedure and test. Terminal sterilization provides a higher level of sterility assurance and easier validation and documentation process. However, because of their sensitivity to heat and package integrity, a large number of drugs are manufactured by aseptic processing. Aseptic processing provides a reliable process for manufacturing of heat-labile compounds. However, process control of aseptic processing is more rigorous and complicated than terminal sterilization. Process control optimization comprises a continuous and reliable environmental monitoring program along with sterility testing of finished products. Sterility testing is an important component in the process control of sterile manufacturing. On the basis of the sample chemical composition and antimicrobial nature, validation studies qualify the media and conditions for optimal microbial recovery. Process control and optimization of sterile processes and testing rely on the proper validation, training, and documentation of all procedures to comply with GMP.

REFERENCES

1. United States Pharmacopeial Convention. Sterilization and sterility assurance of compendial articles. US Pharmacopoeia. Vol. 25. Rockville, MD: United States Pharmacopeial Convention, 2002:2250–2255.
2. European Pharmacopoeial Convention. Sterility. European Pharmacopoeia. 3rd ed. Strasbourg, France: Council of Europe, 2001:63–67.
3. Parenteral Drug Association. Aseptic processing; how good science and good manufacturing practices can prevent contamination. PDA Lett 2002; 38:10–11.
4. FDC Reports. Quality control reports "The Gold Sheet." 1998; 32(1).
5. FDC Reports. Quality control reports "The Gold Sheet." 1997; 31(1).
6. FDC Reports. Quality control reports "The Gold Sheet." 1999; 33(8).
7. FDC Reports. Quality control reports "The Gold Sheet." 2000; 34(2).
8. FDC Reports. Quality control reports "The Gold Sheet." 2001; 35(3).
9. FDC Reports. Quality control reports "The Gold Sheet." 2002; 36(3).
10. FDC Reports. Quality control reports "The Gold Sheet." 2003; 37(3).
11. Parenteral Drug Association. PDA testifies before FDA pharmaceutical advisory committee. PDA Lett 2002; 38(1):12–15.
12. Anderson RL, Bland LA, Favero MS, McNeil MM, Davis BJ, Mackel DC, Gravelle CR. Factors associated with *Pseudomonas picketti* intrinsic contamination of commercially respiratory therapy solutions marketed as sterile. Appl Environ Microbiol 1985; 50:1343–1348.
13. Roberts LA, Collignon PJ, Cramp VB, Alexander S, McFarlane AE, Graham E, Fuller A, Sinickas V, Hellyar A. An Australia-wide epidemic of *Pseudomonas picketti* bacteraemia due to contaminated "sterile" water for injection. Med J Aust 1990; 152:652–655.
14. McNeil MM, Solomon SL, Anderson RL, Davis BJ, Spengler RF, Reisberg BE, Thornsberry C, Martone WJ. Nosocomial *Pseudomonas picketti* colonization associated with a contaminated respiratory therapy solution in a special care nursery. J Clin Microbiol 1985; 22:903–907.
15. Akers JE, Agalloco JP. Aseptic processing, elephants, blind men, and sterility. PDA J Sci Technol 2002; 56:231–234.
16. Pflug IJ. Microbiology and Engineering of Sterilization Processes. Minneapolis, MN: University of Minnesota, 1995.
17. Hugo WB. Bacteria. In: Hugo WB, Russell AB, eds. Pharmaceutical Microbiology. 6th ed. Oxford, England: Blackwell Science, 1998:3–34.

18. United States Pharmacopeial Convention. Bacterial endotoxin test. US Pharmacopoeia. Vol. 26. Rockville, MD: United States Pharmacopeial Convention, 2003:2023–2026.
19. Levy RV. Sterilizing filtration of liquids. In: Prince R, ed. Microbiology in Pharmaceutical Manufacturing. 1st ed. Baltimore, MD, USA/Surrey, UK: PDA/Davis-Horwood International Publishing Limited, 2001:399–412.
20. Sundaram S, Eisenhuth J, Howard G, Brandwein H. Method for qualifying microbial removal performance of 0.1 micron rated filters: Part I. Characterization of water isolates for potential use as standard challenge organisms to qualify 0.1 micron rated filters. PDA J Pharm Sci Technol 2001; 55:346–372.
21. Sundaram S, Mallick S, Eisenhuth J, Howard G, Brandwein H. Retention of water-borne bacteria by membrane filters: Part II. Scanning electron microscopy (SEM) and fatty acid methyl ester (FAME) characterization of bacterial species recovered downstream of 0.2/0.22 micron rated filters. PDA J Pharm Sci Technol 2001; 55:87–113.
22. Sundaram S, Eisenhuth J, Lewis M, Howard G, Brandwein H. Method for qualifying microbial removal performance of 0.1 micron rated filters: Part III. Bacterial challenge tests on 0.2/0.22 and 0.1 micron rated filter cartridges with *Hydrogenophaga* (formerly *Pseudomonas*) *pseudoflava*. PDA J Pharm Sci Technol 2001; 55:393–416.
23. Jornitz MW, Soelkner PG, Meltzer TH. Sterile-filtration—a review of the past and present technologies. PDA J Sci Technol 2002; 56:192–195.
24. Pflug IJ, Evans KD. Carrying out the biological qualification: the control operation of moist-heat (steam sterilization) processes for producing sterile pharmaceuticals and medical devices. PDA J Sci Technol 2000; 54:117–135.
25. Pflug IJ, Odlaug TE. Biological indicators in the pharmaceutical and the medical device industry. J Parenter Sci Technol 1986; 40:249–255.
26. Cristina de Oliveira D, de Jesus Andreoli Pinto T. Study of sterilizing effectivity of different Ethylene Oxide gaseous mixtures using CFCs and HFCs (Oxyfume 12R and 2002R). PDA J Sci Technol 2002; 56:242–247.
27. United States Pharmacopeial Convention. Microbiological evaluation of clean rooms and other controlled environments. US Pharmacopoeia. Vol. 25. Rockville, MD: United States Pharmacopeial Convention, 2002:2206–2212.
28. Underwood E. Ecology of microorganisms as its affects the pharmaceutical industry. In: Hugo WB, Russell AB, eds. Pharmaceutical Microbiology. 6th ed. Oxford, England: Blackwell Science, 1998:339–354.
29. Hyde W. Origin of bacteria in the clean room and their growth requirements. PDA J Sci Technol 1998; 52:154–164.
30. The Japanese Pharmacopoeia. Media Fill Test. 14th ed. Tokyo, Japan: The Society of Japanese Pharmacopoeia, 2001:212–215.
31. Van Doorne H, Van Kampen BJ, Van der Lee RW, Rummenie L, Van der Veen AJ, De Vries WJ. Industrial manufacture of parenteral products in The Netherlands. A survey of eight years of media fills and sterility testing. PDA J Pharm Sci Technol 1998; 52:159–164.
32. Kawamura K, Abe H. Consideration of media fill test for evaluation and

control of aseptic processes: a statistical approach to quality criteria. PDA J Sci Technol 2002; 56:235–241.

33. United States Pharmacopeial Convention. Sterility tests. US Pharmacopoeia. Vol. 25. Rockville, MD: United States Pharmacopeial Convention, 2002:1878–1883.
34. The Japanese Pharmacopoeia. Sterility test. 13th ed. Tokyo, Japan: The Society of Japanese Pharmacopoeia, 1996:69–71.
35. Besajew C. Importance of incubation time in the test for sterility. Pharm Ind 1992; 54:539–542.
36. Cundell AM. Review of the media selection and incubation conditions for the compendial sterility and microbial limit tests. Pharm Forum 2002; 28:2034–2041.
37. Christianson GG, Koski TA. A comparison of a disposable membrane filtration system with a direct inoculation system for sterility testing of veterinary biologics. J Biol Stand 1983; 11:83–89.
38. Proud DW, Sutton SV. Development of a universal diluting fluid for membrane filtration sterility testing. Appl Environ Microbiol 1992; 58:1035–1038.
39. d'Arbelloff N. Improving integrity of pharmaceutical sterility testing: a new robotic approach. Drug Dev Ind Pharm 1988; 14:2733–2740.
40. United States Pharmacopeial Convention. Sterility testing—validation of isolator systems. US Pharmacopoeia. Vol. 25. Rockville, MD: United States Pharmacopeial Convention, 2002:2247–2249.

5

Environmental Monitoring

Luis Jimenez
Genomic Profiling Systems, Inc., Bedford, Massachusetts, U.S.A.

1. INTRODUCTION

The materials, facilities, and personnel where sterile pharmaceutical products are manufactured are major factors to consider in the final product quality. To prevent microbial contamination, these facilities, materials, and personnel should provide an environment that will minimize the survival, growth, and distribution of microorganisms. Environmental monitoring provides the evidence and documentation necessary to determine the efficiency of different systems to prevent microbial contamination [1]. A process must be capable of controlling the presence, distribution, and survival of microorganisms in clean rooms and other controlled environments. This applies to manufacturing environments and testing laboratories.

Optimization of that process requires the development of an environmental monitoring plan that includes:

- Sample sites
- Site maps
- Sampling procedure
- Sampling frequency
- Sample handling and incubation

- Statistical data trending and establishment of alert/action levels
- Personnel training
- Documentation of the different areas by written procedures

Although regulatory agencies and scientific associations have provided industry with guidelines for environmental control of sterile pharmaceutical products, there is a discrepancy between the different documents (Table 1).

For nonsterile products, the problem is that environmental monitoring of production facilities and testing laboratories is not performed as frequent as in sterile environments [2]. Furthermore, there are no specific guidelines for nonsterile production facilities. Several companies have modified and adapted the aseptic processing guidelines for controlled environments and applied to nonsterile manufacturing [3]. The goal of an environmental monitoring program for nonsterile pharmaceuticals is then to prevent the introduction of significant numbers of microorganisms and objectionable microorganisms into the manufacturing process, raw materials, and finished product. The presence of microorganisms in nonsterile manufacturing is not by itself a problem, but the critical part is to determine if the numbers and types of microorganisms represent a risk to the processes and products. High numbers of microorganisms might compromise the efficiency and safety of a nonsterile product.

To ascertain the status of environmental monitoring in nonsterile production areas, a survey has been completed and published to determine the most common practices regarding areas monitored, frequency, test methods, data evaluation, and corrective actions [2]. The results indicate that practices and program goals are based upon the types of products manufactured and facilities design. In some cases, facilities, materials, and personnel are moni-

TABLE 1 Regulatory Guidelines in the United States and Europe for Environmental Monitoring of Pharmaceutical Environments

21 CFR.211.42—Design and Construction Features
21 CFR.211.46—Ventilation, Air Filtration, Air Heating, and Cooling
21 CFR.211.113—Control of Microbiological Contamination
21 CFR.211.22—Responsibilities of the Quality Control Unit
FDA Guideline on Sterile Drug Products Produced by Aseptic Processing, June 1987
FDA Guide to Inspection of Sterile Drug Substance Manufacturers, July 1994
EU Guide to Good Manufacturing Practice. Annex I on the Manufacture of Sterile Medicinal Products, June 1997
USP Chapter 1116. Microbiological evaluation of clean rooms and other controlled environments

tored regularly or sporadically. Identification of microbial isolates ranges from a simple gram strain to complete identification to genera and species.

Environmental monitoring has always been an important part of aseptic processing of sterile pharmaceuticals [4]. The manufacturing environment must be in control to minimize the possibility of microbial contamination. Systems that prevent microbial survival and distribution must be installed, validated, and maintained. However, in the absence of regulatory and compendial guidelines, nonsterile products are manufactured using good manufacturing practices (GMPs) as the primary regulatory requirement. Therefore microbial quality is most effectively controlled through strict adherence to GMP. Common deficiencies in the area of environmental monitoring are:

- Not monitoring in all aseptic process areas
- Not responding in a timely fashion to out-of-limit results
- Inadequate corrective actions
- Not following written procedures
- Inadequate documentation of follow-up
- Inadequate environmental monitoring program
- Failure to validate cleaning and sanitization procedures
- Lack of an environmental monitoring program
- Failure to trend environmental monitoring data
- Inadequate assessment of root cause for deviation
- Failure to identify common microorganisms
- Inadequate laboratory facilities
- Lack of written procedures
- Lack of an identification program for microbial isolates
- Inadequate documentation of deviation
- Failure to finalize investigation reports for deviations

2. FACILITIES

Clean and controlled rooms are built to facilitate the cleaning, disinfection, and sanitization of materials and surfaces [5]. They are spacious areas where walls, floors, ceilings, and cabinets are smooth, nonporous, and nonshedding. The surfaces are easy to clean and disinfect. These facilities allow the smooth flow of personnel and equipment. Surfaces are resistant to sanitizers and disinfectants such as ultraviolet radiation, alcohol, etc. Corners and edges are curved to prevent buildup of contaminating agents.

Airflow, humidity, and temperature are controlled by humidity ventilation air-conditioning units (HVAC) units. High-efficiency particulate air (HEPA) filter systems are used to remove particulates from the air to comply with the different room classification systems, numbers of air changes per hour, and velocity [6]. Room classification is based upon the activities and work performed within the rooms. Airflow is controlled by pressure differential gradients between rooms with air flowing out of the most controlled room, e.g., Class 100 to 1000 to 10,000, and into a lower class. Calibration studies are conducted to verify the consistent and uninterrupted airflow. Neither equipment nor personnel must disrupt or affect the air patterns. Gowning rooms allow sufficient space for personnel to dress without contaminating their clean room garb. All the systems such as HEPA, laminar flow cabinets, if present, and pressure gauges are calibrated on a 6-month or yearly basis.

3. CLEANING PROCEDURES IN PHARMACEUTICAL ENVIRONMENTS

To reduce the probability of microbial contamination in clean rooms and controlled environments, a cleaning, sanitization, and disinfection program is a critical component in the process control of pharmaceutical manufacturing. The reliability of these written procedures is ascertained by environmental monitoring data analysis.

Sanitizers or disinfectants that are effective against vegetative cells maybe ineffective against spores. Some of the disinfectants utilized by industry are:

- Ecophene II
- Phase
- Lysol
- Pesthole
- Deco phase
- Ethanol
- Sparkling
- Isopropyl alcohol (IPA)

Disinfectants such as sodium hypochlorite and formaldehyde, which are effective against spores, are corrosive to surfaces. Before a specific disinfectant is used, several parameters must be determined, for instance, the concentration to be used, contact time, activity in the presence or absence of organic substrates, surface, toxicity, residual concentration, and delivery systems. Testing is performed on standard microbiological cultures and environmental microorganisms isolated from the facility where the disinfectant is used. A chapter on validation of disinfectants is included in this book.

Validation studies demonstrating the efficacy of the agents, disinfection, and sanitization procedures used in pharmaceutical environments ensure the reproducibility, robustness, and accuracy to support a given cleaning program. In some cases, cleaning validation might allow the use of common pieces of equipment for multiple products. A cleaning and disinfection program must include the dissembling of the equipment, cleaning, drying, assembling procedures, and disinfectants. Sometimes, different detergents are needed for different products and cleaning procedures are product-specific. To determine the cleanliness of a cleaning process, chemical residues, detergent residues, and microbial counts must be ascertained. Acceptance criteria are based upon visual, toxicological, pharmacological, and microbiological analyses. An effective program comprises a maintenance schedule, personnel training, and changes in equipment aging and repair, product, detergents, disinfectants, equipment, and manufacturing process. Disinfection and sanitization testing is performed on both commercially available and environmental microorganisms. Standard microbiological cultures such as:

- *Salmonella choleraesuis* ATCC 10708
- *Staphylococcus aureus* ATCC 6538
- *Pseudomonas aeruginosa* ATCC 15442

are used to test against different types of disinfectants. Testing is performed in the presence of organic materials such as bovine serum albumin (BSA) or fetal calf serum (FCS). Testing is performed according to current regulatory guidelines. Good disinfection and sanitization studies include testing of environmental isolates from the facility in question. This provides a measure of the capacity of "house" microorganisms to resist sanitization and disinfection procedures. Some environmental isolates found in manufacturing environments exhibit a higher resistant to disinfectants and sanitizers [7]. Therefore industrial practices are currently rotating disinfectants on a weekly and monthly basis. However, there is no scientific study published to support that practice.

4. ENVIRONMENTAL MONITORING SYSTEM

What are the critical areas in a manufacturing environment susceptible to microbial contamination?

4.1. Water

Water is the most common raw material in pharmaceutical formulations and processes and a major source of microbial contamination when GMP standards are not followed [8]. Water is also used in different process for the cleaning and rinsing of equipment. During process validation and production, water samples are analyzed to determine the microbiological quality of the facilities water. In general, sample frequency relies on the type of water and the use of it. There are several categories of water in a pharmaceutical environment. These are:

- Potable water
- Purified water
- Water for injection

For instance, monthly sample of potable water for total microbial count and coliforms is usually performed. Purified and water for injection (WFI) lines are sampled daily, weekly, biweekly, or as specified by the product's monograph. For potable, purified, and WFI water, sample volume ranges from 1 to 100 mL. However, microbial densities in WFI and purified water are usually low. Sample concentration by membrane filtration of 100 mL can provide more accurate information on the microbiological quality of the systems. In some cases, even 1-L volumes are filtered. The following procedures are used to monitor the microbiological quality of pharmaceutical waters.

4.1.1. Membrane Filtration

Membrane filtration is used for enumerating total microbial count for potable water, pure water, and water for injection (WFI) lines. Growth media such as R2A and plate count agar (PCA) are used for bacterial enumeration. These media provide a low-nutrient environment for microorganisms to grow. Low-nutrient media exhibit higher recovery of water microorganisms than regular media such as soybean-casein digest agar (SCDA). Sabouraud dextrose agar (SDA) is a selected media for yeast and mold (Table 2).

4.1.2. Pour Plate

In some cases, when sample volume is 1 or 5 mL, pour plating is performed. However, it is not recommended for larger volumes.

4.1.3. Coliform Detection

To determine the presence of enteric bacteria in water systems, coliform counts are performed using m-ENDO or most probable number (MPN) counts using lauryl tryptose and brilliant green lactose bile broth, or the Colilert system.

4.1.4. *Pseudomonas* spp. Detection

To determine the presence of *P. aeruginosa* and other *Pseudomonas* species, membrane filtration or pour plates can be performed using *Pseudomonas* isolation agar (PIA) or Cetrimide agar.

TABLE 2 Microbiological Tests for Water Analysis

Method	Media
A. Heterotrophic microorganisms	
Membrane filtration	R2A
	Plate count agar (PCA)
	Sabouraud dextrose agar (SDA)
	Soybean-casein digest agar (SCDA)
Pour plate	R2A
	Plate count agar (PCA)
	SDA
	SCDA
B. Coliforms	
Most probable number	Lauryl tryptose broth
	Brilliant green lactose bile broth
Membrane filtration	Endo agar
Colilert	Coli broth
C. *Pseudomonas* species	
Membrane filtration	*Pseudomonas* isolation agar (PIA)
	Cetrimide
Pour plate	PIA
	Cetrimide

4.2. Compressed Gases

The use of compressed gases such as helium, carbon dioxide, hydrogen, and nitrogen can become sources of microbial contamination if proper procedures for testing and control are not developed [6]. For instance, in some situations, these gases are expelled into a laminar flow environment for testing. This can be performed monthly or on a quarterly basis. The methods to sample compressed gases are discussed in the next section.

4.3. Air

Air can be a major source of microbial contamination. Air sampling comprises the routine monitoring of:

- Viable airborne particulates
- Nonviable airborne particulates

Viable particulates are major sources of contamination in sterile and nonsterile manufacturing [2,4,9]. However, for sterile products, nonviable particulates are required to be frequently monitored. The major sources of both kinds of particulates are laboratory personnel [10]. To reduce the levels of particulates from pharmaceutical manufacturing rooms, the use of HEPA filters is widely implemented, although studies demonstrating the correlation between the levels of viable and nonviable particulates are contradictory. The general belief among regulatory agencies is that the lower the levels of particulates, the lower the number of microorganisms present in a given controlled environment. Airflow pattern and velocity are measured to demonstrate that the appropriate conditions continue to exist within the controlled environment. Equipment design and placement along with personnel intervention during processes must not be disruptive.

The level of nonviable particulates in the air determines the classification of production areas. For instance, the lower the classification of the room, e.g., class 1000, the lower the levels of particulates allowed. The airborne cleanliness is defined by the concentration of airborne particles. In the United States, this will include viable and nonviable particulates retained on a 0.5-μm filter [1] (Table 3). The United States classifies by class, critical area, or controlled area. However, the European Community (EC) uses the term grade for clean areas such as A, B, C, and D [11] (Tables 4 and 5).

However, in the European Union, regulations also require the monitoring of nonviable particulates larger than 5 μm. Particulate requirements are also based upon whether the clean room is at rest or in operation (Tables 4 and 5). Therefore EC regulations show dynamic and static monitoring requirements not shown in the United States Pharmacopeia (USP). Evidently, the EC determines whether or not the process is in control when the ventilation systems are functional and equipment is present but not used by any personnel.

Microbial monitoring of air is used to determine the microbial bioburden surrounding the manufacturing operations. Air sampling can be

TABLE 3 Classification of Clean Rooms Commonly Used in Pharmaceutical Facilities

	Particles equal to and larger than 0.5 μm	
Classification	(m^3)	(ft^3)
Class 100	3530	100
Class 1000	35,300	1000
Class 10,000	353,000	10,000
Class 100,000	3,530,000	100,000

m = meters; ft = feet.
Source: Ref. 1.

TABLE 4 At Rest EP Classification Requirements for Clean Room Environments Commonly Used in Pharmaceutical Facilities

	Particles equal to and larger than			
	0.5 μm		5.0 μm	
Classification	(m^3)	(ft^3)	(m^3)	(ft^3)
Class A	3530	100	0	0
Class B	35,300	1000	0	0
Class C	353,000	10,000	2000	57
Class D	3,530,000	100,000	20,000	570

m = meters; ft = feet.

performed using different methods. Table 6 describes the methods used for monitoring of air and compressed gases. These methods are:

- Slit-to-agar sampler
- Sieve impactor
- Centrifugal sampler
- Sterilizable microbiological atrium
- Surface air system sampler
- Gelatin filter sampler
- Settling plates

4.4. Surfaces

Other critical areas of environmental monitoring in pharmaceutical facilities are surfaces. Surface monitoring of floors and walls is used to determine the

TABLE 5 In Operation EP Classification Requirements for Clean Room Environments Commonly Used in Pharmaceutical Facilities

	Particles equal to and larger than			
	0.5 μm		5.0 μm	
Classification	(m^3)	(ft^3)	(m^3)	(ft^3)
Class A	3530	100	0	0
Class B	350,000	10,000	2000	57
Class C	3,500,000	100,000	20,000	570
Class D	Not defined		Not defined	

m = meters; ft = feet.

TABLE 6 Microbiological Methods for Sampling Airborne Microorganisms

Method	Action
Slit-to-agar sampler	Impaction
Sieve impactor	Impaction
Centrifugal sampler	Centrifugal
Sterilizable microbiological atrium	Impaction
Surface air system sampler	Centrifugal
Gelatin filter sampler	Centrifugal
Settling plates	Impaction

bioburden of surfaces in controlled environments. Furthermore, equipment and product-contact surfaces are also tested to determine the presence of microorganisms that may impact the quality of the processes, raw materials, and finished products. These are the areas that come in contact with the product or any adjacent areas. The surface area sampled is approximately 25 cm^2. Surface monitoring can provide quantitative and qualitative information. Whether the data will be quantitative or qualitative will depend upon the method used. Microbial recovery depends on the growth media. For instance, media with neutralizers recover higher number of microorganisms from surfaces treated with antimicrobial agents. There are three surface sampling methods:

- Contact plates. Replicate organism detection and counting (RODAC) plates are 6 cm in diameter with an agar layer creating a high convex meniscus. The cap is removed and the agar surface is applied to the test surface. Once sampling is completed, the cap is replaced and the RODAC plates are incubated for 2 days at 30–35°C followed by an additional incubation of 3–5 days at 20–25°C.
- Swabs. Swab sticks are made or purchased with different types of material such as cotton or calcium alginate. Sterile swabs are rubbed against the surface to be analyzed and placed in different types of media. Dilutions are performed and plated on growth media for quantitation of microorganisms. Plates are incubated as described above.
- Surface rinses. Surfaces are washed with buffer or media followed by dilution and plating on different media. Plates are incubated as described above.

Contact plates are used for sampling regular or flat surfaces such as ceilings, walls, floors, and uniforms. However, swabbing is used when irregular surfaces are in contact with the product or adjacent to production areas. They are also useful when pipes or equipment parts are sampled. Following swabbing, the swab is placed into a diluent then vortexed to release all microorganisms into solution. After vortexing, the sample is streaked or plated onto solid media. Membrane filtration can also be performed. Surface rinses are applied to irregular surfaces when swabbing or contact plates are difficult to use.

4.5. Personnel

It has been extensively documented that human personnel shed and spread microorganisms and nonviable particles. They are the primary sources of contamination in controlled environments [8,10]. Gowning of personnel prevents the shedding of human microbial flora into products, surfaces, air, and samples. Furthermore, other sources of particles in clean rooms and controlled environments are pollen, smoke, and dust.

An example of the different gowning requirements to work in the different rooms is shown in Table 7. In class 100,000 rooms, all personnel must wear hair nets and laboratory coats, with the cover of facial hair as an option. However, in class 10,000, additional requirements are the mandatory use of gloves and cover of facial hair.

Therefore training of personnel in aseptic techniques and proper gowning must be a priority. Routine microbiological monitoring of garments and finger impressions must be completed to determine general aseptic techniques. In general, microbiological sampling of the personnel includes contact plate samples of:

- Right chest
- Left chest
- Forehead
- Right sleeve

TABLE 7 Gowning Requirements for Aseptic Processing Areas by Room Classifications

Room	Requirements
100,000	Hair net, shoe covers, lab coat, (optional cover of facial hair)
10,000	Same as 100,000, but gloves and facial hair cover required
1000	Same as 10,000 but with coverall
100	Same as 1000 but with facemask, boots, hood with three seals, neck, wrist, and ankles

- Left sleeve
- Right-hand glove fingers
- Left-hand glove fingers

A certification program must be developed to evaluate the effectiveness of gowning training and support procedures. Documentation of analyst's bio-burden indicates the potential risk of the laboratory personnel to impact product quality and process control. Sanitization of hands before and after every working day reduces the possibility of microbial contamination. Similar practices are performed during the use of gloves in clean room environments. Common practices in industry range from sanitizing gloves every time a new sample is handled to wearing a new set of gloves for every new sample tested. The use of laboratory coats must be restricted to laboratory areas and hall-ways. Wearing laboratory coats in bathrooms, break rooms, or dinning rooms must be prohibited. Hair and body must be cleaned daily. The use of cosmetics and jewelry must be kept to a minimum. Personnel with a contagious disease such as cold, flu, and pink eye must stay away from controlled environments.

5. SELECTION OF SAMPLING SITES

Environmental monitoring for sterile and nonsterile pharmaceutical manu-facturing requires the selection of sampling sites to determine the microbial bioburden of the manufacturing facility and process. Processes and rooms with activities such as blending, compression, filtration, heating, encapsula-tion, shearing, tableting, granulation, coating, and drying must be evaluated. Furthermore, rooms where equipment is cleaned, assembled, and dis-assembled are also critical. The questions to ask are: how do these sites can contribute to the potential microbial contamination of a given product? For how long will the product, raw material, or equipment will be exposed to a noncontrolled area? Sites with direct contact with product and equipment must be sampled frequently. Some companies sample these sites every time they are in use, while others rely on the activity inside the room to determine sampling frequency. There are cases when there is no activity in a room for 1 month and sampling frequency continues on a weekly basis. However, in other cases, sampling is discontinued until activity resumes. A list and map indicating the location of selected environmental monitoring sites ensure the consistency and proper documentation of data analysis. Table 8 shows an example of a list of all environmental systems at a given manufacturing fa-cility that can be sampled to monitor process control. Some common envi-ronmental sites are:

- Compounding rooms
- Filling rooms
- Mixing rooms

TABLE 8 Environmental Monitoring Sampling Sites

Potable water
Purified water
Water for injection (WFI)
Air-compressed
Air
Personnel—chest, gloves, forehead
Equipment-
Product-contact surfaces-
Nonproduct-contact surfaces-

- Component preparations
- Stoppering rooms
- Air ventilation systems
- Water lines

For nonsterile products, sampling should include those areas most likely to cause contamination, such as processing equipment, product-contact surfaces, ventilation systems, process gases, purified water systems, non-product-contact surfaces in processing, and packaging areas.

6. FREQUENCY OF SAMPLING SITES

How often a pharmaceutical site is supposed to be sampled? The frequency of sampling can go from daily to weekly for sterile products, to monthly or quarterly sampling for nonsterile products. However, it is based upon the room classification and activity. For instance, a series of environmental sample in class 100 room is taken at every shift, while class 100,000 rooms are sampled twice a week (Table 9).

TABLE 9 Suggested Frequency of Sampling on the Basis of Criticality of Controlled Environments

Sampling areas	Frequency of sampling
Class 100	Each operational shift
Class 10,000	Each operational shift
Class 100,000	Twice a week
Product/container contact areas	Twice a week
Other support areas to aseptic processing areas but nonproduct contact (Class 100,000 or lower)	Once a week

Source: Ref. 1.

When it comes to nonsterile products, the frequency depends upon the production process, companies' compliance record, formulation chemistry, history of product, and controlled area design. Furthermore, other important factors are the amount of human intervention in the process, environmental monitoring history of the facility, and whether the product is aseptically filled or terminally sterilized.

For instance, daily monitoring of critical areas during aseptic manufacturing is common practice in industry [5]. However, nonsterile monitoring does not include daily monitoring of the environment, process, equipment, and personnel.

Corrective actions when limits are exceeded must be properly documented and defined [11]. Trending will indicate the pattern or status of the program for the optimization of process control and identification of adverse trends. Trends indicate that counts are increasing or decreasing over time, a change in the microbial composition due to failures in the processes protecting the environment against microbial insult. Why changes occurred must be determined to determine if there is a significant impact on the process affecting the quality and integrity of the product.

For both sterile and nonsterile products, the monitoring must be dictated by circumstances and classification of the products manufactured. Priority should be given to products that are susceptible to microbial contamination or that support microbial growth. For instance, nonsterile liquid and topical formulations may require special attention, while solid dosages might have lower priorities. Products most susceptible to microbial contamination might require daily, weekly, monthly, or lot-by-lot environmental sampling. As a minimum, quarterly sampling of the environment to establish a historical database appears to be current industrial practice [6]. Cundell [13] has indicated that the priority, from high to low, must be based upon the route of administration and risk of infection such as:

- Parenteral and ophthalmic solutions
- Inhalation solutions
- Aerosol inhalants
- Nasal sprays
- Vaginal and rectal suppositories
- Topicals
- Oral liquids
- Oral tablets and capsules

In these cases, the risk of infection decreases from products injected into the body and ophthalmic solutions to oral tablets and capsules. A written protocol based upon the accurate assessment for every product will optimize the quality evaluation and decision-making process.

7. ALERT AND ACTION LEVELS

Alert and action limits are established after sampling, analyzing, and trending the values obtained during at least 3 months of intensive microbiological testing of facilities and personnel. Once the trends are determined, then limits are set upon historical, regulatory, and industry guidelines. Alert levels are values that, when exceeded, indicate a potential deviation of the system from normal operating conditions [11]. However, action levels are values that, when exceeded, indicate that the system is not in compliance and an investigation report and corresponding corrective actions must be performed and implemented. Some companies establish an action level after 2 or 3 alert level notifications.

There are cases where different levels are set for rooms within the same facilities. For instance, it might be that a particular operation such as compounding takes place in room A and a filling line down the hallway requires a completely different alert and action levels. Nevertheless, once the values are established, they must be implemented and enforced. If an investigation is needed to investigate any deviations from the established values, proper documentation of the excursion, investigation, and corrective action must be completed within reasonable time. These investigations are usually completed within 1–3 months. Improper or late closure of an investigation is one of the major reasons for noncompliance with GMP regulations. Tables 10 and 11 show the different recommended alert levels for air and surface samples including personnel.

However, action and alert level values do not have to be static. They can also be reviewed to reflect changes in the facility and production processes. However, proper studies are performed to support and document any changes to the limits.

Sterile manufacturing sites are not frequently reviewed as much as nonsterile manufacturing sites. Conditions for sterile manufacturing are more stringent than nonsterile manufacturing. Therefore action and alert limit values do not change as frequent.

TABLE 10 United States Pharmacopeia Microbial Levels for Air Sampling

Classification	Zone	Levels CFU/ft^3	Levels CFU/m^3
Class 100	M1	<0.1	<3
Class 1000–10,000	M2	<0.5	<20
Class 100,000	M3	<2.5	<100

m = meters; ft = feet.
Source: Ref. 1.

TABLE 11 United States Pharmacopeia Microbial Levels for Surface Sampling

Classification	Zone	Surface CFU per contact plate	Personnel CFU per contact plate
Class 100	M1	3	3-gloves 5-masks/gown
Class 1000–10,000	M2	5 (10 floors)	10-gloves 20-masks/gown
Class 100,000	M3	20 (30 floors)	15-gloves 30-masks/gown

Source: Ref. 1.

In nonsterile facilities, the most typical responses to action-level excursion are reporting result (100%), additional cleaning (94%), historical data review (83%), review of cleaning procedures (75%), investigation of environmental control systems (72%), additional training (66%), and additional product testing (64%) [13].

What type of corrective actions can be implemented when an action level has been exceeded? For instance, sampling and testing are almost immediately repeated if conditions indicate that the product quality has been compromised. Sanitization procedures are reviewed and repeated. Retraining of personnel is performed if the investigation report indicates analyst error. Review of controlled environment certifications might indicate system breakdown during manufacturing, sampling, and testing. Basically, all the systems and validation procedures are reviewed to determine the root cause of the action level. If there is an indication that the product manufactured has been compromised, the batches are placed on hold until the investigation is completed and the product is cleared for release.

8. MICROBIOLOGICAL METHODS FOR ENVIRONMENTAL MONITORING

The presence of microorganisms in air can impact the quality of the processes and products manufactured in pharmaceutical environments. Although quantitation of the airborne microbial flora depends upon the sensitivity and accuracy of the methods used, several methods are recommended for air monitoring [14,15]. The most common methods are based upon active procedures such as impaction and centrifugal samplers (Table 6). The slit-to-agar air sampler (STA) is an example of a method based upon impaction. STA uses an agar plate which is revolving under a slit type orifice. Air goes through the orifice directly on a collecting agar. Settling plates are based upon the exposition of open agar plates to collect particles by gravity from the environment that settle on the agar surface [16].

Centrifugal sampler functions on the impaction principle. The air sample is sucked into the sampler by an impeller. The air goes through the impeller drum in a concentric rotating way. Particles in the air are impacted by centrifugal force onto a plastic trip containing an agar media.

Gelatin filter sampler uses a vacuum pump with an extension hose terminating in a filter holder that can be located remotely in the critical space. The system consists of random filters of gelatin capable of retaining airborne microorganisms. After a specified exposure time, the filter is aseptically removed and dissolved in an appropriate diluent and then plated on an appropriate agar medium for microbial content enumeration. The microbial level in the air of a controlled environment is expected to contain not more than 3 CFU per cubic meter [1].

Another important component of the environmental control program in pharmaceutical environments is surface sampling of equipment, facilities, personnel, and personnel gear used in laboratories. To minimize disruptions to critical operations, surface sampling is performed at the conclusion of operations. Surface sampling by contact plates, swabbing, and/or surface rinses is performed on areas that come in contact with the product or adjacent to those contact areas (Table 12). Contact plates (RODAC) filled with different agar media are used when sampling regular or flat surfaces and are incubated at the appropriate time for a given incubation time. Different types

TABLE 12 Microbiological Monitoring for Surface Monitoring

A. Contact Plates (RODAC):
- Soybean-casein digest agar (SCDA)
- SCDA with 1.5% Tween 80
- Letheen agar
- Sabouraud dextrose agar (SDA)
- D/E agar
- R2A agar

B. Swabbing:
- Saline
- D/E broth
- Letheen broth
- SCD broth

C. Rinses:
- Saline
- Phosphate-buffered saline

of media such as SDA can also be used to enhance the detection of mold. The following media are commonly used for contact plates testing:

- Soybean-casein digest agar with or without neutralizers
- R2A
- D/E agar
- SDA

The swabbing method may be used for sampling of irregular surfaces such as equipment and pipes. The swab is then placed in an appropriate diluent and serially diluting the samples to obtain a microbial count. The areas to be swabbed are defined using a sterile template. In general, the diameter range is 24 to 30 cm^2. Microbial counts are reported per contact plate of swab. The type of medium, liquid or solid, that is used for sampling or quantitation of microorganisms in controlled environments depends on the procedure and equipment used. Table 13 shows a list of enrichment media and diluents used for recovering microorganisms from environmental samples in pharmaceutical environment. Some of the most commonly used all-purpose media are:

- Soybean-casein digest agar
- Tryptone glucose extract agar
- Lecithin agar
- Brain heart infusion agar
- D/E neutralizing agar
- Letheen agar

TABLE 13 Enrichment Media and Diluents Used in Environmental Monitoring Studies

Nutrient agar
Lecithin agar
Letheen agar
Dey/Engle (D/E) neutralizing agar
Sabouraud dextrose agar
Brain heart infusion agar
Tryptone glucose extract agar
Soybean-casein digest agar
Peptone water
Buffered saline
D/E broth
Soybean-casein digest broth
R2A agar

The liquid media can be peptone water, buffered saline, brain heart infusion broth, and soybean-casein digest broth. When disinfectants or antibiotics are used in the controlled area, inactivating agents such as polysorbate 20 and 80, sodium thiosulfate, and D/E broth are used. Addition of penicillase to the media neutralizes the antimicrobial activity of penicillin derivative compounds.

The 1997 survey conducted by the PDA ($N = 53$) found that the frequency of use of different monitoring methods in sterile facilities was RODAC plates (98%), STA sampler (60%), centrifugal sampler (55%), settling plates (55%), swabs (49%), sieve-type samplers (9%), and gelatin filters (8%) [5].

For nonsterile, the most common air-sampling methods were centrifugal air samplers (76%), settling plates (52%), and slit-to-agar (33%). Product-contact surfaces are more likely to be monitored using swabs (76%) than RODAC plates (24%), while with nonproduct-contact surfaces, RODAC are used more frequently (77%).

9. CHARACTERIZATION OF ENVIRONMENTAL ISOLATES

When microorganisms are isolated from environmental sites such as equipment, excipients, raw materials, finished products, air, and water, they must be identified to at least the genus level. This is important for sterile products and especially critical for nonsterile products since nonsterile products do contain some minimal microbial bioburden [17–23]. Therefore nonsterile samples containing low microbial numbers and absence of pathogenic microorganisms might be perfectly safe for quality control release and consumer use. For sterile products, microbial characterization indicates the possible sources of contamination. The presence of microorganisms in a sterile product is by itself a reason to reject the product and not to release to the market.

Identification of microbial contamination provides information for the possible sources of contamination. When samples are contaminated with microorganisms such *Staphylococcus epidermidis* and *Staphylococcus hominis*, that indicates the possibility of human contamination during manufacturing or testing, while bacterial species such as *Burkholderia cepacia*, *P. aeruginosa*, and *Pseudomonas* spp. indicate lack of process control in water distribution systems. Other gram-negative bacteria such as *Enterobacter* spp. and *Escherichia coli* indicate fecal contamination by raw materials.

For example, in sterile manufacturing, weekly environmental monitoring of facilities and personnel typically yield almost 99.9% of gram-positive cocci and rods [8]. Characterization of these isolates is not performed by some companies. Identification is then limited to a Gram stain reaction. However, several companies pursuit major characterization of every environmental microorganism isolated from controlled environments. Bacterial and yeast identification is usually performed using biochemical systems such as the API, Biolog, and Vitek [23].

Characterization of the manufacturing facility's microbial flora provides an understanding of the microbial ecology allowing a better understanding of the distribution, activity, and numbers of bacteria, yeast, and mold. Better understanding of the microbial community in manufacturing sites and processes allows the development of proper procedures to control microbial survival, distribution, and proliferation. Optimization of processes relies on the development, validation, and maintenance of critical environmental parameters to minimize microbial populations.

Microbial ecology in pharmaceutical environments is controlled by environmental factors such as temperature, pH, nutrient availability, pressure, and water availability. Microbial flora in clean room environments can be effectively controlled by adjusting different parameters [10,12].

Microorganisms recovered from production environments are stressed due to the fluctuations of parameters during manufacturing processes, lack of nutrients, low water activity, contact with chemicals, and temperature changes. Pharmaceutical manufacturing comprises physical processes such as blending, compression, filtration, heating, encapsulation, shearing, tableting, granulation, coating, and drying. These processes expose microbial cells to extensive environmental stresses.

Microorganisms respond to the lack of nutrients and other environmental fluctuations by undertaking different survival strategies [24]. Microorganisms are not always metabolically active and reproducing. For instance, gram-positive bacteria such as *Bacillus* spp. and *Clostridium* spp. develop dormant structures called spores [8]. On the other hand, gram-negative bacteria such as *E. coli*, *Salmonella typhimurium*, and other gram-negative rods undergo a viable but not culturable stage [24]. Furthermore, bacterial cells that do not grow on plate media but retain their viability going through the viable but culturable stage are still capable of causing severe infections to humans. Several studies have shown that microbial cells in pharmaceutical environments have changed the cell size, enzymatic, and physiological profiles as a response to environmental fluctuations [25–30]. These responses are named stress-induced which allow the microbes to repair the damage caused. Similar responses have been reported by bacteria exposed to drug solutions where significant morphological and size changes are observed [27]. Bacterial cells spiked into different types of injectable products have shown different changes in their metabolism, enzymatic profiles, and structural changes that interfered with their identification using standard biochemical assays [26]. Furthermore, bacteria undergoing starvation survival periods are capable of penetrating 0.2/0.22 μm rated filters which are supposed to retain all bacterial species [25].

Therefore using enzymatic and carbon assimilation profiles, e.g., biochemical identification, along with colony and cell morphology to discriminate and identify microorganisms from environmental samples might, in some cases, yield unknown profiles that will not provide any significant information on the microbial genera and species. Standard identification systems rely on the detection of proteins and enzymes to characterize clinical isolates. Furthermore, these identification systems are based upon the characterization by enzyme production, substrate utilization, and phenotypic analysis. Because of the stress, environmental isolates develop different sets of proteins and enzymes [25–27]. Therefore when analyses are completed, a different profile is obtained. Approximately 20–45% of the environmental isolates in a quality control laboratory are misidentified or are given an unidentified profile result (Jimenez, personal communication). In pharmaceutical environments, information on the genera and species of a microbial contaminant will provide valuable information on the possible sources of the contamination allowing the implementation of effective corrective actions.

Environmental samples, e.g., raw materials, finished products, air, water, equipment swabs, and contact plates, taken from production facilities are not rich in nutrients (oligotrophic), and temperature fluctuates below and above ambient temperature. Low water activity, low-nutrient concentration, and dramatic changes in pH also contribute to microbial stress. It has been also shown that the recovery of microorganisms from environmental samples including clean room environments is enhanced by using low-nutrient media [16]. The recovery of microorganisms from pharmaceutical water samples has been shown to be increased by the use of a low-nutrient media, R2A [28,29]. A recent study has also shown that the majority of bacteria present in a pharmaceutical clean room environment are recovered and counted by using a low-nutrient media and longer incubation times [16]. Oligotrophic bacteria counts in clean rooms have been shown to be up to 2 orders of magnitude higher than the number found on SCDA [16]. The microbial composition of the SCDA plates comprised micrococci, staphylococci, and spore-forming gram-positive rods. However, the low nutrient demonstrated a more diverse microbial flora composed of the microorganisms mentioned above along with gram-negative rods, gram-positive coccobacilli, and gram-positive nonsporeforming rods. Of 25 samples with zero counts on SCDA, 12 exhibited growth. The need for a stress recovery phase is demonstrated by longer incubation times and low-nutrient media [31]. In the case of heat-damaged bacterial spores, recovery and growth are based upon media composition, pH, incubation temperature, and incubation time.

However, genus/species identification can be accurately and reproducibly obtained using new genetic identification methods [28,30,32]. Table 14 shows the identification of environmental isolates from several pharmaceu-

TABLE 14 Microbial Identification of Common Microbial Contaminants in Pharmaceutical Environments Using Lipid Analysis and DNA-Based Tests

Species	Lipid	DNA fingerprinting	DNA sequencing
Ralstonia spp.	*R. pickettii*	*R. pickettii*	*Ralstonia* spp.
Kokuria rosea	Unidentified	Unidentified	*K. rosea*
Bacillus pumilus	*B. pumilus*	*B. pumilus*	*B. pumilus*
Bacillus pumilus	*B. pumilus*	Unidentified	*B. pumilus*
Bacillus pumilus	*B. pumilus*	*B. pumilus*	*B. pumilus*
Bacillus pumilus	Unidentified	*B. pumilus*	*B. pumilus*
Bacillus pumilus	*B. pumilus*	*B. pumilus*	*B. pumilus*
Bacillus pumilus	*B. pumilus*	*B. pumilus*	*B. pumilus*
Bacillus pumilus	*B. pumilus*	*B. pumilus*	*B. pumilus*
Ralstonia pickettii	*R. pickettii*	*R. pickettii*	*R. pickettii*
Staphylococcus hominis	*S. hominis*	*S. epidermidis*	*S. hominis*
Ralstonia pickettii	Unidentified	*R. pickettii*	*R. pickettii*
Corynebacterium spp.	Unidentified	*C. amycolatum*	*Corynebacterium* spp.
Stenotrophomonas maltophilia	*S. maltophilia*	*S. maltophilia*	*S. maltophilia*
Enterobacter cancerogenous	*E. cancerogenous*	*E. cloacae*	*E. cancerogenous*
Aeromonas hydrophila	Unidentified	Unidentified	*A. hydrophila*
Pantoea spp.	*Cedecea lapagei*	Unidentified	*Pantoea* spp.
Moraxella osloensis	*M. osloensis*	Unidentified	*M. osloensis*
Staphylococcus warneri	*S. warneri*	*S. aureus*	*S. warneri*
Stenotrophomonas spp.	*S. maltophilia*	*S. maltophilia*	Stenotrophomonas spp.
Staphylococcus aureus	Unidentified	*S. aureus*	*S. aureus*
Microbacterium sp.	Unidentified	Unidentified	*Microbacterium* sp.
Bacillus circulans	*Cellulomonas turbata*	Unidentified	*B. circulans*
Bacillus megaterium	*B. megaterium*	*B. megaterium*	*B. megaterium*
Bacillus amyloliquefaciens	*B. subtilis*	*B. subtilis*	*B. amyloliquefaciens*
Bacillus sp.	*Bacillus* sp.	Unidentified	*Bacillus* sp.
Staphylococcus epidermidis	Unidentified	*S. epidermidis*	*S. epidermidis*
Burkholderia cepacia	Unidentified	*B. cepacia*	*B. cepacia*
Micrococcus luteus	Unidentified	*M. lylae*	*M. luteus*
Paenibacillus glucanolyticus	*P. polymyxa*	*P. glucanolyticus*	*P. glucanolyticus*

TABLE 14 Continued

Species	Lipid	DNA fingerprinting	DNA sequencing
Stenotrophomonas maltophilia	*S. maltophilia*	*S. maltophilia*	*S. maltophilia*
Burkholderia cepacia	Unidentified	Unidentified	*B. cepacia*
Burkholderia cepacia	*B. gladiolli*	Unidentified	*B. cepacia*
Pseudomonas veronii	Unidentified	*P. fluorescens*	*P. veronii*
Yokenella regensburgel	*S. typhimurium*	*P. putida*	*Y. regensburgel*
Pseudomonas putida	*P. putida*	*P. putida*	*P. putida*
Pseudomonas stutzeri	Unidentified	*P. stutzeri*	*P. stutzeri*
Chryseomonas luteola	Unidentified	*C. luteola*	*C. luteola*
Micrococcus luteus	Unidentified	Unidentified	*M. luteus*
Staphylococcus haemolyticus	*S. aureus*	*S. haemolyticus*	*S. haemolyticus*
Micrococcus luteus	*M. lylae*	*M. lylae*	*M. luteus*
Micrococcus luteus	Unidentified	*M. luteus*	*M. luteus*
Micrococcus lylae	Unidentified	NT	*M. lylae*

Courtesy of Accugenix.

tical facilities using DNA fingerprinting, DNA sequencing, and lipid analysis. DNA-based methods are more accurate and provide a higher degree of characterization allowing the tracking of contamination sources. Accurate identification of environmental isolates using DNA sequencing demonstrates the accuracy and resolution of this technology [28,30]. In some cases, nonculturable species are detected by direct DNA extraction of pharmaceutical water samples [28]. Therefore alternative microbiological methods can complement standard methods to determine the microbiological quality of pharmaceutical products and processes [33].

However, standard identification methods are commonly performed in quality control laboratories. For sterile facilities, the extent that isolates are identified is morphology (6%), Gram stain (15%), genus (11%), and species (83%) [13]. In nonsterile pharmaceutical environments, identification can be useful in determining the source of environmental contamination or in detecting organisms known to be deleterious to a product and therefore periodically compared with product bioburden isolates. Of an industrial survey on common laboratory practices in QC laboratories, 50% identified all isolates to the genus level [2]. The most widely used identification is based upon Gram staining of bacterial isolates and mold characterization by colony morphology and color. Most firms had both alert and action limits in place (75%) with microbial isolates being identified at the action level (85%), alert level (55%), or all isolates identified (45%).

10. PRODUCT TESTING PROGRAM

After all the air, water, personnel, and surfaces are sampled and results were analyzed, testing of finished products and raw materials becomes the last test prior to product release and testing. Therefore final raw material and product testing are important components of an environmental monitoring system.

The methods to perform the microbiological testing of nonsterile and sterile pharmaceutical product are specified by the different regulatory agencies [17–22]. For nonsterile products, microbiological testing comprises the enumeration of the bacteria, yeast, and mold in raw materials and finished products. As per USP, further testing requires the enrichment of samples to determine the absence of *E. coli*, *P. aeruginosa*, *Salmonella* spp., and *S. aureus*. The European Pharmacopeia (EP) requires the additional testing of the bacterial family, Enterobacteriaceae. A chapter in this book discusses in detail the different requirements for nonsterile products. These tests are described as microbial limits since the numbers and presence of microorganisms by themselves do not make a product unsafe. The test requirements are time-consuming and labor-intensive requiring the inoculation and transfer of aliquots from 12 different types of media [17–19]. The definition of the limits is based upon the nature of the product, route of application, intended use, etc [2]. Tables 15 and 16 show the microbial limit testing of raw materials and

TABLE 15 Distribution of Objectionable Microorganisms in Pharmaceutical Raw Materials and Products over a 3-Year Period

Product	Microorganism	Isolation frequency
A	*Enterobacter agglomerans*	1
	Chromobacterium violaceum	1
	Stenotrophomonas maltophilia	1
B	*Chryseomonas luteola*	1
	Sphingomonas paucimobilis	1
	Stenotrophomonas maltophilia	1
C	*Escherichia vulneris*	1
	Enterobacter sakazakii	1
D	*Enterobacter sakazakii*	1
E	*Enterobacter agglomerans*	21
	Enterobacter sakazakii	17
	Klebsiella pneumoniae	1
	Enterobacter cloacae	6
	Serratia rubidae	5
	Serratia phymuthica	1
Raw material	*Chryseomonas luteola*	2
	Enterobacter agglomerans	2

TABLE 16 Distribution of Objectionable Microorganisms in Pharmaceutical Raw Materials and Products over a 3-Year Period

Product	Microorganism	Isolation frequency
A	*Enterobacter sakazakii*	1
B	*Enterobacter cloacae*	2
	Pseudomonas putida	2
	Acinetobacter baumannii	4
	Serratia fonticola	1
	Flavobacterium oryzihabitans	3
	Enterobacter sakazakii	1
	Acinetobacter spp.	1
	Escherichia vulneris	1
	Leclercia adecarboxylata	1
C	*Pseudomonas stutzeri*	4
	Enterobacter agglomerans	4
	Flavobacterium oryzihabitans	2
	Acinetobacter lwolffii	7
D	*Enterobacter sakazakii*	4
	Enterobacter agglomerans	1
E	*Enterobacter agglomerans*	2
F	*Pseudomonas stutzeri*	1
G	*Pseudomonas stutzeri*	1

finished products in two different facilities over a 3-year period. The frequency of isolation from product to product changes based upon the type of processes used for manufacturing and chemical composition of the product. For instance, product E (Table 15) demonstrates a higher level of contamination. This contamination is based upon the nature of the product which is composed of natural ingredients. Therefore to optimize the elimination of the microbial flora, additional manufacturing steps must be implemented. When compared with all products in Table 15, product D exhibits the lowest incidence of microbial contamination. The manufacturing process and the chemical composition of product D provide the conditions necessary to minimize microbial insult.

Sterile pharmaceutical products must not contain any bacteria, yeast, and mold. Therefore the presence of microorganisms disqualifies the use of the product for human applications. Testing is simpler than for nonsterile products since the test comprises the inoculation of the products into two different types of enrichment media. These media are soybean-casein digest broth (SCDB) and fluid thioglycollate broth (FTB) for detecting aerobic and anaerobic microorganisms, respectively. They test the presence or absence of the above microorganisms. Therefore the assays are not quantitative. A positive result in any of these media indicates a serious breakdown in the process control during manufacturing, quality control testing, or both.

11. ENVIRONMENTAL MONITORING OF THE QUALITY CONTROL MICROBIOLOGY LABORATORY

The facilities and personnel where nonsterile pharmaceutical products are tested are also major factors to consider in the final product quality. To prevent microbial contamination, these facilities should provide an environment that will minimize the survival, growth, and proliferation of microorganisms. The microbiology laboratories are usually classified as 10,000 or 100,000 rooms with laminar flow cabinets classified as class 100.

Disinfecting and cleaning laboratory areas in a QC microbiology laboratory are common current good manufacturing practices (cGMP) practices, which are based uponcleaning of hands, laboratory benches, floors, and hoods during the beginning and ending of the working day. To further prevent microbial contamination by the analysts, good aseptic techniques must be performed during sample analysis. Laboratory facilities must be spacious allowing the smooth flow of personnel and equipment.

However, there are no regulatory guidelines for monitoring viable contamination in microbiology laboratories. In the absence of regulatory guidelines, there are no consensus industrial practices. One monitoring scheme that has been reviewed by regulators without comment includes monthly monitoring of the QC microbiology laboratory with the following sampling sites:

- Laboratory benches
- Air vents
- Water testing
- Laminar airflow systems

Air samples are collected by using settling plates and a centrifugal sampler. Surface monitoring is based upon contact plates and swabs. Data trends are performed on a quarterly basis.

Minimal isolate identification by Gram stain, colony morphology, and color will give an indication of the bacterial flora in the QC laboratory. However, if gram-negative bacterial species are present, biochemical identification must be performed.

Mold identification is based upon colony morphology and color. Characterization of the QC microbiology laboratory microbial flora will provide valuable information for the optimization of the environmental monitoring program leading to a better tracking and understanding of potential sources of microbial contamination.

12. ENVIRONMENTAL MONITORING DATA SYSTEMS

The environmental data generated during the environmental monitoring program allow the analysts to ascertain the functionality of all the systems in place to provide aseptic conditions during the pharmaceutical production. Data are analyzed to determine whether the systems are in control. Manual collection of data requires the generation of worksheets describing sample site, date, analyst signature, and sample type, e.g., air, water, contact. Statistical analysis of test results is trended by using a database computer program or laboratory information system (LIMS).

Several commercial computer systems are available. A thorough discussion on the capabilities of computerized systems for supporting data management and analysis in environmental monitoring by Moldenhauer [34] has recently been published. A reliable software system includes:

- Environmental sites to be sampled
- Types of samples, e.g., air, water, surface
- Data collection
- Reporting
- Automated generation of worksheets
- Automated generation of labels
- Automated alert limit notification
- Automated action limit notification
- Automatic generation of deviation notification
- Record tests to be performed
- Record specifications
- Methods
- Monitoring frequencies
- Capability to input microorganism identification
- Automatic objectionable microorganism notification with review of the previous microbial data
- Trending and statistical analysis
- Computer security to prevent data modification
- Computer security to restrict access only to authorized personnel

Software validation requirements must be determined before routine use for product testing and release. It is important that the software chosen complies with 21CFR part 11 regarding issues such as security, audit trail, and restoration of lost data.

13. CONCLUSION

Environmental monitoring programs for sterile and nonsterile pharmaceutical facilities comprise the analysis of personnel, processes, raw materials, and finished products. Critical areas during pharmaceutical manufacturing must always be in control to minimize the distribution, viability, and proliferation of microorganisms. When an environmental monitoring program is in place, environmental monitoring data are evaluated to determine whether or not the series of environmental controls continue to operate as intended. Statistical analysis is used to evaluate an environmental monitoring program. A gradual increase or decrease in microbial counts over time, or a change in microbial flora or counts on several plates of a particular area on a given day, would constitute a trend. Environmental fluctuations are intrinsic of an environmental monitoring system. This is because clean rooms and controlled environments are not supposed to be sterile, and constant intervention by personnel and materials represents continuous challenge to process control and cGMP. Optimization of pharmaceutical manufacturing relies on the integration of different systems and processes to minimize microbial insult resulting in safe and efficacious products.

REFERENCES

1. United States Pharmacopeial Convention. Microbiological evaluation of clean rooms and other controlled environments. U.S. Pharmacopoeia. Rockville, Maryland: United States Pharmacopeial Convention, 2003; 26:2381–2385.
2. Mestrandrea LW. Microbiological monitoring of environmental conditions for non-sterile pharmaceutical manufacturing. Pharm Technol 1997; 21:58–74.
3. Wilson JD. Aseptic process monitoring—A better strategy. J Parenter Sci Technol 1997; 51:111–114.
4. Reich RR, Miller MJ, Paterson H. Developing a viable environmental program for nonsterile pharmaceutical operations. Pharm Technol 2003; 27:92–100.
5. Parenteral Drug Association. Current Industry Practices in the Validation of Aseptic Processing. PDA Technical Report No. 24, 1997.
6. Parenteral Drug Association. Fundamentals of an Environmental Monitoring Program. PDA Technical Report 13, 2001.
7. Zani F, Minutello A, Maggi L, Santi P, Mazza P. Evaluation of preservative effectiveness in pharmaceutical products: The use of a wild strain of *Pseudomonas cepacia*. J Appl Microbiol 1997; 43:208–212.
8. Underwood E. Ecology of microorganisms as its affects the pharmaceutical industry. In: Hugo WB, Russell AB, eds. Pharmaceutical Microbiology. 6th ed. Oxford, England: Blackwell Science, 1998:339–354.
9. Akers J, Agalloco J. Environmental monitoring: myths and misapplications. PDA J Pharm Sci Technol 2001; 55:176–190.

10. Hyde W. Origin of bacteria in the clean room and their growth requirements. PDA J Sci Technol 1998; 52:154–164.
11. EU Guide to Good Manufacturing Practice. Annex I on the Manufacture of Sterile Medicinal Products, June 1997.
12. Wilson JD. Setting alert-action limits for environmental monitoring programs. PDA J Pharm Sci Technol 1997; 51:161–162.
13. Cundell A. . Environmental Monitoring. 8th Annual Pharmaceutical Meeting, Microbiology Seminar, Newark, New Jersey, Apr, 1999.
14. Lhungqvist B, Reinmüller B. Airborne viable particles and total number of airborne particles: comparative studies of active air sampling. PDA J Pharm Sci Technol 2000; 54:112–116.
15. Lhungqvist B, Reinmüller B. The biotest RCS air samplers in unidirectional flow. PDA J Pharm Sci Technol 1994; 48:41–44.
16. Nagarkar P, Ravetkar SD, Watve MG. Oligophilic bacteria as tools to monitor aseptic pharmaceutical production units. Appl Environ Microbiol 2001; 67:1371–1374.
17. United States Pharmacopeial Convention. Microbial limit test. U.S. Pharmacopoeia. Rockville, Maryland: United States Pharmacopeial Convention, 2002; 25:1873–1878.
18. European Pharmacopoeial Convention. Microbiological examination of non-sterile products. European Pharmacopoeia. 3rd ed. Strasbourg, France: Council of Europe, 2001:70–78.
19. The Japanese Pharmacopoeia. Microbial Limit Test. 13th ed. Tokyo, Japan: The Society of Japanese Pharmacopoeia, 1996:49–54.
20. United States Pharmacopeial Convention. Sterility tests. U.S. Pharmacopoeia. Rockville, Maryland: United States Pharmacopeial Convention, 2002; 25:1878–1883.
21. European Pharmacopoeial Convention. Sterility. European Pharmacopoeia. 3rd ed. Strasbourg, France: Council of Europe, 2001:63–67.
22. The Japanese Pharmacopoeia. Sterility Test. 13th ed.Tokyo, Japan: The Society of Japanese Pharmacopoeia, 1996:69–71.
23. Palmieri MJ, Carito SL, Meyer J. Comparison of rapid NFT and API 20E with conventional methods for identification of gram-negative nonfermentative bacilli from pharmaceutical and cosmetics. Appl Environ Microbiol 1988; 54:2838–3241.
24. Roszak DB, Colwell RR. Survival strategies of bacteria in the natural environment. Microbiol Rev 1987; 51:365–379.
25. Sundaram S, Mallick S, Eisenhuth J, Howard G, Brandwein H. Retention of water-borne bacteria by membrane filters. Part II: Scanning electron microscopy (SEM) and fatty acid methyl ester (FAME) characterization of bacterial species recovered downstream of 0.2/0.22 micron rated filters. PDA J Pharm Sci Technol 2001; 55:87–113.
26. Papapetropoulou M, Papageorgakopoulou N. Metabolic and structural changes in *Pseudomonas aeruginosa*, Achromobacter CDC, and *Agrobacterium radiobacter* cells injured in parenteral fluids. PDA J Pharm Sci Technol 1994; 48:299–303.

27. Whyte W, Niven L, Bell ND. Microbial growth in small-volume pharmaceuticals. J Parenter Sci Technol 1989; 43:208–212.
28. Kawai M, Matsutera E, Kanda H, Yamaguchi N, Tani K, Nasu M. 16S ribosomal DNA-based analysis of bacterial diversity in purified water used in pharmaceutical manufacturing processes by PCR and denaturing gradient gel electrophoresis. Appl Environ Microbiol 2002; 68:699–704.
29. Kawai M, Yamaguchi N, Nasu N. Rapid enumeration of physiologically active bacteria in purified water used in the pharmaceutical manufacturing process. J Appl Microbiol 1999; 86:496–504.
30. Venkateswaran K, Hattori N, La Duc MT, Kern R. ATP as a biomarker of viable microorganisms in clean room facilities. J Microbiol Methods 2003; 52:367–377.
31. Reasoner DJ, Geldreich EE. A new medium for the enumeration and subculture of bacteria from potable water. Appl Environ Microbiol 1985; 49:1–7.
32. Jimenez L. Molecular diagnosis of microbial contamination in cosmetic and pharmaceutical products—A review. J AOAC Int 2001; 84:671–675.
33. Jimenez L. Rapid methods for the microbiological surveillance of pharmaceuticals. PDA J Pharm Sci Technol 2001; 55:278–285.
34. Moldenhauer J. Environmental monitoring. In: Prince R, ed. Microbiology in Pharmaceutical Manufacturing. 1st ed. PDA, Baltimore, Maryland, USA. Surrey, United Kingdom: Davis-Horwood International Publishing Limited, 2001:451–483.

6

Biological Indicator Performance Standards and Control

Jeanne Moldenhauer
Vectech Pharmaceutical Consultants, Inc., Farmington Hills, Michigan, U.S.A.

1. INTRODUCTION

Biological Indicators (BIs) are preparations of specific microorganisms that are resistant to a specified sterilization process [1]. To be utilized as a BI, the preparation should be characterized and "calibrated," i.e., the reaction of the organism to the sterilization process should be known and consistent. BIs are used for a variety of purposes, e.g., qualification of sterilizers, qualification of Steam-in-Place systems, monitoring cycle performance, and so forth. In the United States, the regulatory guidance documents for requesting approval to market drug products require that microbiological challenge studies be performed for the Performance Qualification of the sterilization process [2]. Furthermore, there is an expectation that the pharmaceutical manufacturers verify the accuracy of the thermal death time (D-value) and the organism control counts [2]. Due to a lack of sterilization requirements harmonization this expectation is not shared globally [3]. Some typical performance standards for BIs, e.g., verification of organisms suspension counts and survival kill time studies, are relatively easy to perform and do not require specialized equipment. Other tests, e.g., thermal death time (*D*-value) analysis, or z-value analysis, may require a greater level of expertise and specialized equipment. This chapter discusses the BI performance standards and provides useful information on resolving conflicts in verifying whether these standards have been met. It also discusses some of the advances in rapid microbiology that may be used for BI testing.

2. OVERVIEW OF BIOLOGICAL INDICATORS

BIs are available in a variety of configurations including spore suspensions, self-contained test units, including growth media and a visual indicator for whether growth has occurred, spores on a carrier (e.g., paper strip, disk, coupon, thread), or in a test kit (e.g., a unit inoculated with spores within a test package). Each configuration has advantages and disadvantages that must be considered when selecting and qualifying a BI. They may be purchased commercially or prepared at the User's facility.

Different microorganisms for the BI may be used depending upon the sterilization process to be challenged. Some organisms, such as *Bacillus stearothermophilus* or *Geobacillus stearothermophilus* may be used for more than one type of sterilization process, e.g., moist-heat or vaporized hydrogen peroxide. Table 1 provides an overview of some BIs and their associated sterilization processes [1].

TABLE 1 Sterilization Processes and Typical Biological Indicators Used

Sterilization process	Typically used biological indicators
Moist-heat (steam) sterilization	*Bacillus stearothermophilus*, *Geobacillus stearothermophilus*, *Clostridium sporogenes*, *Bacillus subtilis* var. 5230, *Bacillus coagulans* ATCC 51232
Dry heat sterilization	Some processes utilize *Bacillus subtilis* spp.; however, other companies challenge with endotoxin test units. Reduction of one log of endotoxin is comparable to reduction of 10^{100} bacterial spores
Ionizing radiation	*Bacillus pumilus*
Ethylene oxide	*Bacillus subtilis* spp.
Vapor phase hydrogen peroxide (VHP)	*Bacillus stearothermophilus*, *Geobacillus stearothermophilus*, *Bacillus subtilis*, *Clostridium sporogenes*

BI resistance to the sterilization process may also be affected (positively or negatively) by a variety of parameters [4] including:

- Growth media
- Incubation conditions
- Substrate in or on which the BIs are maintained
- Presence of chelating agents
- pH
- Water content
- Size of inoculum
- Phase of spore maturity

In addition, the methods are prone to variability based upon a number of contributing factors [4]. Accordingly, care must be taken to ensure that the procedures used are definitive, operators performing the testing are qualified, and the data recovered with the BIs are meaningful.

3. RESPONSIBILITIES OF BIOLOGICAL INDICATOR MANUFACTURERS

The U.S. Pharmacopeia (USP) defines several responsibilities for the manufacturer of the BI. This includes both commercial manufacturers and companies that prepare their own BIs [1]. Although this is a general information chapter of the USP, the expectations listed are consistent with the expectations of regulatory inspectors.

The manufacturer is responsible for determining the performance characteristics of each BI lot manufactured. Typical characteristics include [1]:

- Population (spore count)
- *D*-value (at a specified sterilization process condition)
- Characterization of the BI (e.g., pure culture of a specific organism. When characterizing BIs, it is important to ensure that the testing system has the appropriate sensitivity to distinguish between closely related strains, e.g., *B. stearothermophilus* vs. *Bacillus coagulans*, which differ by very few biochemical test reactions.)
- Specifying the optimum storage conditions
- Stability of the BI over the shelf-life of the BI (i.e., at specified storage conditions)
- Verification testing for any other label claims made on the package

In the late 1990s, several lots of BIs were the subjects of a FDA recall [5]. The recall was due to failure of the BIs to meet the stated performance characteristics following shipping. It was indicated that shipping validation studies were required as part of the certification/qualification of the BI process. Biological Indicators shipped using the Post Office (ordinary mail) may be subject to extreme temperatures (hot or cold), X-ray, UV radiation, preventative measures established to protect from bio-terrorism, etc. All of these conditions may affect the quality of the BI when received at the User location.

Some other considerations for manufacturers of BIs are whether the facilities used to manufacture the BIs are appropriately validated or qualified to ensure consistency across lots manufactured. It is also important to know how lot numbers are assigned for batches of BIs, as this may be important in assessing the risks that are taken when an end-user reduces testing on incoming shipments of a previously approved lot of BIs.

Manufacturers should also have detailed records that describe the source of the original spore culture, the traceability to the parent spore crop, descriptions of all culturing, subculturing, harvesting of the spore crops performed, media and reagents used on the lot, any observations made regarding the lot, stability testing results for the shelf life of the lot, etc.

4. RESPONSIBILITIES OF END USERS

The end user should have defined, written procedures for the acceptance of BIs. The selected BIs should be appropriate for the sterilization process in which they are to be utilized. Typical performance tests include [1]:

- Verification of morphology/identity of the microorganism
- Verification of the spore population/control counts
- Verification of the resistance to the sterilization process (e.g., survival kill time, *D*-values, *z*-values, etc.) [2]

Methods for performing these tests are described in the appropriate compendia [6]. In addition, data should be available to support the storage time and conditions used at that location. This is very important for the working suspensions and dilutions manufactured and used on-site.

5. RESOLVING ISSUES IN THE VERIFICATION OF PERFORMANCE STANDARDS

Many times it is difficult for the end-user to verify the performance characteristics claimed by the commercial manufacturer of the BI within the tolerances provided by the USP. Sometimes, the difficulty comes from what may seem like very minor and unimportant procedural differences and other times the BI received may not be the same as the BI originally tested.

5.1. Discrepancies in the Enumeration of Population (Control) Counts

Most often, unless the BI was subjected to an adverse condition during the shipping process, the discrepancies are due to procedural issues. It is important to note that some deviations can occur without doing anything that is "wrong." Like many areas of microbiology, there is more than just "good science" at work. There are numerous factors that can affect the variability seen in verifying counts including [4]:

- Use of different temperature for incubation of organisms than specified on the manufacturer's label claim.
- Using "gross" dilutions, e.g., 1:100 or 1:1000 instead of 1:10.
- Inadequate cooling of the plating media, i.e., not to exceed 50 °C.
- Insufficient humidity present in the incubator during incubation.
- Sonication followed by vortexing frequently increases the recovery of spores. Excessive sonication of BIs, e.g., some brands of sonicators indicate that spores are killed at times greater than "*x*" minutes.
- Heat shocks should be performed using boiling water and time should be measured when the appropriate temperature is reached.
- Accuracy of fill volumes in dilution tubes.
- Insufficient media present in the Petri plates, e.g., 15–20 mL works well.
- Failure to qualify the growth media for the recovery of injured organisms.
- Failure to count organisms around the perimeter of the agar plate.
- Overlay agar frequently hinders the recovery of the injured spores on the agar plate.
- Use of old suspension of the BI (typical cut-off periods are 2 weeks or less).

5.2. Discrepancies in Thermal Death Time Analysis (*D*-Value)

There are numerous factors that can contribute to the variations seen when comparing *D*-value analysis testing from one laboratory to the same type of testing being performed in another laboratory [4]. This testing is somewhat more complex than verification of population control counts and requires the use of sophisticated laboratory equipment, e.g., a BI evaluation resistometer (BIER vessel). As such, it is important that both of the laboratories used to perform the testing be qualified to minimize the potential for variability, e.g., consistencies in equipment performance, operator techniques, methodologies, etc.

5.2.1. Validation Master Plan

Ideally, BI testing laboratories should have an established validation plan (or be part of a facility wide plan) that defines the requirements for the qualification of the laboratory, i.e., Validation Master Plan. Typical types of information included in the plan are:

- Laboratory design features, e.g., prevention of contamination of the remainder of the facility from the cultures maintained and tested in the laboratory
- Types of equipment used for BI testing and the associated validation requirements
- Identification of any software used, either as part of the equipment or quality management systems, and the associated validation requirements
- Requirements for qualification of personnel
- Requirements for qualification or validation of methods used
- Change control systems implemented for equipment, software, methods, and procedures
- Preventative maintenance procedures established
- Requirements for requalification

5.2.2. Equipment Validation (Hardware and Software)

When testing and/or cultivating BIs, there are numerous types of equipment that are routinely used, e.g., incubators, refrigerators, BIER vessels, laminar airflow hoods or bio-containment hoods, water baths, centrifuges, microscopes, pipettors, sterilizers, etc. For multistep processes using different pieces of equipment, it is useful to flow chart the actual process and identify all of the testing performed, and the equipment needed to complete each step. This makes it easier to ensure that all of the equipment necessary has been identified. Following identification of the equipment, one should assess whether the equipment needs to be validated/qualified or only calibrated. Most pipettors, for example, are calibrated to ensure that they deliver the desired volume. Other equipment such as incubators, refrigerators, or BIER vessels is subject to formal qualification testing and evaluation. Any equipment utilizing microprocessor controllers is subject to requiring software validation testing also.

Controlled Temperature Storage Units (e.g., Incubators, Refrigerators). Validation of these units should include temperature-mapping studies to show that the desired conditions are maintained, regardless of the loading configuration. In addition, recorder locations should be representative of the temperatures maintained within the unit.

BIER Vessels. Although small in size, a BIER vessel is really a miniaturized sterilizer or retort. Different sterilization processes are available for these units, e.g., steam, Ethylene Oxide (EtO), vaporized Hydrogen Peroxide. As such, it is qualified in a typical fashion for the associated sterilization process.

The Installation Qualification (IQ) verifies that the unit has been received in good condition and was installed in accordance with the manufacturer's and the end user's requirements. It is very important to document manufacturer, model numbers, and part numbers for key components in order to aid in assessing the impact to the validated status of the equipment at a later date. Utilities should be properly connected and any associated safety features verified to be operational. The completed document serves as a technical reference manual for the system as installed at the User's facility.

The Operational Qualification (OQ) verifies that the unit as installed operates as expected. Some of the typical tests included are:

- Verification of adequate supply for all utilities required for operation, e.g., sufficient quality of water and water pressure to supply the boiler of a steam vessel.
- Temperature distribution studies in the empty chamber to ensure that all temperature requirements are met (typically a minimum of three consecutive acceptable studies)
- Verification that the cycle sequences as designed
- Verification that any alarms and/or error messages operate as expected
- Verification that archival and recovery procedures work as designed
- Verification that security features operate as expected
- Verification that operators have been trained to use the vessel
- Verification that Standard Operating Procedures (SOPs) are issued
- Verification that the unit has been incorporated into the change control and preventative maintenance systems.

The Performance Qualification (PQ) testing is performed to ensure that the unit performs as expected. There are standards for many BIER vessels performance, e.g., American National Standard ANSI/AAMI ST45-1992 is issued and applies to the performance standards for steam (moist heat) BIER vessels. This document identifies the design, electrical, steam supply, safety, and performance characteristics necessary for the system. Part of the testing at this time may include various loading configurations, e.g., racks of tubes or stoppers, to ensure that the desired heating conditions are maintained throughout the load during the cycle. Typically, a minimum of three consecutive acceptable studies is required for each cycle/load configuration.

It is also useful to determine lag correction factors, e.g., does the penetration probe temperature lag behind the distribution probe temperature? Is this lag significant? Should it be incorporated or applied to any testing procedure? All of these questions should be answered to establish a final testing policy.

5.2.3. Method Validation

There are several different methods used in a typical BI laboratory, e.g., enumeration of microorganisms, *D*-value determinations, survival/kill time studies, preparation of working suspensions, freezing or harvesting cultures, etc. Testing should be performed to ensure that each method is accurate, reliable, and reproducible. This can become a bit more difficult than traditional chemistry methods as the limits for acceptability as equivalent may be broader.

The USP [1,6] provides guidance on the performance of many of these studies, as well as tolerances for equivalent methods, e.g., values within $\pm 20\%$ are considered acceptable for confirmation of *D*-values. It is important to note that when performing these comparisons, one may need to send samples out to another laboratory for verification of the *D*-value or have multiple analysts perform the testing to be used as controls in the validation. During validation, the parameters being compared must be equivalent, e.g., *D*-value is determined on a paper strip vs. a paper strip, not a suspension. It is also important to compare the equivalent *D*-value methodology, e.g., fraction negative method to fraction negative method, not to survivor curve methodology. Within the method it is also important that the "details" be equivalent, e.g., starting with cultures at 0 °C should be started at 0 °C not at 25 °C. This type of variation in temperature can lead to a significant difference in *D*-value. It is quite useful to have as much data as possible from the vendor of the BI regarding performance testing, so that your studies can mimic their testing exactly.

Survival kill time studies also should be compared like for like, i.e., using the same conditions as the manufacturer of the BI for the testing and evaluation.

Many individuals forget to qualify the enumeration procedures for BIs. Although this may seem like a trivial method not needing to be verified, it is important to ensure that each operator can reliably and consistently enumerate the cultures. In many laboratories, it is common to have certain individuals who enumerate on the low side and others who enumerate on the high side. It is important to understand these biases when investigating potential problems with the BI. Simple features, such as the order of performing dilutions, how long tubes are vortexed, quantity of media in the Petri plate, can significantly affect the results obtained [4].

5.2.4. Media Qualification

Not all media are created equal. Media that may work well with healthy cells may not be the correct choice for recovery of injured microorganisms. One should have a qualification program for acceptance of lots and kinds of media used for BI testing. In addition to the typical media acceptance procedures and tests, e.g., sterility, growth promotion, one should perform specific studies to verify that the media will recover injured organisms. As the sterilization process used is designed to kill the microorganisms, it is important that media used will be able to recover injured or stressed microorganisms. A simple test can be designed for moist-heat BIs. Culture tubes with the same dilution of a culture are placed into a sterilizer with exposure times set at sublethal conditions, e.g., 5 min. Following the sterilization cycle the cultures are enumerated on the test media and a previously approved lot of media. Criteria are then set for how much difference in the counts is considered equivalent.

5.2.5. Media Supplements

It is common for many spore-growing procedures to include media supplements, e.g., calcium, magnesium, manganese. Many of these supplements change over time. It is important to consider the age of these supplements when preparing media. Typically, these supplements may not be as effective in the development of resistance change as they get older, e.g., more than a year old [4]. It is important to keep detailed production records for media and record the age of supplements used.

5.2.6. pH, Osmolality, Osmolarity

All of these factors can affect the recovery of the spores. As such, it is important to try and mimic the BI manufacturer's data, as closely as possible [4].

5.2.7. Spore Organism Type, Strain, Purity

Different organisms have different growth characteristics. Review the procedures and records for lot production of the spore crops to determine if other factors have changed in the process.

5.2.8. Qualification of Personnel

The qualification and training of personnel performing this type of testing is critical to ensure that the testing personnel are able to reproduce counts with a range of variability that is within the compendial limits. If not, how could they reliably verify counts to be within that same range of variability? There are

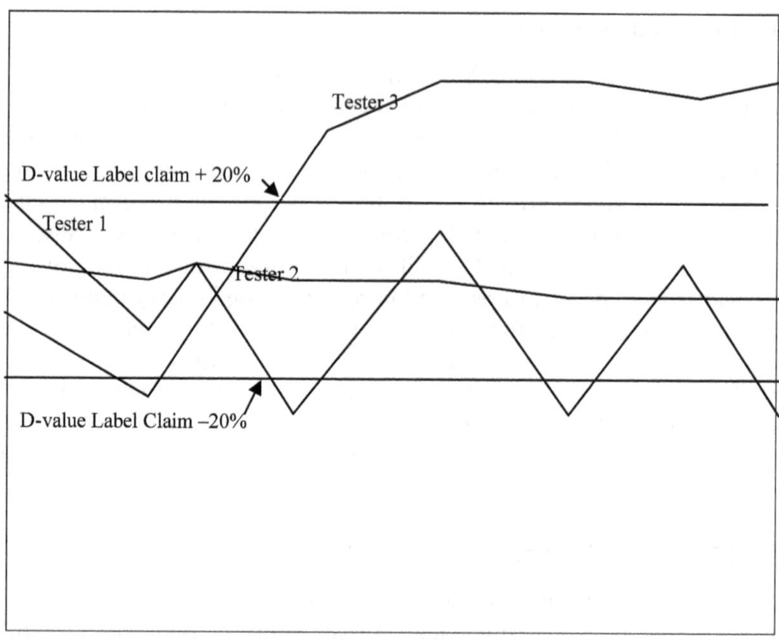

FIGURE 1 Example of operator *D*-value determinations.

numerous methods to qualify new personnel, e.g., sending them to qualify at an outside laboratory with qualified testers, comparing their results for *D*-values to results obtained by another qualified individual, etc. Fig. 1 shows a graph of the *D*-value determinations by three different testers at numerous times, using the same spore crop. Review of this chart shows that tester 1 typically has lower results and is very inconsistent from test to test. Tester 2 has very consistent results for each test. Tester 3 typically tests much higher. This type of information is useful in trying to resolve deviation.

6. IMPACT OF RAPID MICROBIOLOGY ON BIOLOGICAL INDICATOR TESTING

Biological Indicator testing, especially when performed as part of sterilizer validation studies, is frequently a rate limiting factor on returning the sterilizer to production use. Product produced in the sterilizer while awaiting BI results must be placed on hold, pending receipt and acceptability of the results. As storage of product in inventory hold, while awaiting release, can be very expensive many companies are looking at the opportunities provided by rapid microbiology systems to reduce this time period. Additionally, a rapid result may be useful when performing cycle development activities so that the studies may be completed in an expeditious manner.

When selecting a rapid system for BI testing, one must take into account several factors, including:

- Will the system be used for enumeration testing or presence/absence testing?
- Will the system require testing of stressed (e.g., heat injured) organisms?
- Will the system be used for enumeration of control counts prior to testing?
- What BI(s) will be tested?
- Are the expected results total kill of the BI or does one expect levels of spore log reduction (i.e., requiring enumeration of the BI)?
- What time period for obtaining results is deemed to provide an appropriate response?
- Are there any regulatory issues or requirements for testing of BIs?
- Will the use of the rapid system increase the likelihood of counts?
- What is the technology used by the rapid microbiology system?

Different rapid microbiology systems utilize different technologies. Some are based upon the viability on the microorganism and may result in higher counts due to the presence of viable, but not culturable cells. Others automate the reading of results and because machine vision is used, the results are obtained more quickly but are equivalent to traditional methods. Some systems use a totally different technology. It is important to understand what issues are applicable to the system selected for testing and to appropriately validate the system.

6.1. *Bacillus stearothermophilus* or *Geobacillus stearothermophilus*

Several different types of systems may be used with these organisms. For individuals who wish to have a quick read-out of results following the cycle, expect total kill in the cycle, and are willing to use traditional methods for enumeration of BIs prior to use, several inexpensive systems exist that provide results in a few seconds to a few minutes.

An alternative to this type of system is use of systems that utilize enzymes of the BI. Published data indicate that the bacterial enzymes of these organisms can be measured and shown to be deactivated in a known relationship to kill off the BI. Following the cycle, measurements are taken to assess whether the enzyme was appropriately deactivated. This system also works only for tests where total kill is expected.

Systems are available for use with these organisms, when expecting total kill as a result, from Steris, 3M, SGM, and others.

6.2. Systems for Enumeration of Biological Indicator Control Counts, Working with Various Biological Indicator Cultures

Several systems are available for enumeration of BI control counts, i.e., prior to exposure to the sterilization conditions. These systems may work with a variety of BIs and provide some options for personnel using organisms other than *Bacillus stearothermophilus* or *Geobabacillus stearothermophilus*. Papers have been presented at conferences indicating that the RBD system (AATI), the Scan RDI (AES Chemunex), and the Growth Direct™ (Genomic Profiling Systems) can be used for the enumeration of BI control counts.

6.3. Recovery of Biological Indicators Where Total Kill is Not Expected

For companies using cycles that utilize the combined BI bioburden-based approach for their sterilization model, it is not unusual to expect that some BIs may survive the sterilization cycle. As this type of cycle is frequently used with large volume parenterals, is it likely that many of the product formulations may also contain normal saline. Unfortunately, saline can increase the resistance of *Bacillus stearothermophilus* significantly and for this reason many of these companies have chosen to use alternative BIs, e.g., *C. sporogenes* or *B. coagulans*.

Selecting a rapid microbiology system that can enumerate the organisms following the sterilization cycle, and also work with alternative BIs, provides a bit more challenge. Both the Scan RDI (Aes Chemunex) and the Growth Direct™ have been shown to be effective in the recovery of these types of organisms.

Scan RDI utilizes a viability-based technology, i.e., all viable cells are counted. This technology may yield higher counts, because not all viable cells are culturable. The problem occurs when deciding how many additional counts are acceptable. When initially validating this method in my company's laboratory, the counts were almost four logs higher. This could have a devastating effect on the sterility assurance delivered to my product. Subsequent studies were performed at sublethal exposure conditions to assess whether the organisms were being inactivated at a rate comparable to the stated *D*-value of the organism, i.e., approximately 2 min. The enumeration data collected at 2-min intervals showed that the counts were being reduced at the expected rate for the *D*-value. As such, the higher counts persisted to be a problem. Further testing indicated that an additional step was required to eliminate the erroneous counts being identified as viable cells. This alternative was developed in conjunction with the vendor and yielded acceptable, reproducible results for several different BIs. It is critically important to work with the vendor both on the development of the test methodologies and in resolving differences with the various systems.

6.4. Biological Indicators on Paper Strips

The use of paper strips provides some special problems with many of the rapid microbiology systems available today, as paper has the ability to auto-fluoresce and this interferes with the data generated by many rapid microbiology systems.

7. CONCLUSION

Biological Indicators are critical components of sterilization processes, and ensuring that the results are meaningful, manageable, and dependable requires that the steps and practices used be appropriately identified and qualified. End users should have a good working knowledge of the vendor's procedures and practices.

REFERENCES

1. ⟨1035⟩ Biological Indicators for Sterilization. USP25/NF20, United States Pharmacopeia/The National Formulary. Rockville, MD: United States Pharmacopeial Convention, Inc., 2002:2099–2102.
2. FDA Guidance for Industry for the Submission of Documentation for Sterilization Process Validation in Applications for Human and Veterinary Drug Products: Center for Drug Evaluation Research and Center for Veterinary Medicine. Rockville, MD: Government Printing Offices, 1994 (Note: Originally published in the Federal Register Dec. 3, 1993).
3. Moldenhauer JE. Sterilization: steam sterilization and the myth of harmonization. CleanRooms 2000; 14(4).
4. Moldenhauer JE. Contributing factors to variability in biological indicator performance data. PDA J Pharm Sci Technol 1999; 53(4):157–162.
5. Guilfoyle D. Biological Indicators. Steam Sterilization Conference, IIR, Philadelphia, PA, 1999.
6. ⟨55⟩ Biological Indicators—Resistance Performance Tests. USP25/NF20, United States Pharmacopeia/The National Formulary. Rockville, MD: United States Pharmacopeial Convention, Inc., 2002:1871–1873.

7

Rapid Methods for Pharmaceutical Analysis

Luis Jimenez
Genomic Profiling Systems, Inc., Bedford, Massachusetts, U.S.A.

1. INTRODUCTION

The goal of this chapter is to describe the different types of rapid technologies and methods available to pharmaceutical microbiologists working on quality control of pharmaceutical products and develop some awareness among other pharmaceutical scientists. The discussion will be limited to the validation work published in the peer review scientific literature or presented at different scientific conferences.

Although standard microbiological methods are used for routine testing of pharmaceutical products, there are technologies providing rapid quantitative or qualitative information on the microorganisms present in a given pharmaceutical sample, while others are targeting specific microorganisms that might be compromising the product integrity and consumers health. However, validation and implementation of these new methods are not widely adopted by industry. Some of the reasons for the lack of implementation of rapid methods are:

- Absence of validation guidelines
- Uncertain regulatory status

- Lack of validation and technical support
- Lack of understanding of the technologies
- Hesitation from managers and companies to try new methods
- Lack of resources for technology evaluation and implementation
- Systems are expensive
- Underestimation by vendors of regulatory requirements

Fortunately, some of these issues have been corrected. For instance, there is a preliminary informational USP chapter [1] currently in review and a technical report from the Parenteral Drug Association (PDA) [2], addressing the absence of regulatory guidelines to provide guidance on several validation parameters such as:

- Sensitivity
- Accuracy
- Specificity
- Linearity
- Detection limit

TABLE 1 Rapid Microbiological Methods Conferences

Rapid Microbiology User's Group Seminar, Validation Requirements for Rapid Microbiology. Vectec Consultants, April 2003, Baltimore, MD.
Rapid Methods: Strategies for Automation, Detection, and Validation of Microbiology Test Methods for Pharmaceutical, Biotechnology, and Device Applications. Barnett International, February 2003, Philadelphia, PA.
Rapid Screening of Pharmaceutical Samples Using Validated Methods of Technologies: PCR, Immunoassays, and ATP Bioluminescence. Institute of Validation Technology, November 2002, Amsterdam, The Netherlands.
First Rapid Microbiology User's Group Seminar, Validation Requirements for Rapid Microbiology. Vectec Consultants, September 2002, Chicago, IL.
Rapid Methods: Strategies for Automation, Detection, and Validation of Microbiology Test Methods for Pharmaceutical, Biotechnology, and Device Applications. Barnett International, January 2002, Philadelphia, PA.
Rapid Methods and Automation in Microbiology for Pharmaceutical, Biotechnology, and Device Applications. Barnett International, February 2001, San Juan, P.R.
Rapid Methods and Automation in Microbiology for Pharmaceutical, Biotechnology, and Device Applications. Barnett International, November 2000, Brussels, Belgium.
Rapid Methods and Automation in Microbiology for Pharmaceutical, Biotechnology, and Device Applications. Barnett International, April 2000, Washington, DC.

- Robustness
- Ruggedness
- Range
- Precision

Furthermore, vendors are more aware of the need to provide continuous validation and technical support. Because of the interest by industry and regulatory agencies, rapid methods are slowly moving to become part of routine microbiological quality control testing of pharmaceutical products. A reflection of this interest is the number of meetings taking place during the last 5 years (Table 1). These meetings have provided a forum for presentation of innovative technologies, validation work, exchange of ideas, and on how to increase the awareness and successful implementation of rapid methods. As a result, discussion between vendors of rapid methods, users, and regulatory agencies has dramatically improved. Several companies in Europe and the United States are currently pursuing validation work on some of the systems. Furthermore, several publications in peer-reviewed journals and symposium proceedings have demonstrated the applicability and successful validation of some of these technologies [3–5]. As of today, European subsidiaries of major pharmaceutical companies are currently using rapid methods for nonsterile product release and water testing.

2. WHY ARE RAPID METHODS AN ALTERNATIVE TO STANDARD METHODS?

When microorganisms contaminate pharmaceutical products, standard methods are performed to quantify, detect, and identify the numbers and types of microorganisms present in a given pharmaceutical batch [6,7]. Standard methods are based upon the enrichment, incubation, and isolation of microorganisms from pharmaceutical samples. Because of the long incubation times, continuous manipulation, and time-consuming procedures, results are normally obtained within 6–8 days. It has been recently reported that standard methods underestimate the microbial communities present in pharmaceutical environments [8–11]. This has been demonstrated in samples of water, contact plates, and air samples from different pharmaceutical manufacturing facilities and clean room environments. Adenosine triphosphate (ATP) bioluminescence, direct viable counts, deoxyribonucleic acid (DNA), and polymerase chain reaction (PCR) technology have demonstrated that a nonculturable portion of the microbial community in pharmaceutical environments is viable and undetectable by compendial methods. Therefore, these new technologies provide a higher resolution and discrimination between microbial species. Accurate information of the types and numbers of microorganisms in pharmaceutical environments will lead to the implementation of processes that minimize microbial distribution, viability, and proliferation.

Furthermore, identification of several environmental isolates from pharmaceutical environments using standard identification procedures is proven to be incorrect [8,11,12]. When identification is performed by biochemical, lipids, and DNA analyses, DNA analysis provides the best reproducibility, sensitivity, accuracy, and resolution. To develop the proper corrective action when out-of-specification (OOS) results are obtained, accurate microbial identification is needed if the contamination source has to be determined and tracked. A corrective action is not effective if the wrong information is used.

On the basis of these studies, it is evident that in some cases standard methods are not accurate and precise to optimize process control leading to faster releasing time, sample analysis, and high-throughput screening of samples. Although standard methods are valuable and do provide information on the numbers, microbial genera, and species, they were developed for the identification of microorganisms from clinical samples [13]. Most clinical samples originate from human fluids or tissues, which are rich in nutrients and exhibit temperatures of 35–37°C. Environmental samples, e.g., raw materials, finished products, air, water, equipment swabs, and contact plates, taken from production facilities are not rich in nutrients (oligotrophic) and temperature fluctuates below and above ambient temperature. Low water activity and dramatic changes in pH also contribute to microbial stress. Furthermore, manufacturing of pharmaceutical products comprises physical processes such as blending, compression, filtration, heating, encapsulation, shearing, tableting, granulation, coating, and drying [14]. These processes expose microbial cells to extensive environmental stresses.

Microorganisms survive under those conditions by adapting to the lack of nutrients and other environmental fluctuations by undertaking different survival strategies [15]. Microorganisms are not always metabolically active and reproducing. For instance, gram-positive bacteria such as *Bacillus* spp. and *Clostridium* spp. develop dormant structures called spores. On the other hand, gram-negative bacteria such as *Escherichia coli*, *Salmonella typhimurium*, and other gram-negative rods undergo a viable, but not culturable, stage. Furthermore, bacterial cells that do not grow on plate media but retain their viability going through the viable but culturable stage are still capable of causing severe infections to humans. Several studies have shown that microbial cells in pharmaceutical environments have changed the cell size as well as the enzymatic and physiological profiles as a response to environmental fluctuations [16–18]. These responses are considered stress-induced, which allow the microbes to repair the damage caused. Similar responses have been reported by bacteria exposed to drug solutions where significant morphological and size changes are observed. Bacterial cells spiked into different types of injectables products have shown different changes in their metabolism, enzymatic profiles, and structural changes that interfered with their identification using standard biochemical assays. Furthermore, bacteria undergoing starvation survival periods are capable of penetrating 0.2/0.22 μm rated filters which are supposed to retain all bacterial species.

Therefore, using enzymatic and carbon assimilation profiles, e.g., biochemical identification, to discriminate and identify microorganisms from environmental samples might in some cases yield unknown profiles that will not provide any significant information on the microbial genera and species. In pharmaceutical environments, information on the genera and species of a microbial contaminant will provide valuable information on the possible sources of the contamination allowing the implementation of effective corrective actions.

It has been also shown that the recovery of microorganisms from environmental samples including clean room environments is enhanced by using low nutrient media [19]. The recovery of microorganisms from pharmaceutical water samples has been shown to be increased by the use of a low nutrient media, R2A [8,10]. A recent study has also shown that the majority of bacteria present in a pharmaceutical clean room environment are recovered and counted by using a low nutrient media [9]. Similar results are observed for other environmental samples when low-nutrient medium is used [20]. The need for a stress recovery phase is demonstrated by longer incubation times and low nutrient media. In the case of heat-damaged bacterial spores, recovery and growth is based upon media composition, pH, incubation temperature, and incubation time.

Although the development and application of current good manufacturing practices (cGMP) has improved process control in pharmaceutical environments, microbial contamination is still one of the major causes for product recalls worldwide. Some of the reasons for the lack of compliance with cGMP guidelines are:

- Poor sanitization practices
- Lack of personnel
- Lack of training
- Lack of resources
- Inadequate facilities for quality control testing
- Absence of process validation
- Absence of process documentation
- Lack of understanding of basic microbiological principles

When products are contaminated, microbial growth will have a negative impact on product integrity creating a serious health threat to consumers. Therefore, there is a need to develop and implement rapid microbiological methods. Rapid methods have proven to be sensitive, accurate, robust, and provide faster results that might indicate problems in processes and systems used in pharmaceutical environments. Earlier detection of microbial contamination allows rapid implementation of corrective actions resulting in the minimization of manufacturing losses and optimization of risk assessment. Current good manufacturing practices (cGMPs) are a dynamic and ongoing process based on applying the latest technological advances to the manufacturing of pharmaceutical products to provide effective and safe products. Quality control analysis is one of the most important aspects of pharmaceutical process control. Therefore, reducing testing time, increasing throughput, with faster product release optimize process control.

3. ATP BIOLUMINESCENCE

Adenosine triphosphate (ATP) is the most important high-energy phosphate compound present in a microbial cell [21]. ATP carries an important function in the microbial cell by providing the energy source to drive microbial viability and growth. ATP bioluminescence technology is based upon the reaction of the enzyme complex luciferase–luciferin, in the presence of oxygen and magnesium, with ATP released from microbial cells resulting in the production of light (Table 2). The light emitted is proportional to the amount of ATP released. Light emission is measured using a luminometer. Several studies have demonstrated the applicability of ATP bioluminescence to pharmaceutical quality control. The first reported studies have relied on laborious sample preparation for ATP extraction from microbial cells and manual addition of reagents. Once the ATP is extracted and reacted with the enzyme, the samples are added to a luminometer to detect the production of light. These studies were used as an alternative to the visual endpoint used in standard sterility testing by determining the total microbial biomass present in samples in a shorter time period. For instance, standard sterility testing relies on the addition of product samples to different types of enrichment media. Because of the chemical composition of some pharmaceutical products, the addition of the product to the media results in a turbid broth that does not indicate the presence of microbial growth. However, after incuba-

TABLE 2 ATP Bioluminescence Reaction

Firefly Luciferase Magnesium ↓ ATP + D-Luciferin + Oxygen→Light AMP + PPi + Oxyluciferin + CO_2

tion, ATP bioluminescence has indicated that although the broth was turbid there was no microbial growth.

During the 1990s, technological improvements in instrumentation have provided for the complete automation and processing of multiple samples, cell lysis, and reagent addition—allowing minimization of sample handling and time-consuming extraction procedures. Some instruments have developed quantitative information but others only indicated the presence or absence of microbial cells in samples after an incubation step [22,23].

Another case where ATP bioluminescence assays have been used for rapid monitoring of quality is pharmaceutical water systems [23]. Of all raw materials present in a pharmaceutical formulation, water is extremely susceptible to microbial contamination. Therefore, the microbiological analysis of water is a critical parameter in pharmaceutical quality control. Standard methods for water testing comprise membrane filtration and incubation times ranging from 48 hr, with plate count agar (PCA), to 72 hr, with R2A media.

After a 4-month performance evaluation, a quantitative ATP bioluminescence assay has been shown to provide a 24-hr total count of bacteria present in water samples taken from a reverse osmosis/ultra filtration water system, hot water circulating system, and cold tap water [24]. The overall correlation between the assay and standard methods is greater than 82%. After membrane filtration by the analyst, the system simultaneously lyses the microbial cells on the filters, adds the reagents, and quantitatively determines the number of cells in a given sample. Water samples with microbial numbers from 1 to 75 colony forming units (CFU)/100 mL are accurately quantitated. However, accurate quantitation is not possible with water samples containing >75 CFU/100 mL. The linearity between the bioluminescence assay and standard methods is demonstrated when the system is challenged with water samples artificially contaminated with *Pseudomonas aeruginosa* ATCC 9027.

A different quantitative ATP bioluminescence system has been shown to be effective for monitoring purified and water for injection in a pharmaceutical plant. After a 1-month evaluation, comparable counts are obtained with the system and standard methods [23]. Microbial counts are obtained within 24 hr. The system combines a specialized membrane filtration assay with ATP bioluminescence and enhanced image analysis for quantitation purposes. The linearity, accuracy, and reproducibility of the system are demonstrated by analyzing water samples artificially contaminated with *Burkholderia cepacia*. Similar responses are demonstrated with water samples artificially contaminated with *P. aeruginosa* and *Bacillus subtilis* using a second ATP bioluminescence quantitative system [25]. Replica plates of microbial colonies enumerated with the ATP bioluminescence system are identified and compared to the microorganisms found by using standard methods. Bacterial species such as *Ralstonia pickettii*, *Bacillus sphaericus*, *Steno-*

trophomonas maltophilia, and Staphylococcus species have been isolated using both methods.

The studies discussed above have demonstrated the accuracy, reproducibility, and linearity of two different ATP bioluminescence based systems for the monitoring of pharmaceutical water systems. However, for finished product and raw material testing a different ATP bioluminescence assay has been validated. This system does not provide quantitative information on the numbers of microorganisms because it requires enrichment of the samples for assay detection. Therefore, this system provides qualitative information on the presence or absence of microorganisms in samples [22].

A wide variety of pharmaceutical formulations have been validated by using the qualitative ATP bioluminescence assay. Different types of pharmaceutical drug delivery systems such as capsules, tablets, liquid, solids, and emulsions were found to be compatible with the system. To validate the assay, two steps must be performed prior to spiking the samples with different

TABLE 3 Sample Effects of Pharmaceutical Products Using 1% and 10% Sample Suspensions

Product	Enrichment broth	Response to ATP must be within 25–200%	[ATP] picomolar
1 g cream	Letheen	100	10
10 g	Letheen	97	10
1g emulsion	Letheen	107	9
10 g	Letheen	94	13
1 g tablet	Letheen	86	11
10 g	Letheen	52	10
1g ointment	Letheen	97	10
10 g	Letheen	73	11
1 g powder bulk	R broth	97	47
10 g	Letheen	192	6
1 g cellulose	Letheen	109	10
10 g	Letheen	102	11
1 g tablet	Letheen	108	19
10 g			
1 g tablet	Letheen	95	17
10 g	Letheen/Lec	81	133
1 g tablet	Letheen	107	9
10 g	Letheen	85	7
1 g liquid	Letheen	104	9
10 g	Letheen	107	8

concentrations of microorganisms. As enrichment of sample suspensions is needed, the enrichment broth must be free of indigenous ATP to avoid a false positive reaction. Second, the sample suspension in broth should not enhance nor inhibit the bioluminescence reaction to indicate that the light emitted after assay completion is not an artifact but a real signal recorded as positive or negative.

An example of these analyses is shown in Table 3. When 1% and 10% product suspensions in enrichment media are analyzed, no indigenous ATP concentration is found neither the reaction is enhanced or inhibited. Table 3 shows pharmaceutical samples containing 1% and 10% product suspensions exhibiting similar responses when spiked with ATP. Evidently, increasing the product suspension from 1% to 10% does not inhibit the reaction neither adds additional ATP. The product response to ATP ranges from 25% to 200%, which are within the specifications recommended [22].

As soon as it has been demonstrated that the enrichment broth does not contain significant ATP and the sample suspensions neither inhibits nor enhances the reaction, the next step is to spike different levels of microorganisms to demonstrate the sensitivity and accuracy of the assay. The sample sus-

TABLE 4 Detection Times (hr) of Microbial Contamination by ATP Bioluminescence

	Product A		Product B	
Enrichment media	R broth	MR broth	TAT broth	R broth
P. aeruginosa	24	24	24	24
S. aureus	48	24	24	24
E.coli	24	24	24	24
S. typhimurium	24	24	24	24
C. albicans	24	24	24	24
A. niger	24	27	48	27
	Product C		**Product D**	
	R broth	Letheen/lecithin	MR broth	Letheen/lecithin
P. aeruginosa	72	24	48	24
S. aureus	96	24	48	24
E. coli	48	24	48	24
S. typhimurium	72	24	48	24
C. albicans	72	24	48	24
A. niger	72	48	72	48

TABLE 5 Detection of Different Levels of Spiked Microorganisms in a Pharmaceutical Product by ATP Bioluminescence and Standard Methods

P. aeruginosa ATCC 9027			Detection time = 24 hr	
Mean RLU of R Broth	Mean RLU of 1 g sample in R broth	CFU per 10 μL	Mean RLU of 1 g sample in R broth + 10 μL of inoculum	Growth on Agar
1306	1627	25.0	22,500,000	+
		3.0	22,400,000	+
		1.0	22,400,000	+
		0.3	2173	−
S. aureus ATCC 6538			Detection time = 24 hr	
Mean RLU of R broth	Mean RLU of 1 g sample in R broth	CFU per 10 μL	Mean RLU of 1 g sample in R broth + 10 μL of inoculum	Growth on agar
1306	1627	27.0	1,516,276	+
		4.0	7,640,774	+
		1.0	16,321,052	+
		0.3	2440	−
E. coli ATCC 8739			Detection time = 24 hr	
Mean RLU of R broth	Mean RLU of 1 g sample in R broth	CFU per 10 μL	Mean RLU of 1 g sample in R broth + 10 μL of inoculum	Growth on agar
1799	1752	43.0	392,736	+
		5.0	295,350	+
		2.0	1805	−
		0.2	2054	−
S. typhimurium ATCC 13311			Detection time = 24 hr	
Mean RLU of R broth	Mean RLU of 1 g sample in R broth	CFU per 10 μL	Mean RLU of 1 g sample in R broth + 10 μL of inoculum	Growth on agar
1549	1523	31.0	864,608	+
		7.0	1,960,812	+
		2.0	1,628,083	+
		0.1	1802	−

TABLE 5 Continued

C. albicans ATCC 10231		Detection time = 24 hr		
Mean RLU of R broth	Mean RLU of 1 g sample in R broth	CFU per 10 μL	Mean RLU of 1 g sample in R broth + 10 μL of inoculum	Growth on agar
1549	1523	33.0	1,110,004	+
		6.0	254,703	+
		2.00	39,269	+
		0.4	1879	−
A. niger ATCC 16404		Detection time = 27 hr		
Mean RLU of R broth	Mean RLU of 1 g sample in R broth	CFU per 10 μL	Mean RLU of 1 g sample in R broth + 10 μL of inoculum	Growth on agar
1853	1953	12.0	38,369	+
		3.0	49,387	+
		1.0	2341	−
		0.30	2243	−

pensions are inoculated with different levels of *P. aeruginosa*, *Staphylococcus aureus*, *E. coli*, *S. typhimurium*, *C. albicans*, and *Aspergillus niger*. The use of different types of microorganisms demonstrates that the assay is sensitive enough to detect all types of microbial contamination such as bacteria, yeast, and mold.

After samples are spiked with different types of microorganisms, detection time range between 24 and 96 hr (Table 4). The criteria for passing or failing a sample are simple. A positive sample is indicated when the relative light units (RLU) of the contaminated samples in the enrichment broth are two times the values of the sample in the broth. When product suspensions inoculated with different concentrations of microorganisms are incubated, a positive response is detected in all the samples exhibiting two times the values of the control (Table 5). As shown in Table 5, all microorganisms spiked into pharmaceutical product A have been shown to grow on standard media and exhibited bioluminescence values twice the values of the control sample. Therefore, equivalency to the standard method is demonstrated. Bacteria and yeast are easily detected after a 24-hr incubation period while mold detection requires 27 hr. This is because mold exhibit slower growth rate than bacteria and yeast.

This qualitative ATP bioluminescence system has been shown to allow high-throughput screening of more than 180 samples/day. Furthermore, faster detection times for finished product samples range from 24 to 48 hr. Because of the need for an enrichment-incubation step, assay optimization requires the development of different enrichment media to overcome the antimicrobial nature of the different pharmaceutical actives (Tables 4 and 6). For instance, for optimal recovery of bacteria, yeast, and mold, from pharmaceutical products containing halogenated compounds, it was necessary to add sodium thiosulfate to the enrichment media (R, MR, and MR2 broth) (Table 4). Furthermore, different nutrients are also added to optimize recovery for *S. aureus*, e.g., glycine, and mold, e.g., sodium acetate and glycerol (Table 6). Optimization of detection of microbial contamination in some products required the use of Letheen broth with 1.5% Lecithin.

Another ATP assay relies on the differentiation of free extracellular ATP from intracellular ATP to determine viable microbial communities in clean room environments [11]. Extracellular ATP is degraded by using an ATP somase enzyme. Samples from clean room environments exhibit lower levels of ATP when compared with samples obtained from ordinary rooms.

TABLE 6 Enrichment Media for ATP Bioluminescence Analysis

R broth	R2 broth
TAT broth	TAT broth
4% tween 20	4% tween 20
1% dextrose	1% dextrose
1% neopeptone	1% neopeptone
0.25% sodium thiosulfate	0.25% sodium thiosulfate
MR broth	0.5% sodium acetate
TAT broth	1% glycerol
10% tween 20	1% sucrose
1% dextrose	MR2 broth
1% neopeptone	TAT broth
1% glycine	10% tween 20
1% triton X-100	1.2% dextrose
0.5% sodium phosphate dibasic	1.2% neopeptone
0.5% sodium thiosulfate	1% $MgSO_4$
Letheen Broth with 1.5% Lecithin	0.25% KH_2PO_4
TAT Broth with 4% Tween 20	0.25% sodium thiosulfate

However, a large fraction of the samples yield no colony forming units (CFU) on soybean casein digest agar (SCDA) plates but are positive for intracellular ATP. Viable microbial contamination in clean rooms can be detected by using this assay, which might give a better indication of the presence of microbial biomass.

4. DIRECT VIABLE COUNTS (DVC)

Microbial enumeration in pharmaceutical samples can be performed using plate counts and direct microscopy along with viability dyes. Direct counting of individual microbial cells using epifluorescence microscopy has been shown to detect physiologically active bacteria in purified water used in manufacturing processes [10]. The samples have been processed through a 0.45-μm filter to retain the bacteria. The bacteria on the filter are then stained with different types of dyes. The dyes are specific for different types of metabolic reactions in the microbial cell. Fluorescent staining with 5-cyano-2,3-ditolyl tetrazolium chloride (CTC) and 6-carboxyfluorescein diacetate (6CFDA) has detected bacterial cells with respiration and esterase activity, respectively. The CTC and 6CFDA results have indicated that large number of bacteria in purified water retained physiological activity, while a large percentage could not form colonies on conventional media. Therefore, microbial counts using DVC are always higher than standard plate counts. However, epifluorescence microscopy analysis is a time consuming procedure that at the time does not allow the rapid screening of multiple samples.

5. FLOW CYTOMETRY

Several studies have shown the applicability of using "viability markers" and flow cytometry for the rapid enumeration of microorganisms in pharmaceutical grade water [26–28]. The viability maker most commonly used is based upon the reaction of bacteria with the ChemChrome B (CB) dye. Sample preparation consists in filtering the sample through a 0.45-μm membrane followed by cell labeling and laser scanning (Fig. 1). The dye, a fluorescein-type ester, is converted to a fluorescent product, a free fluorescein derivative, by intracellular esterase activity after being taken up by microbial cells previously captured by membrane filtration (Fig. 2). Microbial cells with an intact cell membrane only retain the fluorescein derivative. The bacteria are then enumerated by using a laser scanning instrument, which has been shown to be sensitive down to one cell in a sample within 90 min, and demonstrated a substantially wider linear range than the conventional heterotrophic plate count method. Similar results have been found by fluorescent staining using 4′-6-diamine-2-phenylindole (DAPI), membrane filtration with

◇ Membrane Filtration

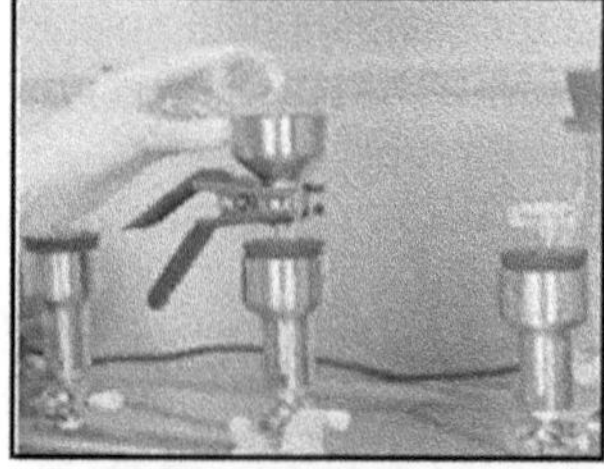

◇ Cell Labeling

◇ Laser Scanning

FIGURE 1 Sample preparation for flow cytometry analysis. Courtesy of AES-Chemunex.

Criteria of Viability:

◇ **Enzyme activity (Cellular esterase activity)**

◇ **Membrane integrity**

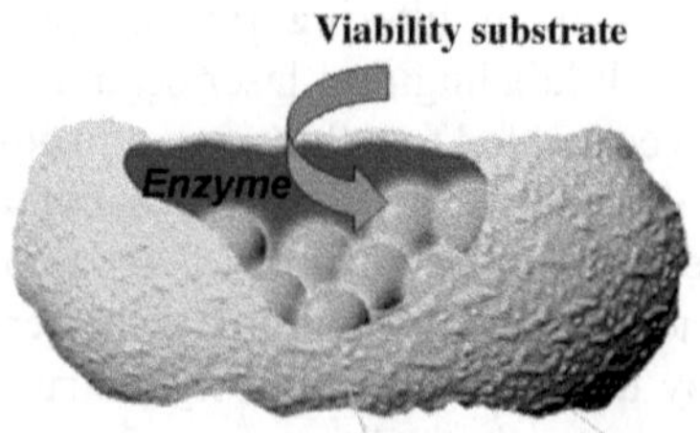

FIGURE 2 Determination of viability by flow cytometry analysis. Courtesy of AES-Chemunex.

tryptic soy agar (TSA) and R2A as growth media, and flow cytometry. An ion-exchange system, reverse osmosis system, and purified water in a hot loop have been sampled and processed. Fluorescence microscopy analysis of water samples using DAPI has resulted in higher microbial counts because DAPI stained all cells containing DNA including dead cells. Of the two growth media used for membrane filtration, R2A has shown higher microbial numbers than TSA because of the longer incubation time. However, flow cytometry has generally demonstrated a cell recovery closer to R2A. Rapid and accurate enumeration of labeled microorganisms is completed within 90 min. Bacterial numbers obtained by the laser scanning instrument appear to be higher than standard plate counts by an order of magnitude. Analysis of tap water, purified water, and water for injection (WFI) at several pharmaceutical sites has also shown that flow cytometry is equivalent to the conventional membrane filtration method. Recovery studies in pure cultures demonstrate a good correlation between methods, with a coefficient of correlation of >0.97 for all organisms tested (vegetative bacteria, spores, yeast, and mold). However, none of the studies reported the multiple processing of water samples. Furthermore, the assay does not provide accurate quantitation when samples exhibit more than 10^4 cells/membrane. The scanning of the filters is interrupted due to the agglomeration of cells resulting in a high fluorescence background. Nevertheless, because of recent modifications to the instrument, a higher accuracy can be achieved with 10^5 cells/membrane for bacteria and 10^4 cells/membrane for yeast and mold [29].

Additional studies have recently been performed on the macrolide antibiotic, Spiramycin, using solid phase cytometry [30]. Artificially contaminated samples of the antibiotic have been analyzed. The solid phase cytometry has been found to detect all microbes regardless of their sensitivity to the bacteriostatic activity of the drug. With the conventional heterotrophic plate method run in parallel, complete recovery has been only obtained for Spiramycin-resistant organisms. The spiked microorganisms that were sensitive to the antibiotic have remained inhibited or stressed by the action of the Spiramycin and do not grow on the plate but are detected by flow cytometry. These results further indicate the inadequacy of standard methods to recover injured microorganisms.

Bioburden of in-process samples of recombinant mammalian cell cultures have also been performed using flow cytometry [31,32]. Instead of the 7-day incubation time required for standard bioburden testing, analyses are completed within 4 hr. The assay is sensitive enough to detect from 5 to 15 CFU/mL after 4 hr. The advantage of rapid analysis of in-process samples is that bioburden results are known before a batch is pooled or processed. In some cases, microbial contamination has been found after the batches are polled and processed resulting in huge financial losses. However, to optimize the detection of bacteria from a background of mammalian cells, different sample preparation procedures and modification of the original protocol are needed. Residual fluorescence appears to be a problem when detection limits go down to 1 cell/filter.

Another current application of flow cytometry to pharmaceutical quality control is the enumeration of biological indicators (BIs) [33]. BIs are used for the validation of sterilization cycles in pharmaceutical environments. Once the BIs are exposed to the sterilizing agent, the level of lethality must be determined. Conventional enumeration testing of BIs is based on the standard plate count of serial dilutions. Because sample incubation is required for growth of visible colonies, results are obtained after 2 or more days. Furthermore, results might vary for different types of BIs based on media and culture conditions. Flow cytometry analysis has demonstrated that spore trips showed interference from paper, counts were lower than plate counts. Modifications of the sample preparation prior to flow cytometry analysis demonstrated that enumeration of BIs is faster, e.g., 2–4 hr, and that results were equivalent to standard plate counts. The advantages of using a rapid method to analyze BIs are a significant reduction in sterilizer holding time, cycle development time, and better understanding of lethality and sterility assurance.

6. IMPEDANCE

When microorganisms grow in enrichment media as a result of microbial metabolism, some of the substrates are converted into highly charged end products. These substrates are generally uncharged or weakly charged but are transformed during microbial growth. Because of their nature, the end products increase the conductivity of the media causing a decrease in impedance. Impedance is the resistance to flow of an alternating current as it passes through a conducting material.

Impedance detection time (T_d) is when the resistance to the flow of an alternating current indicates the growth of a particular microorganism as a result of changes in the growth media. Several studies have shown the applicability of direct impedance for detecting microbial activity in pharmaceutical products. Because impedance is a growth-dependent technology, a medium must be chosen that will support the growth of microorganisms and also to be optimized for electrical signal. Substrates for this kind of media will be uncharged or weakly charged—such as glucose that, when converted to lactic acid, will increase the conductivity of the media. However, a current modification called indirect impedance monitors microbial metabolism by measuring the production of carbon dioxide. The carbon dioxide removed from the growth media results in a decrease in conductivity. The use of indirect impedance allows the use of media that might not generate an optimal electrical response by using the direct method.

A good correlation between direct impedance detection time (T_d) and total colony counts has been obtained for untreated suspensions of *S. aureus* ATCC 6538, *C. albicans* ATCC 10231, *A. niger* ATCC 16404, and *P. aeruginosa* ATCC 9027 in phosphate-buffered saline (PBS) [34]. Similar results have been found with suspensions of test microorganisms treated for varying contact periods with selected concentrations of antimicrobial agents. The only difference found is that the detection time for treated cells is extended. The assay is sensitive enough to detect bacteria, yeast, and mold.

Impedance has been compared to the direct epifluorescence technique (DEFT-MEM) and ATP bioluminescence (ATP-B) for detecting microbial contamination in cells exposed to different antimicrobial agents [35]. ATP-B, impedance, and DEFT-MEM have shown a strong correlation between the rapid method response and total colony counts for bacteria and yeast. However, for mold, impedance has been the only rapid method that showed a strong correlation between colony counts and the rapid method. When chlorhexidine-treated suspensions of *S. aureus* ATCC 6538 and *C. albicans* ATCC 10231 have been analyzed by impedance a good dose–response curve was obtained. Different results have been found with ATP-B and DEFT-MEM methods, which underestimate the kill by the order of 1–6 logs. Impedance application to pharmaceutical screening requires the development of growth curves for different microorganisms. Furthermore, the systems available do not provide high throughput.

7. PCR TECHNOLOGY

Deoxyribonucleic acid (DNA) contains the genetic information that controls the development of a microbial cell. DNA determines the genotypic and phenotypic potential of a microbial cell. With the latest advances in genomics, where more than 25 microbial genomes have been sequenced, the potential to use genetic information for the detection and discrimination of microorganisms is endless. Genetic technologies can increase the resolution and specificity of microbial detection and identification in pharmaceutical environments. DNA-based technologies are used in clinical, food, and environmental samples providing valuable information on the survival, distribution, and function of microorganisms in those habitats [36,37]. One of the technologies based on DNA analysis is the polymerase chain reaction.

Polymerase chain reaction (PCR) amplifies specific DNA sequences along the microbial genome. For example, a set of DNA primers is used to target the specific sequence to be amplified (Table 7). The PCR reaction takes place in three different steps. First, the target sequence is denatured by

TABLE 7 PCR Assay Reaction Steps

```
(1) Double helix denatured by heating
5′                        3′
A T C G C A G G G A T C        95°C       5′                      3′
T A G C G T C C C T A G                   A T C G C A G G G A T C
3′                        5′    →
                                         T A G C G T C C C T A G
                                         3′                      5′

(2) Primers are bound to complementary sequences on template strands
Template Strand
5′                        3′
A T C G C A G G G A T C
T A G
      | > > > > |
      Target Region
             |
      | < < < A T C
T A G C G T C C C T A G
3′                        5′
Template Strand

(3) Primers are extended by DNA polymerase resulting in two
    DNA strands
5′                        3′
A T C G C A G G G A T C
T A G C G T C C C T A G
3′                        5′

5′                        3′
A T C G C A G G G A T C
T A G C G T C C C T A G
3′                        5′
```

heating. Second, the primers anneal to complementary sequences on the target DNA strands. Third, the primers are extended by the DNA polymerase enzyme resulting in two different strands. The three steps are repeated again for a given number of cycles, e.g., 30–35. As soon as the target is amplified, the products are detected by gel electrophoresis. However, new systems have been developed that rely on fluorescence detection of amplified products. PCR based assays are used routinely in the food industry and clinical laboratories to detect and identify bacteria, yeast, and mold [36,37].

In pharmaceutical laboratories, PCR-based assays have been shown to be capable of detecting *S. typhimurium*, *E. coli*, *P. aeruginosa*, *S. aureus*, *B. cepacia*, *A. niger*, and eubacterial sequences after an incubation period [38–43]. Analysts, raw materials, equipment, or water contamination introduces some of these microorganisms into pharmaceutical environments. Furthermore, when analysts do not follow good laboratory practices, they become major sources of microbial contamination in clean rooms and aseptic manufacturing. Rapid detection of objectionable microorganisms results in faster implementation of corrective actions. Detection times using PCR range from 24 to 27 hr (Table 8). This is a significant reduction when compared to the standard 5–7 days detection time. Furthermore, high-throughput screening of samples is possible by using a 96-well format.

The simplification of PCR analysis for pharmaceutical quality control is achieved by using a tablet and PCR bead formats. The PCR reagents, including DNA primers, are combined in a tablet form, while the beads provide the necessary reagents for the PCR reaction but without the DNA primers. Time-consuming preparations and handling of individual PCR reagents are not required due to the tablet and bead formats incorporated in the assay. During assay development, different experiments are performed to determine

TABLE 8 Pharmaceutical Samples Analyzed by PCR [21,23–27]

	Inhibitory reaction	Detection	Dilution	Time (hr)
Neobee Oil	No	Yes	1/10	24–27
Simethicone	No	Yes	1/10	24–27
CMC	No	Yes	1/10	24–27
Sodium alginate	No	Yes	1/10	24–27
Rasberry flavor	No	Yes	1/10	24–27
Hydroxymethylcellulose	No	Yes	1/10	24–27
Xantham gum	No	Yes	1/10	24–27
Silica calcinated	No	Yes	1/10	24–27
Guar gum	No	Yes	1/10	24–27
Starch	No	Yes	1/10	24–27
Lactose monohydrate	No	Yes	1/10	24–27
Diatomaceous earth	No	Yes	1/10	24–27
Tablets	No	Yes	1/10	24–27
Medicated skin cream	No	Yes	1/10	24–27
Ointment	No	Yes	1/10	24–27
Antiflatulent drops	No	Yes	1/10	24–27
Medical device	No	Yes	1/10	24–27
Laxative tablets	No	Yes	1/10	24–27

the optimal numbers of beads. Optimal DNA amplification is found to be obtained with two or three beads (Table 9).

DNA extraction from sample enrichments is performed in single-step assays. For bacteria and yeast, a sample preparation using Tris–EDTA–Tween 20 buffer with proteinase K at 35°C resulted in high-quality DNA, while boiling the samples in sodium dodecyl sulfate (SDS) for 1 hr is required for efficient mold DNA extraction. None of the product suspensions shows PCR inhibition allowing rapid determination of sample quality (Table 8). The amount of DNA needed for detecting the different target sequences ranged from 10 to 50 μl of lysate (Table 9). Higher concentrations of the lysate are found to be inhibitory for successful PCR amplification.

TABLE 9 Optimization of PCR Reactions for Objectionable Microorganisms in Pharmaceutical Products

Microorganism	Beads	Aliquot	PCR band
S. aureus	3	10	+
S. aureus	2	10	–
S. aureus	1	10	–
S. aureus	3	25	–
P. aeruginosa	2	10	+
P. aeruginosa	1	10	–
P. aeruginosa	2	25	–
P. aeruginosa	1	25	–
E. coli	2	10	+
E. coli	1	10	–
E. coli	2	25	–
E. coli	1	25	–
S. typhimurium[a]	1	50	+
B. cepacia	2	10	+
B. cepacia	1	10	–
B. cepacia	2	25	–
B. cepacia	1	25	–
A. niger	2	50	+
A. niger	2	25	–
A. niger	2	10	–
A. niger	1	50	–
C. albicans	2	50	+
C. albicans	2	25	–
C. albicans	2	10	–
C. albicans	1	50	–

[a] Commercial system.

The development of new PCR formats allows for the simplification of PCR protocols where only sample addition and primers are needed to perform the assay. With the latest advances in microbial genomics, the availability of DNA primer sequences are limitless allowing the development of universal primers for bacteria, yeast, and mold. A recent study has shown the applicability of detecting bacterial contamination for sterility testing by using a simple PCR assay. The study is based upon the universal and inclusivity nature of the DNA sequences coding for bacterial ribosomal genes. DNA primers targeting these common bacterial sequences are capable of rapidly screening samples for bacteria contamination.

All the previously discussed studies have been performed using a single PCR amplification format where a specific microorganism DNA sequence was targeted. However, simultaneous detection of bacteria and mold DNA sequences in pharmaceutical samples using a gradient thermocycler has been recently reported [44]. The gradient thermocycler allows the use of primers with annealing temperatures ranging from 54 to 65 °C leading to the detection of different microorganisms in a single PCR run. This allows the immediate screening of a pharmaceutical sample for bacteria, yeast, and mold.

PCR has also been used for the monitoring of pharmaceutical water samples in manufacturing processes [8]. Ribosomal DNA sequences are amplified with universal bacterial primers. After amplification, the samples are loaded onto polyacrylamide gels [denaturing gradient gel electrophoresis (DGEE)] to detect the amplified products. This will allow the separation of DNA fragments of the same length but different pair sequences. After separation, the gels are scanned to generate a densitometric profile. The sequencing of the amplified fragments has revealed that the dominant bacteria in the water samples are not culturable on standard media. Most of the culturable bacterial species have been found to be related to *Brudyrhizobium* spp., *Xanthomonas* spp., and *Stenotrophomonas* spp., while the dominant unculturable bacterial species have not been characterized. These studies further showed the limit capacity of standard methods to determine and characterize the community structure of pharmaceutical environments. Similar results have been found in other environmental conditions.

Similar results are found in pharmaceutical clean room environments [11]. DNA extracted from selected samples have been analyzed by using 16S rDNA sequencing. Results indicate that bacterial isolates do not grow on plate media but are major components of the microbial populations.

8. GENETIC IDENTIFICATION

When microbial contamination is detected in a given pharmaceutical sample, characterization of the types of microorganisms by genera and species is an important criterion to determine the source of the contamination. The first step in the phenotypic identification of microorganisms in pharmaceutical laboratories is performed using the gram strain method [7]. This method is based upon the chemical and structural differences between the membranes and cell walls of gram-negative and gram-positive bacteria.

For bacteria, once the results of the gram reaction have been determined and other simple biochemical tests are completed, e.g., catalase and oxidase test, a standardized pure culture suspension of the isolate is inoculated into strips, cards, or microtiter plates [7]. These systems are based upon the detection of enzymatic activity by different types of enzymes such as oxidases and carbon utilization profiles. However, new genetic tests provide a greater resolution and discrimination for microbial identification. Table 10 shows a comparison of phenotypic and genotypic identification of bacterial species by biochemical, lipids, and genetic methods. The genetic method demonstrated a higher accuracy and reproducibility than lipid and biochemical analysis. Similar results were obtained with environmental isolates from different pharmaceutical environments (Table 11). DNA sequencing analysis provided

TABLE 10 Microbial Characterization of Bacteria Using Different Identification Systems

Species	Vitek	Biolog	Lipids	Genetic
Bacillus cereus	Unidentified	Yes	Yes	Yes
Burkholderia cepacia	Yes	Unidentified	Yes	Yes
Enterobacter cloacae	Yes	Yes	Unidentified	Yes
Escherichia coli	Yes	Yes	Yes	Yes
Micrococcus luteus	Unidentified	Yes	Unidentified	Yes
Pseudomonas aeruginosa	Yes	Yes	Yes	Yes
Shigella flexneri	Unidentified	Yes	Unidentified	Yes
Staphylococcus aureus	Unidentified	Unidentified	Unidentified	Yes
Staphylococcus epidermidis	Yes	Unidentified	Unidentified	Yes
Acinetobacter radioresistens	Unidentified	Yes	Yes	Yes
Macrococcus caseolyticus	Unidentified	Yes	Yes	Yes
Methylobacterium radiotolerans	Unidentified	Unidentified	Unidentified	Yes
Ochrobactrum anithropi	Yes	Yes	Yes	Yes
Ralstonia pickettii	Unidentified	Yes	Unidentified	Yes
Streptococcus salivarius	Unidentified	Unidentified	Unidentified	Yes
Corynebacterium xerosis	Unidentified	Unidentified	Unidentified	Yes
Kokuria rosea	Unidentified	Unidentified	Unidentified	Yes
Paenibacillus glucanolyticus	Unidentified	Unidentified	Unidentified	Yes

Source: Ref. 12.

TABLE 11 Microbial Identification of Common Microbial Contaminants in Pharmaceutical Environments Using Lipid Analysis and DNA-Based Tests

Species	Lipid analysis	Genetic fingerprinting	Genetic sequencing
Ralstonia spp.	*R. pickettii*	*R. pickettii*	*Ralstonia* spp.
Kokuria rosea	Unidentified	Unidentified	*K. rosea*
Bacillus pumilus	*B. pumilus*	*B. pumilus*	*B. pumilus*
Bacillus pumilus	*B. pumilus*	Unidentified	*B. pumilus*
Bacillus pumilus	*B. pumilus*	*B. pumilus*	*B. pumilus*
Bacillus pumilus	Unidentified	*B. pumilus*	*B. pumilus*
Bacillus pumilus	*B. pumilus*	*B. pumilus*	*B. pumilus*
Bacillus pumilus	*B. pumilus*	*B. pumilus*	*B. pumilus*
Bacillus pumilus	*B. pumilus*	*B. pumilus*	*B. pumilus*
Ralstonia pickettii	*R. pickettii*	*R. pickettii*	*R. pickettii*
Staphylococcus hominis	*S. hominis*	*S. epidermidis*	*S. hominis*
Ralstonia pickettii	Unidentified	*R. pickettii*	*R. pickettii*
Corynebacterium spp.	Unidentified	*C. amycolatum*	*Corynebacterium* spp.
Stenotrophomonas maltophila	*S. maltophila*	*S. maltophila*	*S. maltophila*
Enterobacter cancerogenous	*E. cancerogenous*	*E. cloacae*	*E. cancerogenous*
Aeromonas hydrophila	Unidentified	Unidentified	*A. hydrophila*
Pantoea spp.	*Cedecea lapagei*	Unidentified	*Pantoea* spp.
Moraxella osloensis	*M. osloensis*	Unidentified	*M. osloensis*
Staphyloccus warneri	*S. warneri*	*S. aureus*	*S. warneri*
Stenotrophomonas spp.	*S. maltophila*	*S. maltophila*	*Stenotrophomonas* spp.
Staphyloccus aureus	Unidentified	*S. aureus*	*S. aureus*
Microbacterium sp.	Unidentified	Unidentified	*Microbacterium* sp.
Bacillus circulans	*Cellulomonas turbata*	Unidentified	*B. circulans*
Bacillus megaterium	*B. megaterium*	*B. megaterium*	*B. megaterium*
Bacillus amyloliquefaciens	*B. subtilis*	*B. subtilis*	*B. amyloliquefaciens*
Bacillus sp.	*Bacillus* sp.	Unidentified	*Bacillus* sp.
Staphylococcus epidermidis	Unidentified	*S. epidermidis*	*S. epidermidis*
Burkholderia cepacia	Unidentified	*B. cepacia*	*B. cepacia*
Micrococcus luteus	Unidentified	*M. lylae*	*M. luteus*
Paenibacillus glucanolyticus	*P. polymyxa*	*P. glucanolyticus*	*P. glucanolyticus*
Stenotrophomonas maltophila	*S. maltophila*	*S. maltophila*	*S. maltophila*
Burkholderia cepacia	Unidentified	Unidentified	*B. cepacia*

TABLE 11 Continued

Species	Lipid analysis	Genetic fingerprinting	Genetic sequencing
Burkholderia cepacia	*B. gladiolli*	Unidentified	*B. cepacia*
Pseudomonas veronii	Unidentified	*P. fluorescens*	*P. veronii*
Yokenella regensburgel	*S. typhimurium*	*P. putida*	*Y. regensburgel*
Pseudomonas putida	*P. putida*	*P. putida*	*P. putida*
Pseudomonas stutzeri	Unidentified	*P. stutzeri*	*P. stutzeri*
Chryseomonas luteola	Unidentified	*C. luteola*	*C. luteola*
Micrococcus luteus	Unidentified	Unidentified	*M. luteus*
Staphylococcus haemolyticus	*S. aureus*	*S. haemolyticus*	*S. haemolyticus*
Micrococcus luteus	*M. lylae*	*M. lylae*	*M. luteus*
Micrococcus luteus	Unidentified	*M. luteus*	*M. luteus*
Micrococcus lylae	Unidentified	NT	*M. lylae*

Source: Ref. 12.

a higher level of accuracy resolution, and identification than lipid analysis or DNA fingerprinting. Several studies demonstrate that unidentified environmental isolates not characterized by phenotypic analysis are correctly characterized by 16S rRNA and 16S rDNA sequencing [45]. This provides accurate information for the tracking of the contamination source in pharmaceutical environments and microbial community characterization allowing faster corrections actions to be implemented.

9. DNA MICROCHIPS

The miniaturization of genetic analyses such as PCR and DNA hybridization using DNA microchips have been reported to be used to ascertain the microbial composition of environmental, food, and clinical samples [46–48]. DNA microchips provide an automated, accurate, and high-throughput alternative to phenotypical identification. Miniaturized DNA chips are divided into four processes: sample preparation, assay, detection, and analysis. Eggers and Ehrlich [49] have reviewed these four steps in details. In general, oligonucleotides (oligos) probes are short DNA or RNA sequences of different sizes immobilized on a solid support such as glass, gel, or silicon. Binding of the oligos to the solid support requires the design of probe arrays at predetermined locations (Table 12). After binding, a positive reaction is detected by autoradiography or fluorescent dyes (Table 13). A positive reaction is indicated by the development of a black dot on the microchip array (Table 13).

TABLE 12 Microchip Probe Array, Before Sample Binding

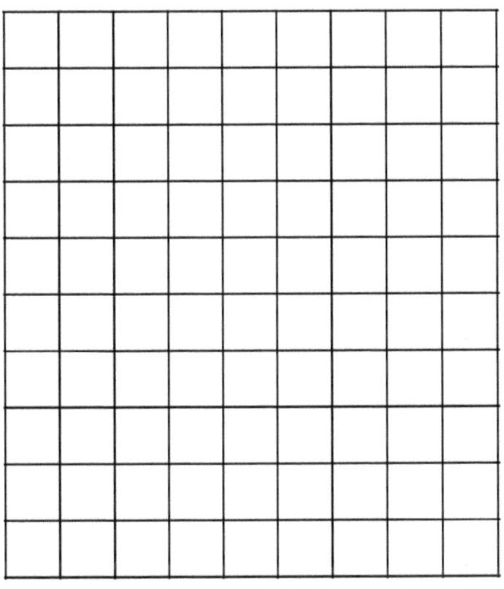

90 samples on a microchip.
Oligonucleotide immobilized on solid support.

A single microchip can be used up to 30 times without noticeable deterioration of the hybridization signal. Thousands of oligonucleotides can be immobilized on a single microchip, allowing for the simultaneous detection of a wide variety of microorganisms in a single sample. Environmental studies have demonstrated the specificity, sensitivity, and accuracy of DNA microchips to detect specific microbial communities in mixed cultures. Oligonucleotides probes on microchips targeting specific rRNA sequences of nitrifying bacteria are shown to be capable of detecting nitrifying bacteria with high resolution and sensitivity. Total RNA of *Geobacter chapellei* and *Desulfovibrio desulfuricans* have been shown to hybridize to oligonucleotide arrays of universal and species-specific 16S rRNA probes [50]. *G. chapellei* is also detected with total RNA extracted from soil. *E. coli* and *E. coli* O15:7 are reported to be detected using DNA microchips in clinical and food samples, respectively [46]. Furthermore, a microarray assay is capable of detecting and discriminating six species of the *Listeria* genus. Detection is based upon the amplification of six virulence factor genes and hybridization with multiple oligonucleotides probes specific for each species [51]. The potential for using DNA microchips for pharmaceutical quality control is enormous. In theory,

TABLE 13 Microchip Probe Array, After Sample Binding

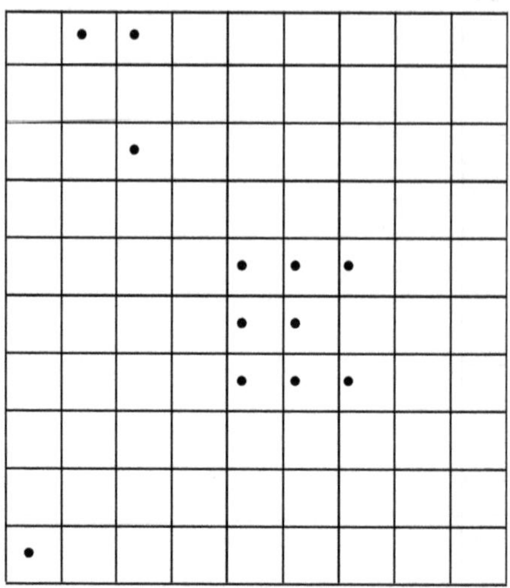

	•	•						
		•						
				•	•	•		
				•	•			
				•	•	•		
•								

Oligonucleotides with hybridized DNA sequence detected 12 positive samples.

on one single chip, the presence of all different types of microorganisms can be ascertained.

10. IMMUNOASSAYS

Although enzyme-linked immunosorbent assays (ELISA) are widely used in clinical and food analyses, it was not until recently that these methods were applied to pharmaceutical quality control. ELISA tests are performed using different formats. The most common format to pharmaceutical quality control analysis is based upon the immobilization of high affinity antibodies, specific for different types of microorganisms, on the surface of microtiter wells. The sample is then applied to the well and incubated. If there is a microorganism in the sample, it is captured by the immobilized antibody (Fig. 3).

An enzyme-conjugate antibody is then added to react with the captured microorganism. This will result in the formation of an antibody–microorganisms conjugate "sandwich." To develop a detection signal, a chemical substrate is added to react with the enzyme in the conjugate. If there is a microorganism in the sample, a color reaction will develop. Absence of a specific microbial target is indicated by the absence of color. The use of a 48–

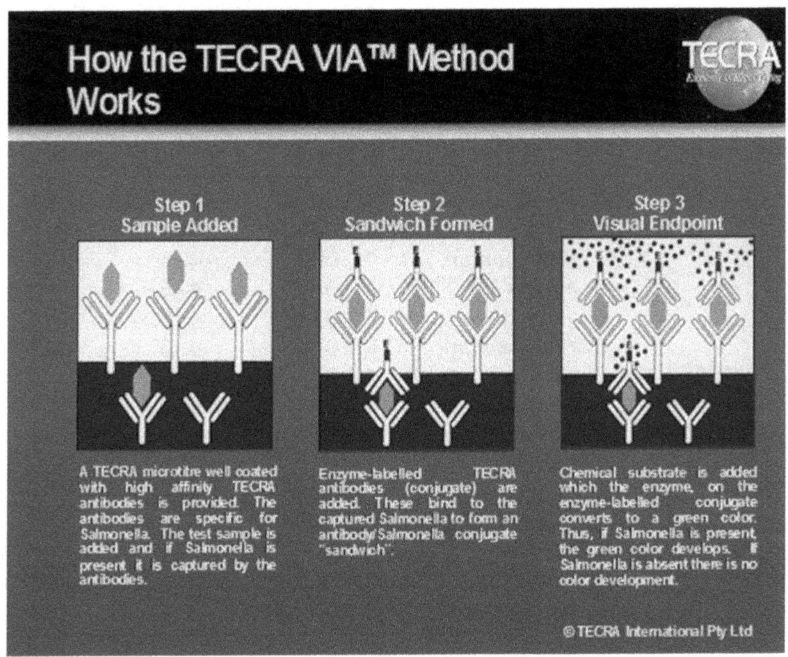

FIGURE 3 Immunoassay format used for detecting microbial contamination in pharmaceutical samples. Courtesy of Tecra International.

96-well microtiter plate format allows the high-throughput screening of pharmaceutical samples.

Pharmaceutical samples contaminated with pure and mixed cultures have been shown to detect microbial contamination by *S. aureus* within 24 hr [52]. These results indicated that the assays are specific enough to detect the target microorganisms in the presence of other microbial species. When compared to the 4–5 days detection time using standard methods, the ELISA method is found to be more effective reducing detection time and labor. Furthermore, multiple processing and analysis of samples has been possible due to the 48–96-well microtiter format. Another validation study has been undertaken to compare ELISA assays with standard methods [53]. Other products tested included a range of pharmaceuticals such as cough mixtures, laxatives, ulcer treatments, infant formulae, antiseptic cream, as well as some pharmaceutical ingredients.

A recent study has ascertained the applicability of three different types of ELISA assays for rapid detection of pathogens. Product suspensions are inoculated with 10 colony forming units (CFU)/mL of *P. aeruginosa, S. au-*

TABLE 14 Detection of Microbial Contamination Using Immunoassays

Pharmaceutical product	Bacteria	Dilution	Detection time (days) Standard method	ELISA
A	*S. aureus*	1:100	4–5	1
	P. aeruginosa	1:100	4–5	1
	S. typhimurium	1:100	4–5	1
B	*S. aureus*	1:10	4–5	1
	P. aeruginosa	1:10	4–5	1
	S. typhimurium	1:10	4–5	1
C	*S. aureus*	1:10	4–5	1
	P. aeruginosa	1:10	4–5	1
	S. typhimurium	1:10	4–5	1
D	*S. aureus*	1:10	4–5	1
	P. aeruginosa	1:10	4–5	1
	S. typhimurium	1:10	4–5	1
E	*S. aureus*	1:10	4–5	1
	P. aeruginosa	1:10	4–5	1
	S. typhimurium	1:10	4–5	1

reus, and *S. typhimurium*. Samples are then incubated for 24 hr at 35°C. After incubation, samples are analyzed as described in Fig. 3. Table 14 shows the results of the analysis of pharmaceutical products by using three different types of ELISA methods. Results demonstrated that standard methods and the immunoassays exhibit a 100% correlation. No interferences, false negatives, or false positives were found by any of the products. However, the immunoassays detected the bacteria in 24 hr while standard methods required from 4 to 5 days. Using the 96-well plate format, sample output is 48 samples every 2 hr counting 2 positive and 2 negative controls simultaneously run with each plate. In an 8-hr laboratory shift a total of approximately 176 samples can be screened for *P. aeruginosa*, *S. aureus*, and *S. typhimurium*.

11. GROWTH DIRECT™

Quantitative analysis using flow cytometry and ATP bioluminescence do provide rapid enumeration of microorganisms in pharmaceutical samples. However, both assays are destructive and do not provide for high throughput screening. Furthermore, in the ATP bioluminescence assay accurate and reliable quantitation beyond 75 CFU/100 mL is not possible. Flow cytometry provides a higher quantitation range. However, membrane filtration through a 0.45-μm, 25-mm-diameter filter is slow. If microbial identification is needed, sample analysis must rely on the standard enrichment plate count for isolation and identification of microbial colonies needed to be characterized.

A nondestructive quantitative analysis of pharmaceutical samples can be performed using the Growth Direct™ system [54]. The system uses non-magnified large-area digital imaging to detect growing microbial cells in water, products, raw materials, air samples, and contact plate samples (Fig. 4). The test uses standard growth media and membrane filtration. The assay requires no reagent addition. Samples are applied to membranes, placed on growth media, incubated, and imaged using a charge couple device (CCD) detector. Microbial colonies are detected by using the intrinsic auto-fluorescence of the microbial cells. Environmental samples exhibited time savings of 50–80% for microbial testing (Fig. 5). Furthermore, when compared to traditional methods, the Growth Direct™ allows the rapid enumeration of microbial cells (Fig. 6). For instance, *E. coli* is enumerated within 12 hr using standard methods while the Growth Direct™ system is completed within 3.5 hr. Time savings for quantitation of yeast and mold are 80% and 70%, respectively.

The system preserves key advantages of traditional testing such as nondestructive, broad range of applications, and facilitates validation. Figure 7 shows that the system does not kill the microbes it detects, the microcolonies in the left panel could continue to divide after early detection to form the

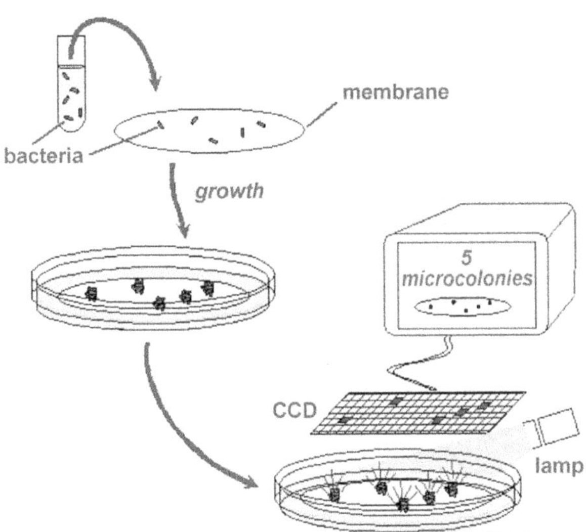

FIGURE 4 Growth Direct™ System.

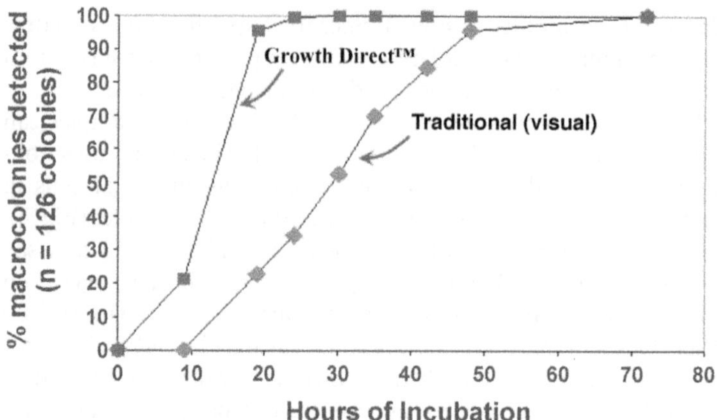

- **99% of the colonies that were detected visually at 72 hr could be detected by the Growth Direct™ test at 24 hr.**

FIGURE 5 Growth Direct™ System detection of bacterial colonies in an environmental water sample.

	Time to Detection (h)		
	Growth Direct	Traditional	Time Saved (%)
Escherichia coli	3.5	12	70%
Pseudomonas aeruginosa	9	24	65%
Staphylococcus aureus	8	24	65%
Bacillus subtilis spores	11	24	55%
Candida albicans	10	48	80%
Aspergillus niger spores	23	72	70%
Penicillium notatum spores	12	96	85%

FIGURE 6 Improved time to detection using the Growth Direct™ System.

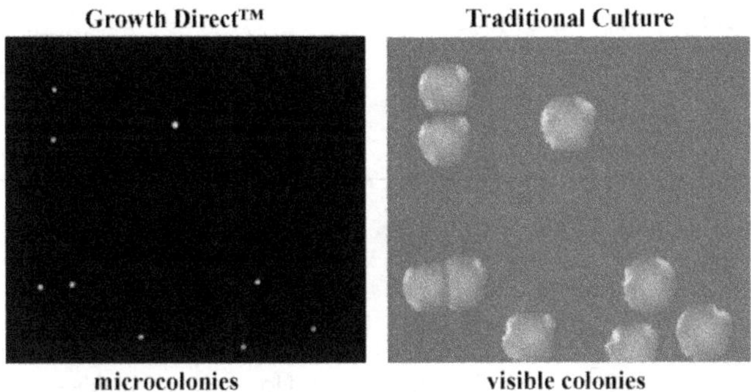

FIGURE 7 Demonstrating equivalence to traditional test is facilitated because the test is nondestructive.

macrocolonies in the right panel. That facilitates the further identification of the colonies isolated on the media.

12. CONCLUSION

On the basis of published scientific studies and presentations, there are several available new technologies that can replace or complement standard microbiological methods for quality control testing of pharmaceutical samples. Rapid methods are proven to be effective, reliable, sensitive, and equivalent to standard microbiological assays. Furthermore, because of the recent demonstration of unculturable bacteria in pharmaceutical environments and the different types of physiological responses to environmental fluctuations, rapid methods provide a more complete description of the microbial community present in products, environment, personnel, and raw materials.

However, rapid methods application must be based upon the needs of a given company and in a case-by-case basis. For instance, in some situations, microbial enumeration is required, while in others the presence or absence of microorganisms results in rapid quality analysis (Table 15). A quantitative ATP bioluminescence system or flow cytometry can be applied to water monitoring while PCR technology and qualitative ATP bioluminescence are applicable to microbial limits.

Because microbial contamination is a sporadic event in pharmaceutical environments, rapid screening of batches using alternative microbiological testing provides a rapid release for approximately 99% of samples tested. When microbial contamination is found, rapid methods such as immunoassays or PCR technology can analyze the sample for the presence of objectionable or pathogenic microorganisms using high-throughput screening

TABLE 15 Comparison of Rapid Methods

Method	Sensitivity (cells/mL)	Detection time (hr)	High quantitation throughput	Quantitation
ATP				
Qualitative	10^4	24–48	Yes	No
Quantitative	1	24	No	Yes
PCR	10^5	24–30	Yes	No
Flow cytometry	1	2	No	Yes
Impedance	10^6	24–30	No	Yes
ImmuNoassays	10^4	24–27	Yes	No
Direct viable counts	1	24	No	Yes
Growth Direct™	45–50	3	Yes	Yes

(Table 15). However, quantitative systems, to this date, do not have high-throughput screening capabilities.

As demonstrated by published scientific reports, validation studies showing equivalency between compendial and rapid methods must be performed before implementation. Some of the rapid technologies are more accurate than standard microbiological methods (Table 15). For example, enumeration and detection of bacteria that did not grow on standard media will create a situation where changes in specifications will be required. However, changes in specifications can be documented if there is a significant advantage in the use of a rapid method. Several terms such as microbial viability will be redefined as per specific data supporting the changes indicating that a microorganism can be viable but not able to grow in enrichment media. For instance, in flow cytometry, DVC, and PCR studies, several microbial species have been found to be predominant members of the microbial community but has not been isolated or detected using standard methods [7,9,24]. However, this should not discourage the use of these technologies but, on the contrary, create an environment where their use will develop additional information where process validation and control can be significantly improved.

Future optimization of pharmaceutical manufacturing and quality control requires faster microbiological analysis than standard conventional methods. Rapid methods identify microbial contamination with detection times ranging from 90 min to 30 hr allowing the monitoring of critical control points, reducing losses, and optimizing resources (Table 15). In the twenty-first century, with advances in computer sciences, automation, combinatorial chemistry, genomics, and medicine, quality control microbiology requires faster turnover times, higher resolution, and sensitivity without compromising efficacy. Rapid technologies enhance the ability of a quality control system for risk assessment and process control. Among other features, an ideal rapid microbiology system will comprise high throughput, rapid identification, ease of use, nondestructive, and easy validation against compendial methods. However, a future quality control microbiology laboratory might have an instrument for water testing and others for microbial limits, environmental monitoring, and microbial identification. Rapid methods can complement standard microbiological testing to provide a greater resolution and analysis of the microbial communities present in pharmaceutical environments.

REFERENCES

1. Pharmacopeial Reviews. Validation of alternative microbiological methods. Pharmacop Forum 2002; 28:154–160.
2. PDA Technical Report Number 33. Evaluation, validation, and implementation of new microbiological testing methods. J Parenter Sci Technol 2000; 54(3).
3. Jimenez, L. Rapid Methods for the Microbiological Surveillance of Pharmaceuticals. PDA J Pharm Sci Technol 2001; 55:278–285.
4. Jimenez L, Smalls S, Ignar R. Use of PCR analysis for rapid detection of low levels of bacterial and mold contamination in pharmaceutical samples. J Microbiol Methods 2000; 41:259–265.
5. Jimenez L. Molecular diagnosis of microbial contamination in cosmetic and pharmaceutical products: a review. J AOAC Int 2001; 84:671–675.
6. Casey W, Muth H, Kirby J, Allen P. Use of nonselective preenrichment media for the recovery of enteric bacteria from pharmaceutical products. Pharm Technol 1998; 22:114–117.
7. Palmieri MJ, Carito SL, Meyer J. Comparison of rapid NFT and API 20E with conventional methods for identification of gram-negative nonfermentative bacilli from pharmaceutical and cosmetics. Appl Environ Microbiol 1988; 54:2838–3241.
8. Kawai M, Matsutera E, Kanda H, Yamaguchi N, Tani K, Nasu M. 16S ribosomal DNA-based analysis of bacterial diversity in purified water used in pharmaceutical manufacturing processes by PCR and denaturing gradient gel electrophoresis. Appl Environ Microbiol 2002; 68:699–704.
9. Nagarkar P, Ravetkar SD, Watve MG. Oligophilic bacteria as tools to monitor aseptic pharmaceutical production units. Appl Environ Microbiol 2001; 67:1371–1374.
10. Kawai M, Yamaguchi N, Nasu N. Rapid enumeration of physiologically active bacteria in purified water used in the pharmaceutical manufacturing process. J Appl Microbiol 1999; 86:496–504.
11. Venkateswaran K, Hattori N, La Duc MT, Kern R. ATP as a biomarker of viable microorganisms in clean room facilities. J Microbiol Methods 2003; 52:367–377.

12. Montgomery S. A comparison of methods for identification of microorganisms in the pharmaceutical manufacturing environment. First Annual Rapid Micro Users Group, Validation requirements for rapid microbiology. Chicago, IL, September 1–7, 2002.
13. Cundell AM. Review of the media selection and incubation conditions for the compendial sterility and microbial limit tests. Pharmacop Forum 2002; 28:2034–2041.
14. Underwood E. Ecology of microorganisms as its affects the pharmaceutical industry. In: Hugo WB, Russell AB, eds. Pharmaceutical Microbiology. 6th ed. Oxford, England: Blackwell Science, 1998:339–354.
15. Roszak DB, Colwell RR. Survival strategies of bacteria in the natural environment. Microbiol Rev 1987; 51:365–379.
16. Sundaram S, Mallick S, Eisenhuth J, Howard G, Brandwein H. Retention of water-borne bacteria by membrane filters: Part II. Scanning electron microscopy (SEM) and fatty acid methyl ester (FAME) characterization of bacterial species recovered downstream of 0.2/0.22 micron rated filters. PDA J Pharm Sci Technol 2001; 55:87–113.
17. Papapetropoulou M, Papageorgakopoulou N. Metabolic and structural changes in *Pseudomonas aeruginosa*, *Achromobacter* CDC, and *Agrobacterium radiobacter* cells injured in parenteral fluids. PDA J Pharm Sci Technol 1994; 48:299–303.
18. Whyte W, Niven L, Bell ND. Microbial growth in small-volume pharmaceuticals. J Parenteral Sci Technol 1989; 43:208–212.
19. Reasoner DJ, Geldreich EE. A new medium for the enumeration and subculture of bacteria from potable water. Appl Environ Microbiol 1985; 49:1–7.
20. Hazen TC, Jimenez L, Lopez de Victoria G, Fliermans CB. Comparison of bacteria from deep subsurface sediment and adjacent groundwater. Microb Ecol 1991; 22:293–304.
21. Hugo WB. Bacteria. In: Hugo WB, Russell AB, eds. Pharmaceutical Microbiology. 6th ed. Oxford, England: Blackwell Science, 1998:3–34.
22. Ignar R, English D, Jimenez L. Rapid detection of microbial contamination in Triclosan and high fluoride dentifrices using an ATP bioluminescence assay. J Rapid Methods Autom Microbiol 1998; 6:51–58.
23. Marino G, Maier C, Cundell AM. A comparison of the MicroCount Digital System to plate count and membrane filtration methods for the enumeration of microorganisms in water for pharmaceutical purposes. PDA J Pharm Sci Technol 2000; 54:172–192.
24. Scalici C, Smalls S, Blumberg S, English D, Jimenez L. Comparison of Millipore Digital Total Count System and standard membrane filtration procedure to enumerate microorganisms in water samples from cosmetic/pharmaceutical environments. J Rapid Methods Autom Microbiol 1998; 7:199–209.
25. Hauschild J. Applying rapid enumeration and identification of microorganisms using the Millipore Microstar™ system. Rapid methods and automation in microbiology for pharmaceutical, biotechnology, and devices applications. San Juan, Puerto Rico, February 1–2, 2001.

26. Reynolds DT, Fricker CR. Application of lasers scanning for the rapid and automated detection of bacteria in water samples. J Appl Microbiol 1999; 86:785–796.
27. Wallner G, Tillmann D, Haberer K. Evaluation of the ChemScan system for rapid microbiological analysis of pharmaceutical water. PDA J Pharm Sci Technol 1999; 53:70–74.
28. Gapp G, Guoymard S, Nabet P, Scouvart J. Evaluation of the applications of a system for real-time microbial analysis of pharmaceutical water systems. Eur J Parenteral Sci 1999; 4:131–136.
29. Costanzo SP, Borazjani RN, McCormick PJ. Validation of the Scan RDI for routine microbiological analysis of process water. PDA J Pharm Sci Technol 2002; 56:206–219.
30. Ramond B, Rolland X, Planchez C, Cornet P, Antoni C, Drocourt JL. Enumeration of total viable microorganisms in an antibiotic raw material using ChemScan solid phase cytometer. PDA J Pharm Sci Technol 2000; 54:320–331.
31. McColgan J. Rapid detection of bacterial contamination in recombinant mammalian cell culture. Rapid methods and automation in microbiology for pharmaceutical, biotechnology, and devices applications. San Juan, Puerto Rico, February 1–2, 2001.
32. Onadipe A, Ulvedal K. A method for the rapid detection of microbial contaminants in animal cell culture processes. PDA J Pharm Sci Technol 2001; 55:337–345.
33. Moldenhauer J, Noverini P, Vukanic N. Feasibility of using solid phase laser scanning cytometry (scan RDI) for the enumeration or biological indicators. Rapid methods and automation in microbiology for pharmaceutical, biotechnology, and devices applications. San Juan, Puerto Rico, February 1–2, 2001.
34. Connolly P, Bloomfield SF, Denyer SP. The use of impedance for preservative efficacy testing of pharmaceuticals and cosmetics. J Appl Bacteriol 1994; 76:68–74.
35. Connolly P, Bloomfield SF, Denyer SP. A study of the use of rapid methods for preservative efficacy testing of pharmaceuticals and cosmetics. J Appl Bacteriol 1993; 75:456–462.
36. Hill WE. The polymerase chain reaction: application for the detection of food-borne pathogens. Crit Rev Food Sci Nutr 1996; 36:123–173.
37. Ieven M, Goosens H. Relevance of nucleic acid amplification techniques for diagnosis of respiratory tract infections in the clinical laboratory. Clin Microbiol Rev 1997; 10:242–256.
38. Jimenez L, Smalls S, Scalici C, Bosko Y, Ignar R, English D. Detection of *Salmonella* spp. contamination in raw materials and cosmetic/pharmaceutical products using the BAX™ system, a PCR-based assay. J Rapid Methods Autom Microbiol 1998; 7:67–76.
39. Jimenez L, Smalls S, Grech P, Bosko Y, Ignar R, English D. Molecular detection of bacterial indicators in cosmetic/pharmaceuticals and raw materials. J Ind Microbiol Biotechnol 1999; 21:93–95.

40. Jimenez L, Smalls S. Molecular detection of *Burkholderia cepacia* in toiletries, cosmetic and pharmaceutical raw materials and finished products. J AOAC Int 2000; 83:963–966.
41. Jimenez L, Bosko Y, Smalls S, Ignar R, English D. Molecular detection of *Aspergillus niger* contamination in cosmetic/pharmaceutical raw materials and finished products. J Rapid Methods Autom Microbiol 1999; 7:49–56.
42. Jimenez L, Ignar R, D'Aiello R, Grech P. Use of PCR analysis for sterility testing in pharmaceutical environments. J Rapid Methods Autom Microbiol 2000; 8:11–20.
43. Jimenez L, Smalls S, Scalici C, Bosko Y, Ignar R. PCR detection of *Salmonella typhimurium* in pharmaceutical raw materials and products contaminated with a mixed bacterial culture using the BAX™ system. PDA J Pharm Sci Technol 2001; 55:286–289.
44. Jimenez L. Simultaneous PCR detection of bacteria and mold DNA sequences in pharmaceutical samples by using a gradient thermocycler. J Rapid Methods Autom Microbiol 2001; 9:263–270.
45. Drancourt M, Bollet C, Carlioz A, Martelin R, Gayral JP, Raoult D. 16S ribosomal DNA sequence analysis of a large collection of environmental and clinical unidentifiable bacterial isolates. J Clin Microbiol 2000; 38:3623–3630.
46. Small J, Call DR, Brockman FJ, Straub TM, Chandler DP. Direct detection of 16S rRNA in soil extracts by using oligonucleotide microarrays. Appl Environ Microbiol 2001; 67:4708–4716.
47. Cheng J, Sheldon EL, Wu L, Uribe A, Gerrue LO, Carrino J, Heller MJ, O'Connell JP. Preparation and hybridization analysis of DNA/RNA from *E. coli* on microfabricated bioelectronic chips. Nat Biotechnol 1998; 16:541–546.
48. Call DR, Brockman FJ, Chandler DP. Genotyping *Escherichia coli* O157:H7 using multiplexed PCR and low-density microarrays. Int J Food Microbiol 2001; 67:71–80.
49. Eggers M, Ehrlich D. A review of microfabricated devices for gene-based diagnostics. Hematol Pathol 1995; 9:1–15.
50. Guschin DY, Mobarry BK, Proudnikov D, Stahl DA, Rittman BE, Mirzabekov AD. Oligonucleotide microchips as genosensors for determinative and environmental studies in microbiology. Appl Environ Microbiol 1997; 63:2397–2402.
51. Volokhov D, Rasooly A, Chumakov K, Chizhikov V. Identification of *Listeria* species by microarray-based assay. J Clin Microbiol 2002; 40:4720–4728.
52. English D, Scalici C, Hamilton J, Jimenez L. Evaluation of the TECRA™ visual immunoassay for detecting *Staphylococcus aureus* in cosmetic/pharmaceutical raw materials and finished products. J Rapid Methods Autom Microbiol 1999; 7:193–203.
53. Hughes D, Dailianis A, Hill L. An immunoassay method for rapid detection of *Staphylococcus aureus* in cosmetics, pharmaceutical products, and raw materials. J AOAC Int 1999; 82:1171–1174.
54. Straus D. The Growth direct™, a novel, rapid, and non destructive method for microbial enumeration. Rapid Microbiology User's Group Seminar, Validation Requirements for Rapid Microbiology. Baltimore, Maryland, February 2003.

8

Endotoxin: Relevance and Control in Parenteral Manufacturing*

Kevin L. Williams
Eli Lilly and Company, Indianapolis, Indiana, U.S.A.

1. INTRODUCTION

If ever a material seemed ill suited for use in analytical assays, it is endotoxin. As a standard, it has been domesticated, but not entirely tamed, captured from the wild, grown in captivity on rich media, chemically groomed (by solvent extraction), and trained to behave in a somewhat civilized manner in modern assays. But, still, it prances like a caged lion, back and forth, unable to escape its dual amphiphilic nature—unable to decide on the direction it should go in aqueous solution. The hydrophobic end would much rather aggregate with ends of its own kind, or stick to the plastic or glass of a test tube or container in which it resides (or parenteral closure to which it has been applied for depyrogenation validation), rather than mingle with water. Furthermore, the biological activity of endogenous endotoxin derived from different bacteria runs the gamut from apyrogenic to highly pyrogenic (the extremes in variability hold true for endotoxicity also). Indeed, laboratories

*Much of this chapter is derived from *Endotoxins: Pyrogens, LAL Testing, and Depyrogenation* by Kevin L. Williams, 2nd ed. Marcel Dekker, 2001 (www.dekker.com).

select different endotoxins for different purposes (i.e., product testing standards vs. depyrogenation validation applications) given varying empirical recovery experiences. This chapter seeks to provide an overview for endotoxin as both a parenteral contaminant and as a standard used in modern assays.

2. ENDOTOXIN NOMENCLATURE AND CLASSIFICATION AS A PYROGEN

Although used interchangeably, Hitchcock et al. have proposed reserving the term "lipopolysaccharide" (LPS) for "purified bacterial extracts which are reasonably free of detectable contaminants, particularly protein" and the term "endotoxin" for "products of extraction procedures which result in macromolecular complexes of LPS, protein, and phospholipid." Any study of endotoxin requires definition as to relative position as one of many pyrogens. Pyrogens include any substance capable of eliciting a febrile (or fever) response on injection or infection (as in endotoxin released in vivo by infecting gram-negative bacteria (GNB). Endotoxin is a subset of pyrogens that are strictly of GNB origin; they occur (virtually) nowhere else in nature. The definition of endotoxin as "lipopolysaccharide–protein complexes contained in cell walls of GNB, including noninfectious gram negatives" has also been used to denote its heterogenous nature [2].

Exogenous pyrogens include any substance foreign to the body that are capable of inducing a febrile response on injection or infection and, of course, include microbial pyrogen—the most potent and predominant of which is endotoxin. Nonmicrobial exogenous pyrogen includes certain pharmacological agents or, for a sensitized host, antigens such as human serum albumen [3]. The exactness of the term "pyrogen" has been eroded by (1) the replacement of the pyrogen assay with the Limulus amebocyte lysate (LAL) test; (2) the characterization of a number of analogous microbial host-active by-products; (3) the identification of deleterious host responses that do not include fever; (4) the discovery of LAL-reactive materials, some of which may be host-reactive but nonpyrogenic; and, (e) perhaps most significantly, the modern focus on cellular and molecular mechanisms, which are not particularly concerned with fever as a measure of biological response. Fever is now known to be only one of a host of physiologically significant aspects of proinflammatory events occurring in response to infection, trauma, and disease progression. Many forms of infection and inflammation progress without the occurrence of fever.

Dozens of microbial compounds have been found to either induce fever or activate host events that may lead to fever, some in combination with endotoxin, but may do so only weakly by themselves or at high doses [see Table 1 for a list of significant host-active microbial components (contaminants)]. The figure does not distinguish the levels of each pyrogen required to

TABLE 1 Bacterial Factors Capable of Stimulating Cytokine Synthesis

Components of gram-positive species	Components of gram-negative species
Lipoarabinomannan	Lipopolysaccharide
Lipomannans	Lipid A/lipid A-associated proteins (LAP)
Phosphatidylinositol mannosides	Outer membrane proteins (OMP)
Proteins (purified protein derivative, mycobacterial heat shock proteins, protein A)	Porins/chaperonins
Lipoteichoic acid	
Cell wall components of gram-positive and gram-negative species	Extracellular products of gram-positive and gram-negative species
Cell surface proteins	Toxins
Fimbriae and pili	Superantigens
Lipopeptides/lipoproteins	
Muramyl dipeptide/peptidoglycan	
Polysaccharides	

Source: Ref. 29.

bring about a host response, or the type of response. LAL activation is considered analogous to the response considered to be pyrogenic, but is specific for endotoxin and is capable of detecting host defense activation at subsystemic levels.

3. STRUCTURE OVERVIEW

The outer membrane of the GNB cell wall is an asymmetrical distribution of various lipids interspersed with proteins (Fig. 1). The membrane is "asymmetrical" in that the outer layer has an inner and outer leaf made up of different constituents. The outer layer contains LPS and the inner leaf contains phospholipids and no LPS. The outer face is highly charged and interactive with cations, so much so that the anionic groups can bind fine-grained minerals in natural environments [3]. LPS contains more charge per unit of surface area than any other phospholipid and is anionic at neutral physiological pH because of exposed ionizable phosphoryl and carboxyl groups [4].

The basic architecture of endotoxin (LPS) is that of a polysaccharide covalently bound to a lipid component, called lipid A. Lipid A is embedded in the outer membrane of the bacterial cell, whereas the highly variable poly-

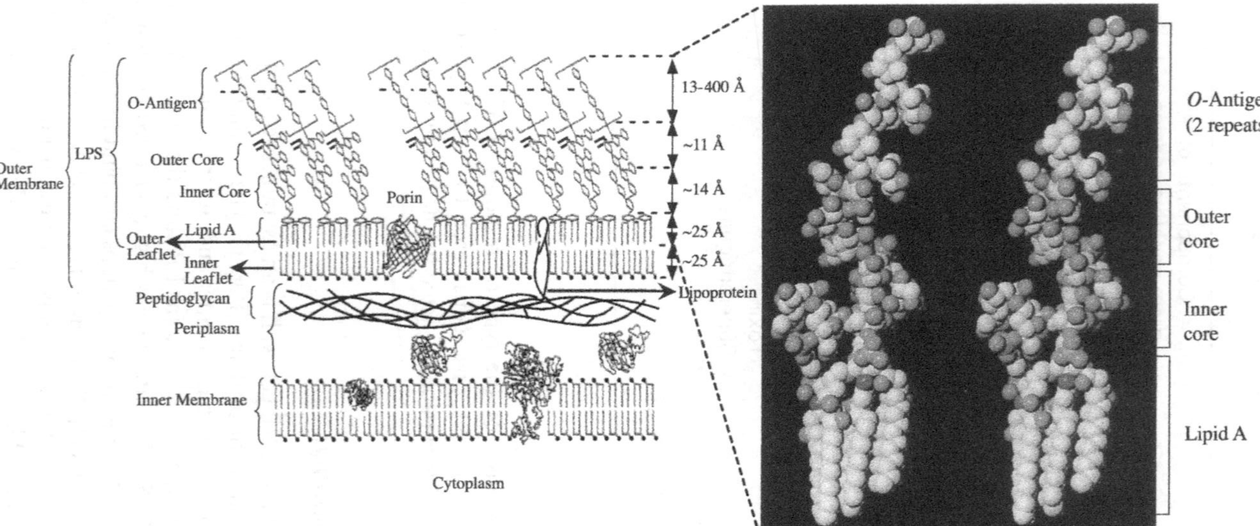

FIGURE 1 Schematic diagram of a portion of the cell envelope in gram-negative bacteria. Stereo view of the model for a single LPS molecule with two repeats of O-antigen is displayed to the right. The model of the single LPS molecule was built using the Sybyl molecular modeling software (Tripos Associates, St. Louis, Mo.), and was energy-minimizing using the Tripos force field available in Sybyl. (From Ref. 7.)

saccharide extends into the cell's environment. The long, hairlike, protruding polysaccharide chain is responsible for the GNB cell's immunological activity and is known as "O-specific side chain" (O = oligosaccharide), or "O-antigen," or "somatic antigen chain." Endogenous endotoxin (as well as purified LPS, depending on the method of extraction) contains cell membrane-associated phospholipids and proteins as well as nucleic acids and glucans [5]. Rietschel and Brade [6] have likened the structure of LPS to that of a set of windchimes. The fatty acids resemble the musical pipes and are embedded in the outer membrane parallel to one another and perpendicular to the cellular wall and to the pair of phosphorylated glucosamine sugars, which form the plate from which they dangle. The "plate" is skewed at a 45° angle relative to the membrane. Connected to the plate is the O-specific chain, which, in this analogy, is the long filament from which the windchime hangs (if, in fact, it did hang, rather than protrude, from the core sugar plate attached to the lipid A fatty acid "pipes" embedded in the outer cell layer).

The O-specific side chain consists of a polymer of repeating sugars and determines the O-specificity of the parent bacterial strain. The O-chain can be highly variable even within a given GNB species and is responsible for the LPS molecule's ability to escape an effective mammalian antigenic response because of the number of different sugars and combinations of sugars that are presented by different strains. Serological identification of members of the family Enterobacteriaceae utilizes the variation inherent in this region of LPS and is the only means of identifying certain pathogenic strains of *Esherichia coli* [8] such as *E. coli* O157, which has been implicated in recent outbreaks of food-borne illness [9]. The O-chain generally (for the most highly studied family, Enterobacteriaceae) contains from 20 to 40 repeating saccharide units that may include up to eight different six-carbon sugars per repeating unit and may occur in rings and other structures. Whereas there are in excess of 2000 O-chain variants in *Salmonella* and 100 in *E. coli*, there are only two closely related core types in the former [10] and five in the latter [11]. Strains with identical sugar assembly patterns may be antigenically different because of different polysaccharide linkages [12]. For this reason, an immune response evoked for one variant of *Salmonella* may produce antibodies oblivious to 2000 other *Salmonella* invaders.

The O-antigen side chain connects to the core oligosaccharide, which is made up of an outer core (proximal to the O-chain) and an inner core (proximal to lipid A). The outer core contains common sugars: D-glucose, D-galactose, *N*-acetyl-D-glucosamine, and *N*-acetyl-D-galactosamine (in *E. coli* and *Salmonella*). The inner core contains two uncommon sugars: a seven-carbon heptose and 2-keto-3-deoxy-D-manno-octulosonic acid (Kdo, systematically called 3-deoxy-D-manno-2-octulosonic acid) [12]. These residues are usually substituted by charged groups such as phosphate and pyrophosphate, giving the LPS complex an overall negative charge that binds bivalent cations such as Ca^{2+} and Mg^{2+}. Kdo very rarely occurs in nature outside of the LPS molecule. Kdo as a polysaccharide acts to solubilize the lipid portion of LPS in aqueous systems (as does O-antigen when it remains attached).

Nowotny [13] and Morrison et al. [14] first precipitated the lipid-rich hydrolytic fragment of LPS and named it "lipid A" (and the other more easily separated portion lipid B). Lipid A is a disaccharide of glucosamine, which is highly substituted with amide and ester-linked long-chain fatty acids. Lipid A is highly conserved across GNB LPS and varies mainly in the fatty acid types (acyl groups) and numbers attached to the glucosamine backbone. The molecular mass of lipid A has been determined to be approximately 2000 Da as a monomer, but largely exists in aggregates of 300,000–1,000,000 Da in aqueous (physiological) solutions [16]. The structure of lipid A demonstrates the general form of lipid A as seen in the *E. coli* structure and natural variants that occur in the fatty acid part of the molecule. Bacterial LPS inside the family Enterobacteriaceae share the prototypical asymmetrical structure with *E. coli* and *Salmonella*, but other GNB organisms may or may not share the structure. The fatty acid groups (acyl groups) may be in either an asymmetrical or symmetrical repeating series, and occur almost exclusively with even-numbered carbon chains. Endotoxic lipid A structures are invariably asymmetrical [15]. It is still unknown whether the endotoxic conformation "relates to a single endotoxin molecule or to a particular aggregation state..." [11].

4. WHY THE PARENTERAL FOCUS ON ENDOTOXIN?

The importance of endotoxin contamination control in parenteral manufacturing becomes apparent when confronted with four aspects of its existence. The first is its *ubiquity in nature*, the second is the *potent toxicity* it displays relative to other pyrogens, the third is its *stability* or ability to retain its endotoxic properties after being subjected to extreme conditions, and the fourth is *the relative likelihood of its occurrence* in parenteral solutions. The concern for endotoxin from a parenteral manufacturing contamination control perspective has superseded concerns for guarding against "all pyrogens" that predominated the first half of almost a century of parenteral testing. The paradigm shift of concern from pyrogens, in general, to endotoxins, in particular,* began with the testing of pharmaceutical waters and in-

---

* And perhaps full circle in the future to include more host-active bacterial and fungal artifacts.

process materials and culminated in the availability of the LAL test for most end-product items as an alternative to the United States Pharmacopeia (USP) pyrogen test in 1980 [17].

The structure of the endotoxin complex has a number of unique properties tied inseparably to its potent ability to elicit host defense mechanisms. A single bacterial cell has been estimated to contain about 3.5 million LPS molecules occupying an area of 4.9 μm^2 of an estimated 6.7 μm^2 of total outer surface area [4]. The outer membrane consists of three quarters LPS and one quarter protein. Endotoxin molecules are crucial to the survival of the GNB, providing structural integrity, and physiological, pathogenic, immunological, and nutrient transport functions. No GNB lacking LPS entirely has been found to survive in nature [6]. Endotoxin molecules are freed from bacteria by the multiplication, death, and lysis of whole cells and from the constant sloughing off of endotoxin, in a manner analogous to the body shedding skin or hair. It builds up in solution as the viable cells and skeletons of dead bacteria accumulate. When such solutions rich in GNB cellular residues find their way into mammalian blood, they retain their ability to activate host defense mechanisms in nanogram per kilogram amounts. GNB organisms occur in virtually every environment on Earth, thus making endotoxin one of the most prevalent complex organic molecules in nature. GNB have been isolated (and are being isolated still) [18] wherever man has gone—in soil, fresh and salt water, frigid oceans, and hot springs, as well as in significant amounts in ocean sediment. Some GNB organisms are able to grow in the coldest regions known (<10°C) [19]. The GNB count of sea water was taken at Woods Hole Oceanographic Institute and found to be in excess of 1 million organisms per milliliter and the sand from the shore contained almost a billion organisms per gram [20].

Given its ubiquity, one wonders at the mammalian host's exaggerated response to endotoxin. It is as though mammalian (and virtually all multicellular organisms) [21,22] and prokaryotic systems are waging war with the mammals—always on the defensive, living in fear, and shouting "barbarian at the gates" at the shadow of this invader. It is as though something larger loomed—as if the body fears another plague, or typhoid (GNB invaders) lies ready to threaten the larger society, with the body reacting accordingly. Viewed in this context, the host response to endotoxin is not as exaggerated as it would seem at first glance. The spectrum of organisms induced to fever by endotoxin is extensive, including reptiles, amphibians, fish, and even insects such as cockroaches, grasshoppers, and beetles [23]. Some animals that were initially believed to be insensitive to LPS such as rodents have subsequently been shown to respond [24].

Endotoxin achieves greater leverage in eliciting deleterious host effects than any other microbial pyrogen as is seen in the relative amount of endotoxin needed to provoke a response, which is in the nanogram per kilogram range. If endotoxin is an alarm marker for hosts in recognizing microbial invasion [25], then it elicits the loudest and most variable response. The leverage of endotoxin can be seen in the wide variety of endogenous mediators elicited, which are active in the picogram (even femtogram) per kilogram range. Therefore, a miniscule amount of endotoxin generates a huge host response in terms of both severity and variety. The complexity of the host response has frustrated efforts to devise treatments. The complexity arises from the interplay of the various mediators (cytokines) produced, which may have proinflammatory and anti-inflammatory host effects as well as synergistic effects on their own kind. A few nanograms of endotoxin translate into the production of a myriad of extremely bioactive manufactured endogenous pyrogens.

In the early use of the pyrogen assay, no attempt was made to quantitate the amount of endotoxin needed to produce a pyrogenic response in rabbits. *E. coli* and *Salmonella* were later chosen, as among the most endotoxic of families of bacteria (Enterobacteriaceae), to determine and quantify the amount of endotoxin by weight considered to be pyrogenic. In 1969, Greesman and Hornick [26] performed a study using healthy male inmates (volunteers) and found the threshold pyrogenic response (TPR) level to be about 1 ng/kg for *E. coli* and *Salmonella typhosa* (approximately 0.1–1.0) and 50–70 ng/kg for *Pseudomonas*. The same study revealed that the rabbit and human threshold pyrogenic responses are approximately the same. Therefore, the amount of purified *E. coli* needed to initiate pyrogenicity in both humans and rabbits is approximately 1 ng/kg, which represents about 25,000 *E. coli* bacterial cells [27]. In terms of whole cells, the injection of an estimated 1000 organisms per milliliter (10,000 per kilogram) of *E. coli* causes a pyrogenic reaction in rabbits, compared with 10^7–10^8 organisms per kilogram of gram-positive or fungal organisms [28]. The fact that many non-LPS products have been recently identified as macrophage activators and that many are associated with devastating diseases supports an underlying theme that there is a wide variety of potential modulators of adverse host effects (including fever) that are not endotoxins but that may proceed by endotoxin-like mechanisms and with endotoxin-like potencies when presented by infecting organisms (although not necessarily relevant from a parenteral manufacturing perspective) (Table 1).

Peptidoglycan (PGN) is usually described only in association with gram-positive bacterial (GPB) infection, but PGN has been found to be released into hosts in several instances of GNB infection [31]. PGN is released (by GPB) during infection and can reach the systemic circulation [32]. Sensitive methods of quantifying PGN and its subunits in a clinical setting have yet to be developed,* leaving the levels associated with GPB sepsis largely unknown.† The incidence of GPB sepsis in the hospital setting is known to equate to that caused by GNB organisms, although studies have proposed that PGN and LTA act synergistically [34,35].

Given the plethora of evidence for nonendotoxin pyrogens (albeit less potent than endotoxin), it remains to be seen which components will be excluded and which will remain classified as "pyrogens." It does seem intuitive that given the range of prokaryotic cellular debris, endotoxins will not be the only significant pyrogenic (or bioactive) harbinger of bacterial origin.

A relevant note concerning the lack of attention given to nonendotoxin cellular components in parenteral manufacturing is the degree of difficulty researchers encounter in obtaining the materials in a pure state devoid of endotoxin. The presence of endotoxin overrides many efforts to study non-endotoxin components because of its potency and can affect research study endpoints at almost undetectable background levels (fg/mL) compared with the levels necessarily used in the study of non-LPS substances (typically in μg mg/mL) (Table 2).

Beverage [3] describes the enduring nature of the GNB cell wall as "strong enough to withstand ~3 atm of turgor pressure, tough enough to endure extreme temperatures and pHs (e.g., *Thiobacillus ferrooxidans* grows at a pH of ~1.5), and elastic enough to be capable of expanding several times their normal surface area. Strong, tough, and elastic..." Endotoxin is extremely heat-stable and remains viable after ordinary steam sterilization and normal desiccation, and easily passes through filters intended to remove whole bacteria from parenteral solutions. Only at dry temperatures exceeding 200°C for up to an hour do they relent.

The amphiphilic nature of the LPS molecule also serves as a resilient structure in solution, with the hydrophobic lipid ends adhering tenaciously to hydrophobic surfaces such as glass, plastic, and charcoal [27], as well as to one another. Many of the most basic properties of LPS are those shared with lipid bilayers in general, which form the universal basis for all cell membrane structures [36]. In aqueous solutions, LPS spontaneously forms bilayers in which the hydrophobic lipid A ends with fatty acid tails that are hidden in the interior of the supramolecular aggregate as the opposite hydrophilic poly-

---

*Or at least widely accepted as the silkworm larvae plasma (SLP) method is a sensitive detection method for PGN.

†Although muramic acid has been used as a sensitive marker for gas chromatography mass spectrometry (GC-MS) detection of GP cellular residues in clinical specimens (septic synovial fluids) at levels of ≥30 ng/mL [33].

TABLE 2 The Relative Biological Activity of Cytokine-Inducing Microbial Components Compared to LPS

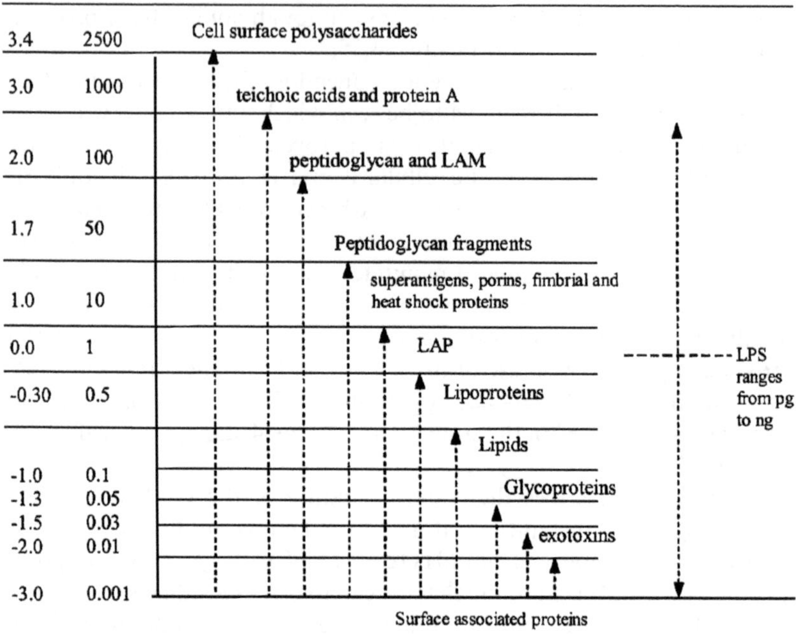

Source: Ref. 30.

saccharide ends are exposed to and subject to solubilization in the aqueous environment. A property adding to the stability of LPS as a lipid bilayer is its propensity to reseal when disrupted, thus preserving the structure's defense against the environment.

A central question that arose with the proposal to replace the rabbit pyrogen test with the Limulus amebocyte lysate test was (and still is): How can one be sure in testing only for endotoxin that other microbial pyrogens will not be allowed to go undetected in the parenteral manufacturing process? In part, we have answered the question by considering the ubiquity, stability, and potency (based on severity of host response), combined with the relative likelihood of endotoxin-bearing GNB as parenteral contaminants. The minimal growth requirements of GNB allow their growth in the cleanest of water. Conversely, the answer can be found by disqualifying from undue concern (1) the environmental predisposition of non-GNB organisms that prevent them from proliferating in largely water-based parenteral manufacturing processes; (2) the relative ease of degradation of their by-products (except heat-stable GPB exotoxins that derive from microbes having significant growth requirements); and (3) modern aseptic manufacturing procedures required by current good manufacturing practices (cGMPs).

5. CONTAMINATION CONTROL PHILOSOPHY IN PARENTERAL MANUFACTURING

Endotoxin is a concern for people only when it comes into contact with the circulatory system. The two relevant mechanisms for such contact involve infection and medically invasive techniques, including injection or infusion of parenteral solutions. A notable exception to limiting the concern for endotoxin to blood contact is the effect that minute, almost undetectable, quantities of endotoxin may have on cell cultures used in pharmaceutical manufacturing. The manufacture of biologics makes use of complex cell culture media including the addition of fetal bovine serum (FBS) as a growth factor (which has been associated with microbial contamination)* to grow mammalian cells used in recombinant and monoclonal expression systems. Serum has presented manufacturers (and clinicians) difficulties in quantifying and reproducing endotoxin levels because of little-understood interference factors. The regulatory precautions set in place are, in many cases (if not most), because of the poor probabilities associated with finding contamination by quality control (QC) sampling techniques. The generally accepted sterility acceptance level (SAL) has been often repeated to be 10^{-6} (i.e., one possible survivor in a million units), but according to Akers and Agalloco [37], the value was selected as a convenience. They maintain that 10^{-6} is a minimal sterilization expectation and should be linked "to a specific bioburden model and/or particular biological indicator... (otherwise) it is a meaningless number that imparts little knowledge on the actual sterilization process."

Bruch [38] relates that the probability of a survivor per item (PSI) for a can of chicken soup is 10^{-11}, whereas the assurance provided by the USP sterility test alone is not much better than 10^{-2} given a 20-item sampling and is, as Bruch says, because of the rigorous heating cycles developed by the canning industry to prevent the possibility of survival of *Clostridium botulinum*. Bruch maintains that the industry has "never relied on a USP-type finished product sterility test to assess the quality of its canned goods... (because) the statistics of detecting survivors are so poor that the public confidence... would be severely compromised through outbreaks of botulism." Bruch cites the generally accepted sterility assurance for a large volume parenteral item as 10^{-9} and 10^{-4} for a small-volume parenteral that has been aseptically filled and sterile-filtered as opposed to terminally sterilized. The

* Being a bovine blood product subject to temperature abuse and containing GNB.

TABLE 3 Probability of Survivor Estimates for Sterilized Items

Item	Probability of survivor/unit
Canned chicken soup[a]	10^{-11}
Large-volume parenteral fluid	10^{-9}
Intravenous catheter and delivery set[a]	10^{-6}
Syringe and needle[a]	10^{-6}
Urinary catheters[a]	10^{-3}
Surgical drape kit[a]	10^{-3}
Small-volume parenteral drug (sterile fill)	10^{-3}
Laparoscopic instruments (processed with liquid chemical sterilants)[b]	10^{-2}

[a] Dosimetric release: no sterility test.
[b] Limits of USP sterility test: $10^{-1.3}$ (with 95% confidence).
Source: Ref. 38.

apparent contradiction in the necessity of more stringent sterility assurance for a can of soup than for a parenteral drug is because of the ability of organisms to grow in soup as opposed to the likelihood of such growth in the parenteral manufacturing environment (Table 3).

The predominant potential source of endotoxin in a pharmaceutical manufacturing environment is the purified water used as a raw material (also used in component sterile rinse depyrogenation processes). Many different grades of water are used and may be variously labeled according to their origin, the treatment they have undergone, quality, or use, and different groups employ different nomenclatures [39]. The only waters that require endotoxin monitoring are "water for injection" (WFI) and "water for inhalation" are prepared via a validated distillation or reverse osmosis process. Distillation is the preferred method and results in sterile, endotoxin-free condensate. However, any water may become contaminated via a number of subsequent distribution or storage mechanisms including the cooling or heating system, storage container, or distribution method such as hoses [39].

6. DEVELOPING AN ENDOTOXIN CONTROL STRATEGY (ECS) FOR DRUG SUBSTANCES/EXCIPIENTS

Finished products often contain ingredients in addition to the active drug substance. Excipients serve as solvents; solubilizing, suspending, thickening, and chelating agents; antioxidants and reducing agents; antimicrobial preservatives; buffers; pH-adjusting agents; bulking agents; and special additives [40]. Recent endotoxin excipient testing references [41,42] dictate limits for some parenteral excipients and require the establishment of endotoxin quality control tests. However, the majority of parenteral excipients still do not have established endotoxin limits. The Food and Drug Administration (FDA) Guideline on Validation of the LAL Test [43] outlines the determination of limits for "end-product" testing and can be misapplied to drug substance and excipient testing. Relevant activities to be established to gain control over a given drug manufacturing process from an endotoxin control perspective include:

1. Identifying the types of excipients used in various drugs
2. The relative amounts of those excipients in each drug type
3. Relevant tolerance limits (TLs) for drug substances and excipients given (1) and (2).

This exercise should establish that proposed limits are appropriate and that existing excipient and drug substance limits used in the manufacturing process will not allow an associated drug product to fail its end-product testing. As the cost of drugs derived from biotechnology increases, so do the business-related requirements for ensuring that the raw materials that go into making the intermediates of the manufacturing process as well as end-products meet appropriate, relevant, and stringent predetermined specifications.

Every marketed product has a level of endotoxin safely tolerated (i.e., an amount below the tolerance limit), which is defined as $TL = K/M$, where K is the threshold pyrogenic dose (TPD) constant in endotoxin units (EU) per kilogram and M is the maximum human dose in units per kilogram of body weight [70 kg/hr as per FDA Guideline] [43]. The TPD is the level of endotoxin capable of eliciting a pyrogenic response in a patient. The relevant dose is that administered in an hour. The TPD constant (K) differs depending on the route of administration (parenteral or intrathecal/radiopharmaceutical). The formula is straightforward except for the units, which vary from product to product depending on the manner in which the product is administered. For drugs administered by weight, the weight to be used is that of the active drug ingredient in milligrams or in units per milliliter. For drugs administered by volume, the potency is equal to 1.0 mL/mL.*

The formulas adjust for a product's potency based on either the weight of the active ingredient or the volume of the drug administered; they constitute a package for determining "how much the product can be diluted and still

*See Appendix D of the FDA Guideline for exceptions to the general formulas including the use of radiopharmaceutical and intrathecal doses, and the use of pediatric weights.

detect the limit endotoxin concentration" [43]. An ECS is a tool to organize and facilitate the laboratory testing of drug substance and excipients at appropriate tolerance limit (and therefore test dilution) levels [44,45]. An example strategy is shown in Table 7.

The table allows the user to view TPD in terms of total EUs delivered in a dose. This rationale for drug substances (active ingredients) and excipients has not been described in any guideline (in that only tolerance limit calculations for "end-products" are described), but the necessity for relevant testing has become a clear expectation as evidenced by the publication of recent monographs for mannitol and sodium chloride and by ongoing excipient harmonization efforts.

In lieu of using the table, a drug substance tolerance limit adjusted for excipients can be calculated:

$$\text{TL (drug substance with excipients (ds/e))} = \frac{\{350 - ((\text{TLe}_1\text{We}_1) + (\text{TLe}_2\text{We}_2)\ldots)\}}{\text{W}_\text{A}}$$

where

TLe_1 = the tolerance limit of excipient 1
We_1 = the weight of excipient 1 per dose of active drug
W_A = the weight or unit of active drug per dose.

Note that the formula ((...)) indicates that all relevant excipients without an exclusion rationale should be included in the calculation. Compare the calculated value of 7.48 EU/mg to the end-product tolerance limit calculated in the formula: TL = 5.0 EU/kg/(35 mg/70 kg) = 10 EU/mg (and 7.0 as assigned in Table 4).

For the above example, the formula would be filled in as follows:

$$\text{TL(ds/e)} = \frac{\{350\ \text{EU} - ((0.0025\ \text{EU/mg} \times 75\ \text{mg}) + (0.005\ \text{EU/mg} \times 50\ \text{mg}) + (1.0\ \text{EU/mg} \times 87.5\ \text{mg}))\}}{35\ \text{mg}}$$

$$\text{TL(ds/e)} = \frac{\{350\ \text{EU} - ((0.19\ \text{EU}) + (0.25\ \text{EU}) + (87.5\ \text{EU}))\}}{35\ \text{mg}}$$

$$= \frac{\{350\ \text{EU} - 87.94\ \text{EU}\}}{35\ \text{mg}}$$

$$\text{TL(ds/e)} = \frac{262.06\ \text{EU}}{35\ \text{mg}} = 7.48\ \text{EU/mg}$$

TABLE 4 Endotoxin Control Strategy Steps

1.	Drug product constituent and weight	
Obtain the unit formula for a given drug product	API	1.0 mg
	Mannitol	2.14 mg
	NaCl	1.43 mg
	Polysorbate	2.5 mg
2.	**Constituent**	**Weight/dose**
Determine the relative amounts of API and excipients based on the dose of API	API	35 mg
	Mannitol	75 mg
	NaCl[a]	50 mg
	Polysorbate 80[b]	87.5 mg
3.	**Constituent**	**Proposed or existing TL assigned**
Assign existing TLs or propose TLs for the drug substance and excipients	API	nmt[d] 7.0 EU/mg
	Mannitol[c]	nmt 0.0025 EU/mg
	NaCl	nmt 0.005 EU/mg
	Polysorbate 80	nmt 1.0 EU/mg

4.	Constituent	Weight/dose	Proposed TL	EU's
Ensure that the final product cannot exceed the TPD given each assigned TL	API	35 mg	7.0 EU/mg	245 EU
	Mannitol	75 mg	0.0025 EU/mg	0.19 EU
	NaCl	50 mg	0.005 EU/mg	0.25 EU
	Polysorbate 80	87.5 mg	1.0 EU/mg	87.5 EU
			Total EU/dose =	332.94 EU

5. Document both the "control strategy" and any "exclusion rationale(s)" used for excipients deemed not to require endotoxin testing.

[a] See European Pharmacopoeia (3rd Ed. 1997) monograph for Sodium Chloride (p. 1481) (41).
[b] No endotoxin limit in monographs.
[c] See European Pharmacopoeia (3rd Ed. 1997) monograph for Mannitol (p. 1143) (41).
[d] Not more than can be interpreted as less than since a test containing the limit concentration of endotoxin would be positive and hence fail the test.

An ECS is appropriate for drug products containing:

1. Numerous excipients
2. Significant (large amounts of one or more) excipients relative to the actives
3. Excipients with tolerance limits set with relatively high limits (perhaps because of difficult/incompatible laboratory tests or ill-conceived historical method of determining its limit)
4. Drug substances and/or excipients with tolerance limits previously calculated using end-product formulas
5. Excipients of natural (animal or plant) origin.

Conversely, an ECS may be unnecessary for drug products containing:

1. Few or no excipients (drug substance = drug product)
2. Excipients in miniscule amounts relative to the actives
3. Excipients with very low tolerance limits (i.e., those with compendial requirements)
4. Excipients incapable of adding appreciable endotoxin because they are antimicrobial and/or inhospitable to microbes due to their method of manufacture, nature or origin, or as a miniscule constituent.

As an example, Cresol (hydroxytoluene) is an antimicrobial excipient obtained from either sulfonation or oxidation of toluene [46]. Therefore, it is (a) manufactured from materials inhospitable to microbial growth (b) at temperatures that are depyrogenating, and (c) is unlikely to be post-manufacture-contaminatable because of the lack of water needed to support microbial growth.

End-product testing provides a test of the total contents of a given vial (see Table 5 below for a proof of this). The ECS is concerned with providing in-process testing that demonstrates that when the parts are combined, they cannot cause the end-product to fail its specification. The trend in drug development is clearly toward greater complexity. New biologically derived drugs may contain a number of unusual excipients in significant amounts (e.g., new sustained-release parenterals contain excipients not traditionally found in nonsustained released drugs [47] and/or present in large quantities). An endotoxin control strategy can provide a frame of reference to determine appropriate drug substance and excipient limits (as opposed to their arbitrary assignment). Although there are arguably safety factors included in endotoxin limit calculations (see "Understanding and Setting Endotoxin Limits") [48], there are also confounding factors such as multiple parenterals given to patients simultaneously. A complete process to account for a drug's entire potential endotoxin contents will aid manufacturers in gaining greater endotoxin control.

TABLE 5 BET Calculations—Active Vs. Total Solids

Calculated by active drug concentration	Calculated by total solids (TS) method (do not use this method, for illustration only)
If active drug is 200 mg and is reconstituted with 20 mL, then the solution is 10 mg/mL. The potency, TL, and λ constitute a "system" to determine the appropriate limit and subsequent dilution (MVD)	If TS of drug is 1 g (this value is not constant as identical drugs made by different manufactures will differ in excipient use and therefore TS)
TL = K/M = 5.0 EU/kg/(200 mg/70 kg) = nmt 1.75 EU/mg drug	TL = 5.0 EU/kg/(1000 mg/70 kg) = nmt 0.35 EU/mg (TS method)
Because MVD = TL × PP/λ, $\text{MVD} = \dfrac{1.75\ \text{EU/kg} \times 10\ \text{mg/mL}}{0.01\ \text{EU/mL}}$	$\text{MVD} = \dfrac{0.35\ \text{EU/mg} \times 1000\ \text{mg/20 mL}}{0.01\ \text{EU/mL}}$
= 1:1750 dilution	= 1:1750 dilution

7. BACTERIAL ENDOTOXIN TEST (BET) STANDARDIZATION

Tied to the concept of a "standard" endotoxin is the historical determination of a threshold pyrogenic dose for endotoxin. The establishment of a defined, specific threshold pyrogenic response level allowed the concept to be established that a certain amount of endotoxin is allowable and a certain amount of endotoxin should not be delivered into the bloodstream. The advent of LAL allowed the quantitation of endotoxin as a contaminant. In turn, quantitation allowed for the creation of specific and relevant endotoxin limits for manufactured drug products, raw materials, active ingredients, devices, components, depyrogenation processes, and in-process samples that constitute the legal requirement for releasing to market products that are not considered "adulterated" by international regulatory bodies.

Today's user of the LAL test rightly views such concepts as the bread and butter of endotoxin testing, but it is good to appreciate the degree to which today's system of endotoxin quantitation has progressed in that:

1. "Quantitation" in the rabbit assay was limited to a pass/fail response (rabbit response = 0.6°C temperature rise).
2. The pyrogen test was initially established without attempting to quantitate the amount of endotoxin necessary to produce a febrile response.

3. Early LAL testing used the weight of dried bacterial endotoxins in nanograms initially with various GNB organisms and then with a specific *E. coli* strain without accounting for the variable potency of a given weight of endotoxin.

None of the early tests could have been used effectively to develop product-specific tolerance limits as they exist today, much less provide the degree of in-process control needed for modern pharmaceutical manufacturing. In some respects, the 10- to 1000-fold greater sensitivity of the LAL test created the "luxury" of controversy on several fronts. A whole new system of relating the new assay to the existing test had to be developed to avoid unnecessary product test failures because of the greater sensitivity of the LAL assay [49]. The "system" included the formation of, or association with: (1) the EU* as a measure of relative biological activity; (2) the TL (endotoxin limit concentration); (3) the maximum valid dilution (MVD) to relate the product dose to the allowable endotoxin content (realizing that a positive LAL response in any given solution as in the pyrogen assay would be inappropriately stringent); and (4) the lysate sensitivity (lambda (λ)) to standardize the relative reactivity of each LAL to each control standard endotoxin (CSE). Prior to this "system," several of the principals of the early LAL assay expressed concern that the greater sensitivity of the assay would end up becoming an apparent disadvantage used by some to confound industry efforts to develop the assay as a replacement for the rabbit pyrogen test. ("I hope that we do not turn the advantage provided by the greater sensitivity of the *Limulus* test into a problem." Jack Levin [50].)

A number of criticisms were put forward with the use of the first assigned endotoxin standard. The major criticisms included the fact that the standard was not "pure" lipid A for which the chemical formula had been defined and the fact that other more potent endotoxins were available. The criticism concerning the purity of the endotoxin was discounted because of the need for a readily soluble standard (lipid A being insoluble). The goal of obtaining a standard endotoxin largely free of biologically active proteins, peptides, polynucleotides, and polysaccharides had been achieved. As for the potency of the new endotoxin reference standard, it was believed that an "average" potency would be more relevant to the testing of a wide range of endotoxins, with a range of potencies likely to be encountered in real world testing.

As recently as the late 1990s, there have been as many as five different official international standards (IS) active at once [51]. For an international

---

*EU is defined as one-fifth the amount of *E. coli* (EC-2) endotoxin required to bring about the threshold pyrogenic response (as established by Greisman and Hornick as 1 ng/kg).

manufacturer, this meant either the construction of a singe test designed to overlap all the test requirements, including the use of a control standard calibrated against each official reference standard, or the performance of multiple testing of each lot of drug material. An initial IS for endotoxin testing was established by the World Health Organization's (WHO) Expert Committee on Biological Standardization (ECBS) in 1987 [52]. The first international standard was calibrated against the U.S. national standard, EC5. However, the potency assignments for the semiquantitative LAL gel clot and photometric tests did not agree. Most of the collaborative data consisted of gel clot testing; therefore, the ECBS of WHO assigned IS-1 as a gel clot standard [53]. The assigned potency was 14,000 IU/ampule.

In 1994, the ECBS of WHO acknowledged that the use of the photometric tests (endpoint and kinetic chromogenic and turbidimetric) had greatly grown in terms of the number of LAL users since IS-1 was established and recognized the need for a common standard for both gelation and photometric tests [53]. The USP made available 4000 vials of a batch of USP-G/EC-6 for the proposed WHO second international collaborative study. Therefore, the stage was set for a comprehensive study organized by the WHO involving the United States, European, and Japanese Pharmacopeias.

Poole and Das [53] describe the ambitious aims of the study:

1. Calibrate the IS compared with EC5 (USP-F) (although superseded by EC6, it was the primary calibrant for IS-1 and the JP reference standard) and assign a single IS unit for all endotoxin applications
2. Compare the current IS (IS-1), EC5, and the candidate standard (CS) using LAL gelation, and kinetic and endpoint assays (chromogenic and turbidimetric)
3. Determine the relationship of IU to EU
4. Compare the CS to the U.S., European (BRP-2), and Japanese reference standards.

A common lysate (supplied by Associates of Cape Cod, Woods Hole, MA) was used in 24 laboratories using two assays and an "in-house" lysate (i.e., whatever was already being used in that laboratory). In all, the participants performed a total of 108 gel clot assays. A total of 33 assays was performed using endpoint chromogenic (3 laboratories), kinetic chromogenic (13 laboratories), and kinetic turbidimetric (12 laboratories) tests. In the gel clot tests, the geometric mean for the candidate standard sublots (therefore, both sublots were considered as a single lot) did not differ significantly from one another, from laboratory to laboratory, or from LAL to LAL reagent source [53].

The candidate standard geometric mean result for each assay type obtained in terms of EC5 is shown in Table 6 (Fig. 2).

TABLE 6 Results Obtained in WHO IS-2 Collaborative Study

Assay Type	Mean recovery	# Tests (n)
Gelation assay	10,300 EU/vial	103
Kinetic chromogenic assays	11,700 EU/vial	13
Kinetic turbidimetric assays	11,800 EU/vial	11
Chromogenic endpoint assays	11,200 EU/vial	3
All assays (gel and photometric)	10,400 EU/vial	68
IS-2 assigned value	10,000 IU/vial	

Source: Ref. 53.

In January 2001, USP 25 created the first harmonized microbiological test, the BET, concomitant with the formation of IS-2 as an international standard endotoxin. Overall, the newly harmonized test has received high marks industrywide for ease of understanding and practicality when applied to real-world test conditions. Furthermore, to multinational companies that must meet international requirements, the benefits of the harmonized

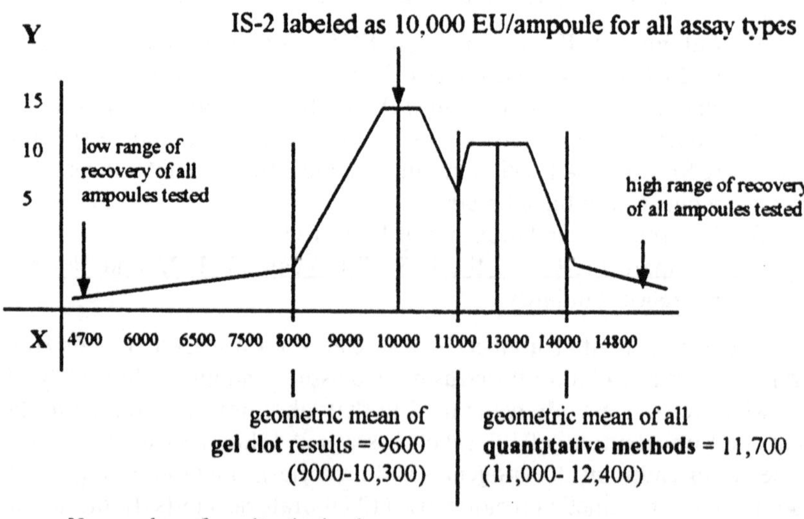

FIGURE 2 Graphical representation of the range of geometric means obtained and the grouping of results for all valid gelation and photometric assays as n (number of assays) vs. EU of EC5 per ampoule of candidate standard. (Derived from 53.)

test cannot be overstated. In a nutshell, the benefits of the harmonized test include:

An elevation of the status of nongel clot tests, including kinetic and endpoint chromogenic and turbidimetric tests, by including them.

The gel clot assay has been split into a limit test or an assay—something that is fairly routine but not specified previously and the limit test no longer requires the confirmation of label claim with each block of tubes tested.

The requisite positive product control standard recovery has been widened from 50–150% to 50–200%, which is in effect the recovery associated with the gel clot assay (one twofold dilution). This change only allows for one's test to overestimate the recovery of endotoxin all the more (200% vs. 150% recovery).

8. ORIGIN AND IMPORTANCE OF LAL

The rabbit pyrogen assay served as the only official pyrogen test for 37 years. However, during the early 1960s, several events occurred, which would eventually lead to the development of a seemingly unlikely replacement: a blood product (lysate) derived from the horseshoe crab, *Limulus polyphemus*. The importance of the changes brought about in the pharmaceutical industry by the switch from the in vivo based rabbit pyrogen test to the in vitro bacterial endotoxin test is often underappreciated for a couple of reasons. First, the labor intensity inherent in the rabbit pyrogen assay served as a lid on the amount of in-process testing that could be realistically be expected to be performed (from a cost and resource perspective) to support the manufacture of parenteral lots (100 rabbit pyrogen tests a day would be a colossal effort). The advent of LAL testing has allowed the broad application of cGMPs as they relate to the detection of endotoxins across the entire manufacturing process. The quality control testing of only the later forms of a parenteral drug provides a greatly reduced probability of detecting a contaminated unit of that material from a statistical standpoint and would make it impossible to preclude the use of contaminated materials prior to manufacture as a means precluding the manufacture (and subsequent destruction) of an expensive biological lot.

Modern pharmaceutical manufacturing processes include sampling and LAL testing of not only the finished (beginning, middle, and end of lot), bulk, and active pharmaceutical ingredient (API) material, but also in-process materials including containers and closures, sterile water, bulk drug materials, and, more recently, excipients. Therefore, the pyrogen assay included the housing of dedicated rabbits and was very expensive, and its expansion was unlikely given cost and other resource constraints. Secondly, the inability to quantify endotoxin associated with pyrogen testing acted as a "blind spot" to restrict the improvement of processes that are now readily monitored given the sensitivity and quantification associated with the LAL test. It is difficult to work toward lower specifications when performing an assay that has an inherent invisible pass/fail result. Modern biopharmaceuticals may indeed contain trace amounts of endotoxin or may have activity (i.e., interferon) mimicking endotoxin and, in such cases, the accurate and reproducible quantification of these minute levels as well as the differentiation of interference and endotoxin content become paramount to demonstrating that allowable levels are present.

The first application of the clotting reaction discovered by Levin and Bang was made by Cooper, Levin, and Wagner in their use of the "pre-gel" to determine the endotoxin content in radiopharmaceuticals in 1970 [54]. According to Hochstein [55], Cooper was a graduate student at Johns Hopkins in 1970 and worked for the Bureau of Radiological Health. That summer, Cooper persuaded the Bureau of Biologics (BoB) group, led by Hochstein, that a lysate from the horseshoe crab's blood would be useful in detecting endotoxin in biological products. Given the short half-life and stringent pyrogen requirements associated with radiopharmaceutical drugs, Cooper believed that LAL could be used to accomplish the improved detection of contaminated products. Though Cooper left the BoB to finish his graduate studies, Hochstein continued the Bureau's efforts to explore the use of LAL in the testing of drug products.

The potential for improvement in the area of pharmaceutical contamination control was evident in Cooper, Hochstein, and Seligman's very first application of the LAL test involving a biological [56]: the results of 26 influenza virus vaccines included as a subset of a 155 sample test using LAL varied from lot to lot by up to 1000-fold and revealed endotoxin in the 1 μg range in the 1972 study. Cooper later pointed out [57] that newer vaccines used in mass inoculation of Americans for A/Swine virus were subsequently required to contain not more than 6 ng/mL of endotoxin, a level that could not be demonstrated with pyrogen testing. Suspected adverse reactions were reported prior to the inception of the LAL assay and were an expected part of some drug reactions such as that associated with L-asparaginase antileukemic treatment as a product of *E. coli* [58]. A third early application (radiopharmaceuticals and biological vaccines mentioned above) involved the detection of endotoxin in intrathecal injections (into the cerebrospinal fluid) of drugs. Cooper and Pearson report [57] that 10 such samples implicated in adverse patient responses were obtained, tested by LAL, and all 10 reacted strongly. The rabbit pyrogen test was negative for all samples when tested on a dose-per-weight basis. They concluded that the rabbit pyrogen test was not sensitive enough for such an application given that endotoxin was determined to be at least 1000 times more toxic when given intrathecally.

9. LAL DISCOVERY

In 1956, Bang, at the Marine Biological Laboratories in Massachusetts, was studying the effects of what he initially believed to be a bacterial disease causing the intravascular coagulation (coagulopathy) of the blood of a horseshoe crab in a group that he was observing. He isolated the bacterium from an ill *Limulus*, believing it to be a marine invertebrate pathogen such as (he cites) the marine bacterium *Gaffkia*, which killed lobsters. He described the basic observation that prompted him to publish the landmark study in *A Bacterial Disease of Limulus polyphemus* [59] as follows:

> Bacteria obtained at random from fresh seawater were injected into a series of horseshoe crabs (*Limulus polyphemus*) of varying sizes. One *Limulus* became sluggish and apparently ill. Blood from its heart did not clot when drawn and placed on glass, and yet instant clotting is a characteristic of normal *Limulus* blood... The bacteria caused an active progressive disease marked by extensive intravascular clotting and death. Injection of a heat-stable derivative of the bacterium also caused intravascular clotting and death. Other gram-negative bacteria or toxins also provoked intravascular clotting in normal limuli. When these same bacteria or toxins were added to sera from normal limuli, a stable gel was formed!

Following Bang's initial observations, he paired up with a hematologist, Levin, at the suggestion of another colleague. Together, they explored the requisite coagulate factors of *Limulus* and published a paper entitled *The Role of Endotoxin in the Extracellular Coagulation of Limulus Blood* [59] in an effort to "study the mechanism by which endotoxin affects coagulation in the *Limulus*, and to elucidate the mechanism by which endotoxin exerts its effect in a biological system that may be less complex than that found in mammals." In this study, they made a number of observations:

1. The amebocyte is necessary for clotting.
2. Clotting factors are located only in the amebocytes (not in the blood plasma).
3. The formation of a gel clot reaction occurs by the conversion of a "pre-gel" material on addition of gram-negative bacteria.

Levin and Bang demonstrated that extracts of the amebocytes gelled in the presence of GNB endotoxin. In the introduction of that early paper, they described the phenomenon that would later become the basis for the LAL assay.

> *Limulus* blood contains only one type of cell called the amebocyte. When whole blood is withdrawn from the *Limulus*, a clot quickly forms. Thereafter, this clot shrinks spontaneously, and a liquid

> phase appears. Under appropriate conditions, this liquid material has the capability of gelling when it is exposed to bacterial endotoxin, and is defined here as *pregel*... The results (of the study that served as the basis for their April 1964 publication) demonstrate that cellular material from the amebocyte is necessary for coagulation of *Limulus* plasma, and that plasma free of all cellular elements does not clot spontaneously nor gel after addition of endotoxin. [60]

Levin and Bang not only used the initial bacterial isolate (they had now identified it as a *Vibrio* species) to bring about gelation, but they also used *E. coli* (Difco, Becton Dickinson and Company, Franklin Lake, NJ) because they now believed that endotoxin common to GNB was bringing about the gelation phenomenon. Their study revealed that agitation of the amebocytes (amebocyte disruption) aided in the production of the pregel (i.e., in the production of gel precursor most susceptible to subsequent endotoxin clotting) and that the rate of gelation of pregel was directly related to the concentration of endotoxin in the mix. In their third paper, Levin and Bang [61] described the "striking similarities between *Limulus* amebocytes and mammalian platelets..." during cellular coagulation on exposure to endotoxin.

10. HEMOLYMPH COAGULATION IN *LIMULUS* AND *TACHYPLEUS*

Invertebrates lack adaptive immune systems and rely on innate immunity to antigens common to pathogenic organisms. Nakamura et al. have extensively studied the hemolymph (blood) system of the Japanese horseshoe crab (*Tachypleus tridentatus*) and found that amebocytes contain two types of granules—large (L) and small (S)—that contain the clotting factors, proteins, and antimicrobials that are released via a process called degranulation into the crab's plasma [62]. Regardless of the relative simplicity of the crab's defense system (the amebocyte), Nakamura et al. consider it to be "a complex amplification process comparable to the mammalian blood coagulation cascade" and "very similar to those of mammalian monocytes and macrophages..." [63]. The ability of *Limulus* and *Tachypleus* blood to clot and form webs of fibrin-like protein serves as a means of entrapping and facilitating the deactivation of both invading organisms and endotoxin by the release of additional antiendotoxin and antimicrobial factors. The clotting action also serves to prevent leakage of hemolymph at external sites of injury.

The "fibrinogen-like" invertebrate protein is called *coagulogen* in its soluble form and *coagulin* in its (postenzyme-activated) gelled form [63]. The conversion of coagulogen to coagulin is mediated by the sequential activation (cascade) of several zymogens arising from the single blood cell of *Limulus* or *Tachypleus* (the amebocyte or granulocyte). The L-granules contain all the clotting factors for hemolymph coagulation, protease inhibitors, and anti-LPS factor, as well as several tacylectins with LPS-binding and bacterial-agglutinating activities (Fig. 3).

On GNB invasion of the hemolymph, hemocytes detect LPS on their surface and release their granule contents (degranulate). The known biosensors consist of coagulation factor C and factor G, which serve as the triggers for the coagulation cascade that converts soluble coagulogen to the insoluble coagulin gel. These two serine protease zymogens are autocatalytically activated by LPS and (1,3)-β-D-glucan, respectively. The LPS-initiated cascade (via activation of the proclotting enzyme) involves three serine protease zymogens: factor B, factor C, and proclotting enzyme. The final step of the clotting reaction involves the creation of coagulin from coagulogen by the excision of the midsection of the protein, called peptide C. Without peptide C, the monomers form AB polymers consisting of the NH_2-terminal A chain and the COOH-terminal B chain covalently linked via two disulfide bridges [65] (Fig. 4).

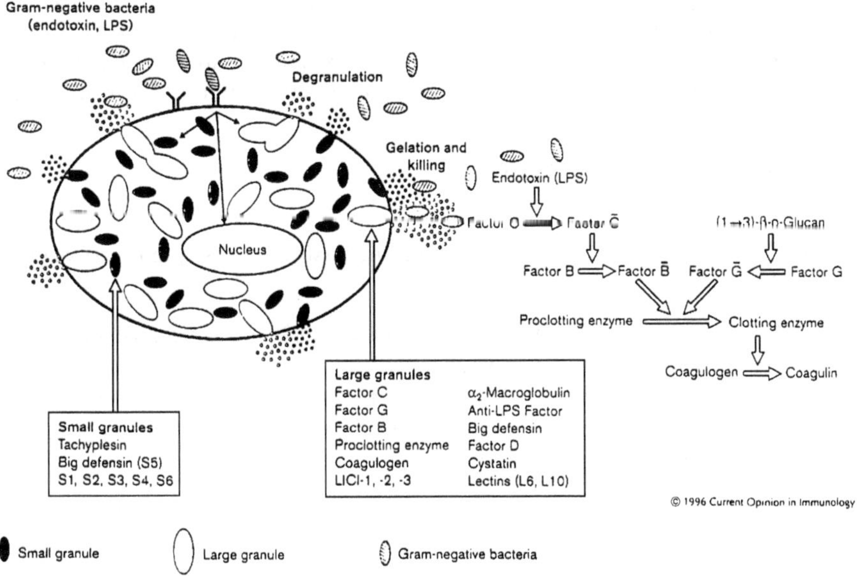

FIGURE 3 The conversion of coagulogen to coagulin is mediated by the sequential activation (cascade) of several zymogens arising from the single blood cell of *Limulus* or *Tachypleus.* (From Ref. 64.)

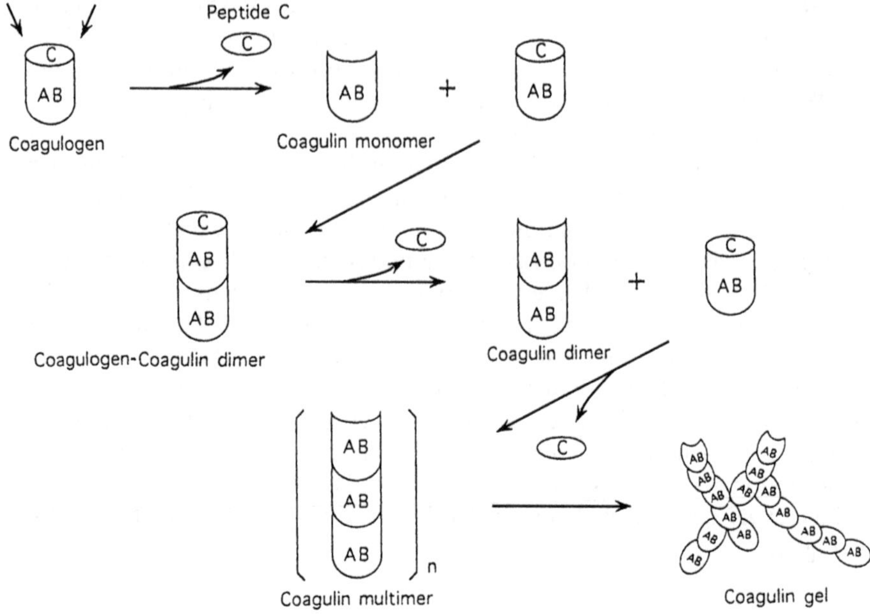

FIGURE 4 Hypothetical mechanism of coagulogen gel formation. Upon gelation of coagulogen by a horseshoe crab clotting enzyme, peptide C is released from the inner portion of the parent molecules. The resulting coagulin monomer may self-assemble to form the dimer, trimer, and multimers. (From Ref. 65.)

11. PROMINENT LAL TESTS

Early on, Levin and Bang described three critical properties of the gelation of LAL in the presence of LPS that formed the basis for subsequent assays [66], including:

1. Increase in OD that accompanies coagulation is because of the increase of clottable protein.
2. The concentration of LPS determines the rate of the OD increase.
3. The reaction occurs in the shape of a sigmoid curve (i.e., a plateau, a rapid rise, and a final plateau).

The total amount of clotted protein formed depends on the *initial LAL concentration*. An excess of LAL is provided for LAL testing and the amount of clotted protein eventually ends up the same, regardless of the amount of endotoxin in the sample. The end result of the enzymatic cascade is the formation of a web of clotted protein. The gel clot and endpoint tests take a single time point reading from the data to determine if the reaction reached an assigned level during the assigned time, whereas the kinetic tests are "watching" (at the appropriate wavelength) throughout the entire course of the reaction. The *endotoxin concentration* determines the *rate* of protein clot formation and thus the optical density change over time as determined by measuring the time to reach an assigned mOD value. The rate of OD formation is then related to the standard curve formed using control standard endotoxin. It can be seen from a plate that sits out that all wells containing endotoxin will eventually form a dark colorimetric or turbidimetric solution regardless of the endotoxin concentration, demonstrating that it is the speed of the reaction that correlates to the endotoxin concentration.

Besides the basic gelation of LAL in the presence of LPS, the two methods of observing the assay include the endpoint and kinetic assays. In the endpoint test, the reaction proceeds until it is stopped by the user by the addition of a stop reagent (such as acetic acid) at which point the optical density readings are recorded for all sample and standard curve points. The drawbacks associated with the endpoint method of observing the reaction are (1) necessity of the user attention at the end of data collection (typically 30 min) and (2) the limited standard curve range (a single log). In the kinetic assay, the spectrophotometer records the optical density reading continuously (as determined by the software settings within the manufacturer's recommendations, typically 1:30- to 2:00-min intervals). Kinetic testing measures the rate of the optical density change by recording the time it takes to reach a preset optical density setting called the "onset" or "threshold" time. The kinetic assay plots the log of the resulting reaction time in seconds against the log of the endotoxin concentration of the known standards and can span several logs (typically 2–4) and proceeds unattended, thus overcoming the two disadvantages presented by the endpoint tests.

The gel clot test is a simple test not far removed from original observations. Until recently, it was the most widely used procedure for the detection of endotoxin in solutions. When equal parts of LAL are combined with a dilution of sample containing endotoxin, one can expect to see gelation in the amount equivalent to the endotoxin sensitivity [called lambda (λ)] of the given lysate. A series of dilutions will reveal the approximate content of a sample—with those samples containing sensitivity equal to, or greater than, the given sensitivity being positive and those below the sensitivity not clotting the mixture. The solutions are incubated at a temperature correlating to a physiological temperature (37°C) for 1 hr and clots are observed by inverting the tubes 180°. In 10 × 75-mm depyrogenated test tubes, the clot must remain in the bottom of the tube when inverted. The method is considered semiquantitative because the true result obtained (indicated by the last gelled sample in the series) is actually somewhere between the two serial dilutions because the result cannot be extrapolated between the (usually twofold) dilution tubes as it is in the kinetic and endpoint assays via the use of a mathematical standard curve extrapolated over the entire range of standards.

Because commercial lysates are available with various standardized endpoints (sensitivities), the assay can be used to quantify the level of endotoxin in a particular solution or product. The level of endotoxin is calculated by multiplying the reciprocal of the highest dilution (the dilution factor) of the test solution, giving a positive endpoint by the sensitivity of the lysate preparation. For example, if the sensitivity of the LAL employed were 0.03 EU/mL and the dilution endpoint were 1:16, then the endotoxin concentration would be $16 \times 0.03 = 0.48$ EU/mL. For products administered by weight, the result (in EU/mL) is divided by the initial test solution potency (as reconstituted, or as per the liquid in the vial) to give a result (in EU/unit) (EU/mg, EU/insulin unit, EU/mL drug, etc.) that can then be compared with the tolerance limit specification. The geometric mean calculation is used for assays as opposed to the pass/fail limit test (that is reported as a "less than" number if there is no activity).

Given that kinetic assays continue to be the overwhelming area of growth in LAL testing (listed as a primary reason for the harmonization of endotoxin standards in IS-2), it is relevant to discuss details of kinetic testing. The development of the chromogenic assay was largely driven by the desire to accurately determine the endotoxin content for bacteremia [67], endotoxemia [68], and bodily fluids such as blood plasma and cerebrospinal fluids [69].

Among the most significant advantages of kinetic and endpoint testing over the gel clot assay is that they allow for the quantitative extrapolation of an unknown result between standard points. In the kinetic test, samples are pipetted into a 96-well microtiter plate, layered with LAL, and read spectrophotometrically at 405 or 340 nm (kinetic chromogenic or turbidimetric). The resulting color or turbidity reaction between LAL and endotoxin is recorded in the form of the time (in sec) that it takes a sample to reach a threshold optical density reading as defined in the reader's software (OD or mOD). The log of the time obtained for each sample is plotted against the standard curve linear or polynomial regression line formed from the log of the endotoxin content obtained for known standards.

The gel clot quantification approach, especially for water and in-process testing, has been largely supplanted by kinetic tests because of the ability of kinetic assays to extrapolate accurate results over a wide range of endotoxin concentration. A positive control consisting of a product sample spiked with a known concentration of endotoxin and a negative control using non-pyrogenic water is used to ensure the lack of interference in the sample matrix. Although a simple clot endpoint may be adequate for routine release testing of various pharmaceuticals, the ability to quantify endotoxin is invaluable for troubleshooting production-related pyrogen problems. Daily monitoring of plant water and in-process testing can alert production personnel to potential pyrogen problems before they become critical. Corrective action can be taken to reduce pyrogen loads and levels of endotoxin at this time. Using the gel clot assay, one would not see the increase in activity until the sample forms a clot. Thus there is little or no warning prior to failing a given lot of water or in-process sample.

The turbidimetric assay gives a quantitative measurement of endotoxin over a range of concentrations. This assay is predicated on the fact that any increase in endotoxin concentration causes a proportional increase in turbidity because of the precipitation of coagulable protein (coagulogen) in lysate (hence forming coagulin). The optical density of various dilutions of the substance to be tested is read against a standard curve obtained, which has been spiked with known quantities of endotoxin in sterile water (Table 7).

The chromogenic assay differs from the gel clot and turbidimetric reactions in that the coagulogen (clotting protein) is partially (or wholly) replaced by a chromogenic substrate, which is a short synthetic peptide containing the amino acid sequence at the point of interaction with the clotting enzyme. The end of this peptide is bound to a chromophore, *para*-nitroanilide (pNA). Japanese workers pioneered the use of chromogenic substrates and lysate (from *Limulus* and *Tachypleus*, the Japanese horseshoe crab) for the detection of endotoxin [70,71]. The chromogenic method takes advantage of the specificity of the endotoxin-activated proclotting enzyme, which exhibits specific amidase activity for carboxyterminal glycine–arginine residues. When such sequences are conjugated to a chromogenic substance, pNA is released in proportion to increasing concentrations of endotoxin. Thus it is possible to measure endotoxin concentration by measuring endotoxin-induced amidase activity as release of chromophore. Release of chromogenic substrate is measured by reading absorbance at 405 nm. Testing is conducted with 100 μL of lysate and an equal amount of sample or diluted sample. The quantitative relationship between the logarithm of the endotoxin

TABLE 7 Standard Curve Values from a Kinetic Chromogenic Assay (λ = 0.05 EU/mL) Using a Commercial Reader/Software System

Coefficient of correlation (r):	−0.999	
Y-intercept:	2.943	
Slope (m):	−0.265	
Blank:	**** (no reaction)	average = ****
Standard 1 (0.05 EU/mL):	1984, 1995, 1996, 1984	average = 1989
Standard 2 (0.5 EU/mL):	1007, 997, 999, 1001	average = 1001
Standard 3 (5.0 EU/mL):	594, 591, 593, 575	average = 588

This is the data from which the kinetic reader software uses a linear (or polynomial) regression standard curve to determine result calculations from sample reaction times.

concentration and amidase activity can be observed between 5×10^{-6} and 5×10^{-2} ng/mL endotoxin [72] and, therefore, can be used for the detection of picogram quantities of endotoxin associated with medical device eluates, immersion rinse solutions, and drug products.

12. METHOD DEVELOPMENT AND VALIDATION: THE IMPORTANCE OF A GOOD TEST

Historically, large-volume parenteral manufacturers have been foremost in developing tests for bacterial endotoxin assays because of the criticality of even minute endotoxin concentrations in solutions administered in large doses. However, many of today's problems revolve around the recovery of control standard endotoxin spike, the difficulty of which is exacerbated by the chemical nature of the small-volume drug materials being validated rather than their dose, which is often small. Small-volume parenteral drugs often contain high drug concentrations, which interfere both with the physiology of rabbits in the pyrogen assay and with spike recoveries in the LAL assay [73]. Some common types of problem compounds encountered in developing endotoxin assays for small-volume parenterals include water-insoluble drugs, drugs containing activity that mimics that of endotoxin, drugs containing endotoxin (that must be removed prior to validation), bulk drugs with variable potencies, multiple drugs in a given container, and potent, highly interfering drugs such as chemotherapy drugs. Now that the science of LAL testing has been firmly established, the challenges that remain often reside in difficult, product-specific applications. Perhaps the last great challenge encountered in each parenteral analytical laboratory is the development of, not just an LAL test, but a rugged, reproducible, and perhaps automatable test that will stand the test of time in routine use (Table 8).

Given all the LAL methods that could be developed, the question may be asked: What characteristics must a good LAL test have? A good LAL test from a legal standpoint must meet the appropriate compendial requirements and need not be quantitative except in its ability to demonstrate the detection of the endotoxin limit concentration (gel clot). However, beyond meeting compendial requirements, the best test is the one that provides the most information on the content of the analyte—endotoxin. The regulatory question that must be answered to put a drug on the market is: "Does it pass the release test?"* The scientific and business questions that remain to be answered

*Significantly, (1) is the manufacturing process used to produce it compliant with cGMP requirements? (2) Do the sampling and testing of precursors to the end-product support the contention that the product is free of endotoxin at the levels required?

TABLE 8 Relative Advantages and Disadvantages of Major LAL Test Types

Kinetic and endpoint tests vs. gel clot method

- Kinetic quantitative extrapolation of an unknown result between standards via linear or polynomial regression
- Less prone to user technique
- Provides "on board" documentation and calculation capabilities for consumables and products used in the test
- The mathematical treatment of data allows for the observance of trends and the setting of numerical system suitability and assay acceptance criteria
- May have different interference profiles than gel clot assays (useful if the gel clot assay will not give a valid result at a sensitive level)
- Assays may be automated
- λ may be varied by changing the bottom value of the standard curve (within the limits of the given LAL), thus allowing the MVD to be extended for difficult-to-test (interfering) products

Kinetic tests vs. endpoint tests

- Quantifies a result over a range of several logs (i.e., the difference between the highest and lowest standard curve points) vs. a single log
- Tests to completion without user intervention after LAL addition precision, speed, and accuracy improved

Chromogenic vs. turbidimetric tests (kinetic and endpoint)

- Calculates a result over a range of several logs (i.e., the difference between the highest and lowest standard curve points) vs. a single log
- Tests to completion without user intervention after LAL addition
- Turbidity determinations are made based upon the physical blocking of transmitted light (like nephlometry)
- Chromogenic methods (endpoint and kinetic) are not limited by particulate constraints associated with Beer's law (absorbance is directly proportional to common parameters such as well depth)
- The chromogenic method may be applied to turbid samples
- The turbidimetric method may be applied to samples with a yellow tint

are: "How much endotoxin does the sample contain?" "How does the result compare to previous lot measurements?" "How close to the endotoxin limit concentration is the result?"

Therefore, characteristics of a good BET *validation* test—in general terms that cover the kinetic, endpoint, and gel clot assays—are as follows:

1. Noninterfering (positives are positive and negatives are negative)
2. Appropriate product solubility if reconstituted and diluted, or as diluted only

3. Demonstration that the method chosen does not reduce (destroy) endotoxin that may be present if harsh conditions or solvents are employed*
4. Performed at the appropriate level as determined by the appropriate drug dose (or as per the USP monograph tolerance limit assigned for existing drugs), potency, lambda, and proposed or dictated specification requirement
5. Not subject to significant reagent batch or laboratory test variability
6. Resolution of a result (well) below the specification to allow manufacturing process contamination problems to be monitored prior to rising to alert levels
7. Demonstration of pH neutrality [6–8] in the inhibition/enhancement (I/E) sample dilution after combination with LAL
8. Appropriate laboratory support testing such as labware qualification (endotoxin-free and noninterfering), reference standard endotoxin/control standard endotoxin (RSE/CSE), limulus amebocyte lysate (LAL) label claim (gel clot) or initial qualification (kinetic and endpoint tests), diluent interference tests (i.e., their effect on LAL sensitivity)
9. Proper documentation of test events
10. Proper supporting documentation: user training, instrument installation qualification/operational qualification (IQ/OQ), preventive maintenance (PMs), computer validation, qualification, data archiving, etc.
11. Appropriate manufacturing support tests, such as component, excipient, and API testing (i.e., appropriate manufacturing process monitoring)

Some basic information must be gathered prior to developing an endotoxin test for a new chemical entity (NEC) or an established product. A list

---

* Validation via a series of sample dilutions in tubes containing spike demonstrates that the sample spikes endures the harsh treatment. However, if a kinetic or endpoint in-plate spike is used at a significant dilution, then the demonstration that the spike has acceptably endured the entire sample preparation method should be performed in the validation testing. For instance, a sample prepared in dimethyl formamide or other suspected harsh treatment then diluted to 1:1000 in water prior to spike in the plate will not demonstrate that the dimethyl formamide (DMF) does not destroy potential endotoxin. This is necessary to mention because of the prevalence today of adding kinetic spikes to only the final dilution of a series in the microtiter plate itself. After all, the goal of validation is to detect, not destroy, endotoxin that may be present in the sample.

of questions for the submitting department or developing scientist(s) may be compiled:

1. The maximum human dose, which will typically allow room for the clinic to increase the dose as needed in safety and efficacy studies. The response should be documented in an e-mail or other mechanism for inclusion in the validation documentation.
2. The formulation should be documented to establish the appropriate excipient tests (as will be discussed) and because it will likely change.
3. The presentation should be recorded as a critical assay parameter and may be subject to change (i.e., the product potency and volume or weight, for a given indication).
4. The approximate scheduling of the manufacture of the (at least) three lots needed for validation testing (if available).
5. A change notification mechanism to notify the laboratory of potency, dose, and/or presentation changes (who is responsible?).
6. Solubility profile (recommended reconstitution diluent(s)). How water-soluble is it? What is it most soluble in?
7. pH profile. What is the expected sample pH range?
8. Interference-related questions:

    - Is it a known chelator (such as EDTA)?
    - Does it possess enzymatic activity (such as trypsin or serine proteases) likely to interfere with LAL?
    - Is the compound likely to be inactivated by heating in a waterbath at 70 °C (an enzyme)?
    - Is it likely to contain cellulosic material?
    - What is the molecular weight of the compound? If there is endogenous endotoxin, it may be advantageous to remove it (via filtration) for validation purposes and the M_w of the sample will determine if it may be filtered and still retain the active compound in the filtrate.

The need for a new bacterial endotoxin test typically begins with a call from a development scientist with a new compound. Perhaps it is a compound prepared for an animal toxicology study, or perhaps it is a lot prepared in the development laboratory (a so-called "lab lot"). The early lots of drug substance or drug product will not be used in people, but there is a need to establish their safety to insure that the studies being performed are not skewed in some manner by the presence of endotoxin. Drug development is a costly endeavor and the generation of misleading results can lead developers down lengthy and costly blind alleys. Typically, compounds have been handed over to a development team from a discovery research effort that has been years in arriving. The compound has been formulated now for parenteral use, perhaps only one of many current or potential formulations, by combining a drug substance (bulk or API), solubilizers, stabilizers, preservatives, emulsifiers, thickening agents, etc. [74]. The compound is in flux and may change several times in its formulation (excipients), presentation (i.e., potency, container, size), and application (i.e., dose and, perhaps, indication). Perhaps, if its prospects seem especially bright, it will spawn a host of sister compounds that vary in the means of drug delivery (i.e., parenteral, for inhalation, time delay parenteral, etc.) and, therefore, in several relevant parameters required to be defined prior to developing additional suitable endotoxin tests.

Assay development for the bacterial endotoxin test for a given compound may be as simple as:

1. Calculating the new product's proposed TL and MVD based on the clinical dose of the material (or USP monograph-listed TL if it is an established drug)
2. Diluting the material in sterile reagent water
3. Testing it by either the gel clot, endpoint, or kinetic (turbidimetric or chromogenic) method at a dilution below the MVD.

However, given that early drugs were much less complex than today's drugs, it seems that the days of simplistic validations that do not require additional sample treatment(s) have passed. Now one would not realistically expect to test most drugs in an undiluted fashion. Many compounds have mitigating factors seemingly designed to frustrate the best assay development efforts as previously described. Additional mitigating sample complications include:

Cost: some product candidates are so expensive that product development scientists are reluctant to supply sufficient quantities for *protracted* method development and validation.

Occurrence of multiple interference properties not overcome by simple dilution, whereby adjusting one causes a deterioration of another.

Poorly characterized products: at an early stage of drug development, one can expect to see drug products that vary greatly from lot to lot (i.e., they are still being adjusted by those charged with establishing their formulation).

The types of testing protocols used in developing a new method may include (1) solubility and pH study protocols, (2) preliminary noninhibitory concentration (pNIC) protocols, and (c) a validation protocol. The tests performed in this sequence are cumulative. In simple terms, the NIC test varies the sample concentration while keeping the endotoxin concentration fixed (none and 2λ for gel clot and the midpoint of the standard curve for kinetic testing), whereas the I/E test varies the endotoxin concentration (to mimic the standard curve) while maintaining a constant product concentration (kinetic I/E uses only the midpoint). The three tests for the gel clot method and subsequent result calculation (which can be applied to the kinetic and endpoint methods with some adjustments) serve to establish parameters on which to base future routine testing:

1. Solubility/pH: One cannot perform the pNIC without having a good idea of the solubility and pH characteristics of the material. To bridge the gap for water-insoluble compounds by dissolving the compound in a suitable solvent that does not destroy endotoxin (dimethyl sulfoxide is such a diluent for many water-insoluble compounds), but that also is readily diluted with water or buffer, the right proportion will have to be found to keep the compound dissolved, but to allow enough dilution in water to overcome potential interference by both the compound and the solvent. The pH characteristics go hand in hand with the solubility. It may be necessary to acidify a given solution before a compound will go into solution.
2. The preliminary NIC determines the dilution at which the full validation test may be performed. Typically, at some point in a series of twofold dilutions of both spiked (2λ) and nonspiked samples, a "breakpoint" will be determined [first positive spike (2λ) recovery of the series coexistent with no recovery in the unspiked sample at the same dilution]. If the unspiked twofold dilution is negative and the positive is positive, then this demonstrates that the observed interference has been overcome by the dilution. Therefore, the noninhibitory concentration is somewhere between the first positive and the negative (2λ) spiked sample test directly preceding it. If it occurs at a level that is compatible with the calculated MVD (MVC for a bulk, excipient, or API sample), then one may proceed to the full validation test.
3. The full validation test typically includes both an NIC confirmation and an I/E curve, which is simply a standard curve performed in sample solution at the concentration of sample that one will not exceed (validated level). The I/E dilution level must not exceed the MVD (or 1/3 MVD for pooled vial tests) and must exceed the minimum valid concentration (MVC) of a sample (or 3 × MVC for pooled vials) needed to detect the endotoxin limit concentration

(the tolerance limit amount of endotoxin). The validation test may include a limit test at the proposed routine test dilution, but it is not necessary because that dilution is contained within the NIC and will be greater than or equal to the I/E dilution being tested.

4. The validation reportable test result will be based on the successful performance of the I/E test. If the I/E test agrees within a two-fold dilution with the labeled LAL label claim (and the included valid CSE curve), then the sample (test result, TR) can be said to contain:

$$\mathrm{TR} = \frac{< \lambda \times \mathrm{DF} \times \mathrm{PF}}{\mathrm{PP}}$$

where

PP = product potency of the active ingredient as reconstituted for a weighed sample or as labeled for a liquid sample containing a predetermined potency
DF = the dilution factor
PF = the pool factor.

A geometric mean is not necessary to determine the result calculation here because the I/E is either valid at the given dilution (sample concentration) or is invalid (i.e., does not confirm the label claim).

13. RESOLVING TEST INTERFERENCES

Given that the LAL assay in its many forms is a water-based assay derived from a sensitive physiological environment (blood of the horseshoe crab), it is not too surprising that as one ventures farther from such an aqueous environment, the results often correspondingly deteriorate. The Catch-22 of such testing resembles the contradiction presented by endotoxin itself (as an amphiphile) in that an increase in water content of a hydrophobic compound in solution will cause the material to precipitate (and endotoxin to aggregate), but, conversely, as the compound gets away from water, the reaction of LAL and endotoxin will be inhibited. Cooper's [75] paper on interference mechanisms encountered during LAL testing is perhaps the most useful on the subject. Cooper lists five major interference mechanisms to be expected when testing various parenteral drugs for BET using the LAL test and points out that often interference mechanisms result from the sample matrix's effect on the aggregation properties of the CSE rather than, or as well as on, the LPS–LAL reaction itself. The broad mechanisms (1–5) listed by Cooper include: (1) suboptimal pH conditions, (2) aggregation or adsorption of control endotoxin spikes, (3) unsuitable cation concentrations, (4) enzyme or protein modification, (5) nonspecific LAL activation, and (6) sometimes an inter-

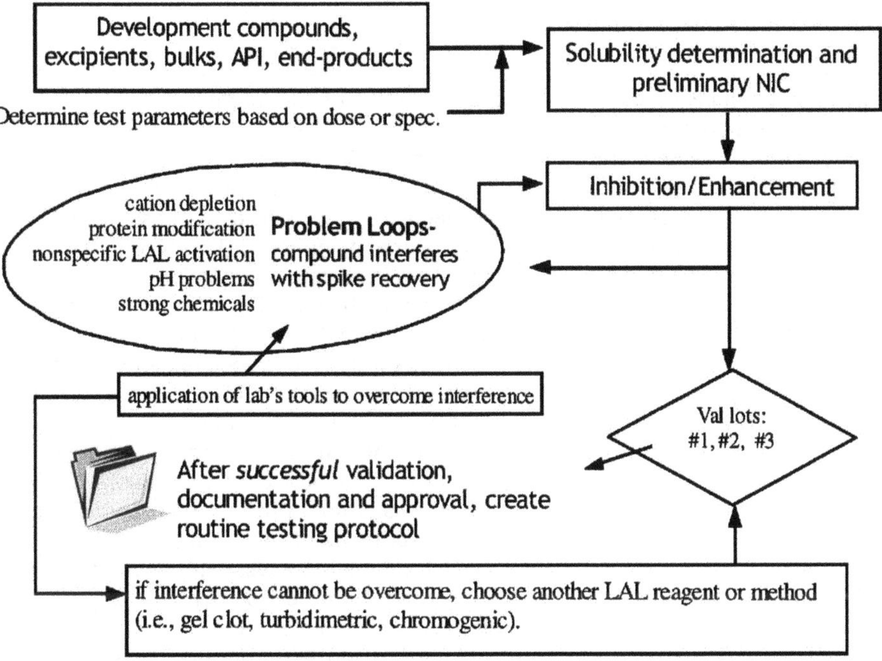

Figure 5 Method development—validation process.

ference mechanism cannot be determined (Fig. 5). Each broad interference mechanism will be briefly explored along with notable (common or unique) means of overcoming the associated interference below in Table 9.

14. SETTING ENDOTOXIN SPECIFICATIONS

The group developing the assay plays a key role in verifying that proposed specifications set are within the appropriate bounds established by the FDA Guideline calculations and pharmacopeial requirements. Practically speaking, the laboratory will determine the informal specification for development testing given the clinician's dose range. At a later date, a specification committee will assign an in-house specification. There appear to be two divergent philosophies on setting specifications. The first is to set the most stringent specification that the laboratory can support (i.e., around the limit of detection). The second is to set the specification around the regulatory limit allowed (i.e., the tolerance limit calculated value), which is the highest legal limit.

TABLE 9 Overcoming Bacterial Endotoxin Assay Interference

Interference/reference	Overcoming interference
(1) Suboptimal pH conditions, LAL is a product of a physiological system and many drugs are not. A pH of 6.4–8.0 is optimal and a pH requirement of 6.0–8.0 taken on a given sample and LAL is referenced by USP [76,77].	Most LAL reagents are buffered either as lyophilized or as reconstituted to overcome minor pH problems. Initial pH adjustment using 0.1 N or lower HCl or NaOH may be needed for more acidic or basic samples. Cooper maintains that pH problems "are the most important biochemical cause of LAL test inhibition". The USP requires the pH of the sample–LAL mixture to fall within the reagent supplier's requirements, which is usually 6.0–8.0. An FDA inspector relates that pH testing is not routinely required for a validated method unless committed to in the firm's new drug application (NDA). He also says that a failure to study the upper and lower limits of the product pH range (in validation) might necessitate routine testing.
(2) Endotoxin modification is a problem involving the amphiphilic properties of the CSE [78–80].	Strong salts and other solutions causing a large increase in test sample ionic strength will cause endotoxin aggregation and poor spike recovery. Dispersing agents such as Pyrosperse™ (Cambrex; BioWhittaker, Walkersville, Maryland) along with dilution (≤MVD) is used to overcome such interference. Adsorption of endotoxin to containers made of polypropylene is avoided in all types of endotoxin testing laboratory ware except pipette tips.
(3) Unsuitable cation concentrations. LAL reaction requires cations [81].	Organic chelators (i.e., EDTA) added for the purpose of complexing heavy metal cations may cause instability in parenteral formulations. A 50 mM $MgCl_2$ is routinely used as a test diluent to provide suitable levels of Ca^{2+} and Mg^{2+}. Reagents vary in cation concentration and buffering capacity among those supplied by LAL manufacturers.

TABLE 9 Continued

Interference/reference	Overcoming interference
(4) Protein or enzyme modification—enzymes needed for LAL gelation reaction[a] are denatured by strong chemicals.	Alcohols, phenols, and oncolytics fall into this category. If the interfering agent is itself an enzyme, it can be denatured by heating a sample or dilution of a sample at 70°C for ~10 min prior to (or post) dilution before testing. Other offenders may be removed by ion or size filtration, although the validation requirements may be onerous.
(5) Nonspecific LAL activation includes the detection of LAL-reactive material and drugs that mimic endotoxin such as those containing serine proteases.	Serine proteases may be heat-inactivated (as above). Products that mimic endotoxin provide a difficult challenge. To show that the activity occurring is not endotoxin, determine the level of activity followed by treatment of the sample to bind endotoxin (if the molecular weight of the product prohibits filtration removal). If the activity is not reduced, then it may not be endotoxin. An alternate test method may be needed or one may lower λ to allow sufficient dilution to overcome ("outrun") the enhancement.
(6) Samples containing endotoxin may present a problem similar to (5).	If the levels are relevant to the required test levels, endotoxin must be removed prior to performing the inhibition/enhancement test (gel clot). Methods of removal include filtration (20,000 Sartorious filter) when the molecular weight of the sample ingredient(s) does not exceed the cutoff rating of the filter.
Insoluble drug products (not on Cooper's list)	The lack of a suitable solvent for poorly water-soluble products is problematic. The LAL assay is a water-based test. DMSO has been used successfully. Mallinckrodt described a method of liquid–liquid extraction capable of pulling endotoxin into the aqueous phase, which leaves an inhibitor or difficult-to-work-with sample in the discarded oily phase. Mallinckrodt (in an old, unreferenceable technical bulletin) detailed that the endotoxin due to its lipid nature tends to remain associated with oils, but by the use of Pyrosperse™ (now a Cambrex product) in the liquid–liquid extraction, endotoxin is coaxed into the aqueous phase.

[a] Serine proteases.

Concerning the first philosophy, setting the specification too tightly may come back to haunt the participants in the form of a test failure and subsequent destruction of an expensive lot of drug that—scientifically and from a regulatory perspective—does not exceed allowable endotoxin levels. Early clinical doses are often several-fold higher than subsequent marketed drug doses, but there often is no communication of the change (downward) to allow specifications to be ratcheted down as doses decrease in the clinic. When products inevitably go to market, they will do so with a dose that is sometimes significantly lower than that used to establish the endotoxin test. The second philosophy is as poor as the first. If the specifications are set too close to the values allowed by law, then the routine examination of the drugs will not detect changes in endotoxin content until they are at failing levels. Ideally, one wants to "see" the endotoxin content well below the specification to serve as a warning that the manufacturing process is beginning to allow contamination well before it reaches a level relevant to the manufacturing process. If the specification is too high, then there will be no time for corrective action preceding a test failure.

Those that are not familiar with endotoxin limit calculations may see a value and gauge whether it is "high" or "low" simply by how large the number is. However, the specification is a function of the dose and any specification that is set appropriately will allow <350 EU/patient dose/hr. Naturally, a several-gram dose may contain less endotoxin on a per-milligram basis than a drug that is delivered in micrograms. The situation may arise in which a limit of nmt 100 EU/mg is set beside another compound with a limit of nmt 0.25 EU/mg, making the 100 EU/mg appear less "stringent" when, in fact, they both allow the same amount of endotoxin delivery as per their associated dose. A committee may scratch their collective heads and determine that the 100 EU/mg specification must be ratcheted down. The proof of this is in the side-by-side calculation:*

$$\text{TL} = \text{K/M} \quad 5.0\ \text{EU/kg}/(3.5\ \text{mg}/70\ \text{kg/hr}) = 100\ \text{EU/mg}$$

$$= 350\ \text{EU/dose}$$

$$\text{TL} = \text{K/M} \quad 5.0\ \text{EU/kg}/(1400\ \text{mg}/70\ \text{kg/hr}) = 0.25\ \text{EU/mg}$$

$$= 350\ \text{EU/dose}$$

The initial process of validation may be as in flux as the compound itself. Factors subject to change include: product potency, presentation, included excipients, interference factors, containers, etc. Factors that are absolutely

*By definition, TL = 350 EU/dose.

critical to establishing a test that will detect the endotoxin limit concentration include: MHD, product potency or concentration (PP), and LAL lambda (λ) to be used in the TL, and MVD (or MVC) calculations. An error in calculation or failure to secure a relevant dose for the TL calculation will nullify subsequent efforts to provide an accurate result. The tolerance limit is equal to the threshold pyrogenic response (K in EU/kg) divided by the dose in the units by which it is administered (mL, units, or mg) per 70-kg person per hour.

Mistakes in this critical calculation may include:

1. Not adjusting for the body weight (conversion from square meters may be necessary)
2. Not clarifying the means of delivery (bolus vs. multiple daily doses, etc.)
3. Basing the dose on a method that is not relevant to the means of administration, or is not based on the units of active ingredient (i.e., using milliliters instead of milligrams, particularly when the reconstitution may vary)
4. Not adjusting the MVD formula calculation for a potency change
5. Having the dose increased in the clinic to a level that exceeds that used as a basis for MVD calculation in the testing laboratory (i.e., poor communication).

The overall process is important in the development of a new LAL assay for a drug to be used in the clinic. Establishing a process that captures all the details is critical to ensuring that the right tasks are performed in the right sequence, the right information is documented, and that the information is correctly applied to the test both in its performance and in the determination of the parameters that govern its proper performance. Such a detailed process may be difficult to capture in a standard operating procedure and extensive experience will be necessary before an analyst is proficient in all the nuances of developing an LAL assay, particularly for a new drug candidate.

The GMP documentation expectation for any analytical test is that of being able to "recreate" the test including all the materials used in a given assay. For the LAL assay, that can be a daunting task if the right systems are not in place. For any given test, there may be dozens of consumables and equipment references (water or other diluent preparation, LAL, CSE, tips, tubes, plates, pipettes, tips, containers, water bath or heating block or kinetic reader, or other equipment, analyst, etc.) for which lot numbers must be recorded. Preventative maintenance records, training records, product validation documentation, certificates of analysis, or other proofs (laboratory test references) that the consumables used are endotoxin-free and do not inhibit or enhance the test, RSE/CSE, and/or COA reagent qualification documents used are all part of the items needed to "back up" any given test.

Printed laboratory notebooks or worksheets are necessary to collect all the pertinent information in an organized fashion.

15. DEPYROGENATION VALIDATION

Integral to the manufacture of sterile and endotoxin-free parenterals is the validation of depyrogenation processes. Endotoxin is notoriously resistant to destruction by heat, desiccation, pH extremes, and chemical treatments. The validation of endotoxin destruction or removal in the manufacture and packaging of parenteral drugs is a critical concern to drug and device manufacturers. LPS requires dry heat treatment of around 250°C for half an hour to achieve destruction, and standard autoclaving will not suffice. Whereas sterilization processes are predictable, depyrogenation procedures are empirical. Many specific instances of applying potent reagents to manufacturing equipment for the purpose of destroying applied endotoxin where one would predict that LPS would be demonstrated to be destroyed have revealed that the LPS has hung on, tenaciously defying preconceived notions of depyrogenation.

Depyrogenation is first thought of as the dry heat incineration of endotoxins from materials able to withstand the protracted dry heat cycle needed to destroy the LPS molecule. Alternatively, the wash/rinse removal of endotoxin from items such as stoppers and plastic vials and alternative vial closures comes readily to mind when heat treatment is not an option. However, there are many additional and hybrid areas of depyrogenation that are less historically entrenched and which are subject to more complex validation support. The two broad classes of depyrogenation processes that may be applied to components, devices, articles coming into contact with parenteral drugs, and drugs are inactivation and removal (see Fig. 6).

The last two decades of biotechnology have brought about the concomitant necessity of removing large populations of endotoxin from products because of their manufacture in microbial expression systems (especially *E. coli*). Selected methods of depyrogenation mentioned in Fig. 6 are employed to remove endotoxin from manufactured materials intended for parenteral use. A few of these methods will be examined. The oldest and simplest method of endotoxin removal from solid surfaces is rinsing with a nonpyrogenic solvent, usually sterile water for injection. Low levels of surface endotoxin contamination can be effectively removed from glassware, device components, and stoppers, for example, with an appropriate washing procedure. Rinse water can be monitored throughout the process with LAL to validate endotoxin removal. An example of such a validation process for large-volume parenteral glass containers was described by Feldstine et al. [82]. Distillation is the oldest method known for effectively removing pyrogens from water.

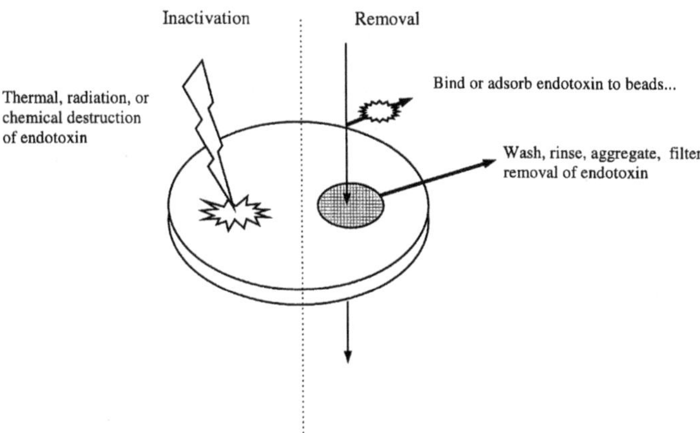

FIGURE 6 Inactivation and removal of bacterial endotoxins. Inactivation: Heat, moist and dry, the use of ionizing radiation of components, chemical inactivation (i.e., strong acid/base solutions), oxidation (i.e., hydrogen peroxide), polymyxin B. Removal: the use of physical size exclusion of endotoxin (ultrafiltration, ion-exchange removal), or aggregation followed by filtration, the use of charge differential (anion exhange), binding treatments (activated charcoal, lipopolysaccharide binding protein products).

Early investigators studying the thermostability of endotoxin concluded that moist heat supplied in conventional autoclaving was ineffective for depyrogenation. Although autoclave conditions for "normal sterilization" of solutions are ineffective for destruction of endotoxin, Banks [83] was able to demonstrate effective depyrogenation by autoclaving at 20 psi for 5 hr at a pH of 8.2, or for 2 hr at a pH of 3.8. Recent studies show that the action of certain depyrogenating agents can be enhanced by autoclaving. Cherkin [84] found that hydrogen peroxide (H_2O_2) was more effective in destroying pyrogen when the solution was autoclaved. Autoclaving also helped to eliminate residual H_2O_2. Similar findings have been reported for other solutions containing acid or base. Novitsky et al. [85] confirmed that autoclaving following conventional methods (121 °C, 15 psi at near-neutral pH for 20 min) was not sufficient to eliminate the pyrogenicity of 100 ng/mL *E. coli* 055:B5. However, autoclaving for longer periods (180 min) successfully reduced endotoxin levels to less than an LAL detectable limit of 0.01 ng/mL. Novitsky et al. also found that activated carbon treatment was more effective in removing endotoxin when solutions containing endotoxin and carbon were autoclaved.

The application of dry heat delivered through convection, conduction, or radiation (infrared) ovens has been the method of choice for depyrogenation of heat-resistant materials, such as glassware, metal equipment, and instruments, and of heat-stable chemicals, waxes, and oils. The standard method described in various national and international compendia and reference texts is an exposure of not less than 250°C for not less than 30 min and is based on the studies of Welch et al. [86] on the thermostability of pyrogens as measured with the rabbit pyrogen test. The mechanism of endotoxin inactivation is incineration. The development of the LAL has provided a more quantitative means of studying dry heat inactivation of endotoxin. Robertson et al. [87], Tsuji and Lewis [88], Tsuji and Harrison [89], and Groves and Groves [90] discovered that the inactivation kinetics of LPS from *E. coli, Salmonella typhosa, Serratia marcescens,* and *Pseudomonas aeruginosa Salmonella* was a nonlinear, second-order process in contrast to the inactivation of bacterial spores, which follow first-order kinetics. They compared the dry heat resistance of intact and purified LPS to that of spores with the greatest heat resistance. Purified LPS was shown to be twice as resistant as the native (whole cell) endotoxin from which it was derived. Of greater importance was the author's convincing evidence that the general practice of increasing exposure time to compensate for lower process temperature is not supportable for LPS destruction, particularly at 175°C or less. Akers et al. [91,92] confirmed these findings and also determined the *F* value requirements for destruction of 10 ng of *E. coli* 055:B5 endotoxin seeded into 50-mL glass vials, using both convection and radiant heat ovens. An *F* value is the equivalent time at a given temperature delivered to a product to achieve sterilization or, in this case, depyrogenation. There were linear relationships between oven temperatures and the logarithms of the *F* values with both treatments.

Before 1978, there were few studies addressing the destruction of endotoxin presumably because of the lack of a suitable quantitative method of measuring endotoxin reductions [93]. Along with the LAL assay and the refinement of LPS standardization came a means of applying (as a biological indicator in a manner analogous to the use of spore-forming *Bacillus* species in sterilization studies) and detecting recovered endotoxin for such studies. Methods and mechanisms of proving the depyrogenation of various items have been largely borrowed from sterilization processes and modified to compensate for the thermal and chemical stability of LPS. The two common types of depyrogenation processes (like sterilization methods) involve (1) the construction of *D* (death or destruction in the case of endotoxin because it is not alive) values and (2) the use of "bioindicators" as an empirical means of demonstrating that a "worst-case" load of applied pyrogenic residue has been removed by a given proposed depyrogenation process. The definition of the death rate (*D* value) in sterilization technology is the "time for a 90% reduction in the microbial population exhibiting first-order reaction kinetics" [38,94]. The number of organisms decreases during sterilization in a log fashion down to one org (log 0) after which it becomes negative where 10^{-1} is the likelihood of a single survivor per 10 items and 10^{-3} is one survivor in 100 items. Therefore, in theory, sterility is never achieved but is reduced to a probability (however remote the likelihood of a survivor). Generic procedures (such as that given in the USP) cannot be assumed to work for a given wash or baking process because of the variety of equipment, loading configurations, times, and temperatures chosen for different process applications. Validation must include "documented evidence" that the process does what it purports to do, namely, provides a three-log reduction of applied endotoxin. Death rate curves in sterility validation (Fig. 7), can be constructed by graphing the number of organisms on the *Y*-axis against the log of either the heating time, exposure time (gas), or radiation dose on the *X*-axis. Similar destruction curves (Fig. 10) can be constructed using endotoxin data.

The tables above show the lack of agreement (and thus empirical nature) of depyrogenation processes and hint at the plethora of conditions that can alter the time and temperature needed to bring about adequate depyrogenation (i.e., load and type of material, oven tunnel speed, etc.). Typical parenteral practice involves moving glass vials on a belt through an oven that blasts it with an excess of heat (~300°C) at speeds of 5–10 min to achieve *F* values equivalent to or exceeding the targeted half hour at 250°C treatment ($F_{250} = 30$) (Table 10).

The requirements for depyrogenation validation processes (from a laboratory perspective) are somewhat vague and interpretive.* A short reference occurs in the USP, Chapter 1211—Sterilization and Sterility Assurance of Compendial Articles, Dry-Heat Sterilization section, as follows:

> Since dry heat is frequently employed to render glassware or other containers free from pyrogens as well as viable microbes, a pyrogen challenge, where necessary, should be an integral part of the validation program, e.g., by inoculating one or more of the articles to be treated with 1000 or more USP units of bacterial endotoxin. The test with Limulus lysate could be used to demonstrate that the endotoxic substance has been inactivated to not more than 1/1000 of the original amount (3 log cycle reduction). For the test to be valid, both the original amount and, after acceptable inactivation, the remaining amount of endotoxin should be measured.

---

*21 CFR Parts 210 and 211 Subpart E—Control of Components and Drug Product Containers and Closures (211.80, 211.82, 211.84, 211.86, 211.87, 211.89, and 211.94) discuss component testing requirements in general terms.

(a)

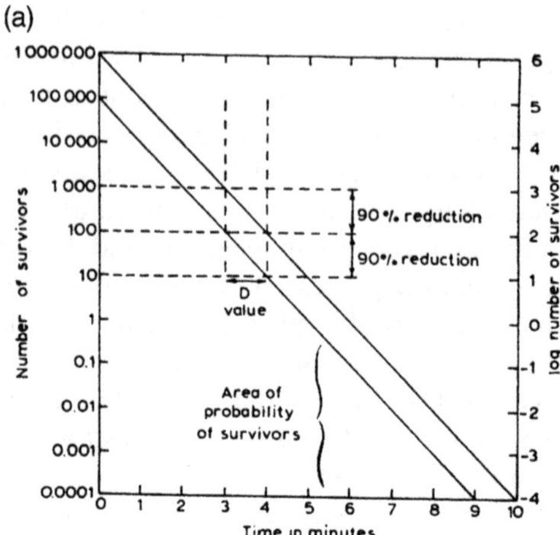

(b)

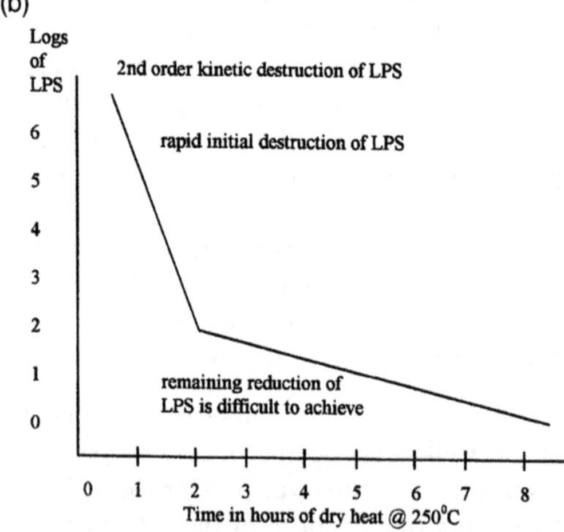

Figure 7 Microbial death rate curves (a) illustrate concept of decimal reduction (D values) and probability of survivors (from Ref. 38) and (b) hypothetically demonstrate the more difficult to achieve reduction of LPS after an initially relatively rapid reduction.

The only other USP references to depyrogenation are in the Bacterial Endotoxins Test chapter (Chapter 85), which states that one should "treat any containers or utensils employed so as to destroy extraneous surface endotoxins that may be present, such as by heating in an oven at 250°C or above for sufficient time" and then references the above paragraph as a means of validating the oven referred to here. Moreover, "render the syringes,

TABLE 10 Time Required to Achieve Multiple Log Reductions Using Different Sources of Endotoxin

Log reduction	Temperature (°C)	Tsuji and Lewis, and Tsuji and Harrison[a] (min)	Ludwig and Avis[b] (min)
3	@ 210	13.6	7
	@ 300	0.089	<0.5
5	@ 210	Infinity[c]	19
	@ 300	0.19[c]	1
6	@ 300	0.27[c]	11

Derived from Ref. 97

Log reduction	Temperature (°C)	BioWhittaker (min)	Difco (min)	ACC (min)
3	@ 225	5	5	5
	@ 250	<0.5	NA	2
5	@ 225	15	45	45
	@ 250	5	NA	19

[a] Used aluminum cups.
[b] Ludwig and Avis used glass.
[c] Extrapolated value.
Source: Refs. 88, 95, and 96.

needles, and glassware (to be used in the pyrogen test) free from pyrogens by heating at 250°C for not less than 30 minutes or by any other suitable method," respectively.

The USP/FDA "Guideline on Sterile Drug Products Produced by Aseptic Processing" [97] provides a review of the requirements for container/closure depyrogenation:

> It is critical to the integrity of the final product that containers and closures be rendered sterile and, in the case of injectable products, pyrogen-free. The type of processes used to sterilize and depyrogenate will depend primarily on the nature of the material which comprises the container/closure. Any properly validated process can be acceptable. Whatever depyrogenation method is used, the validation data should demonstrate that the process will reduce the endotoxin content by 3 logs. One method of assessing the adequacy of a depyrogenation process is to simulate the process using containers having known quantities of standardized endotoxins and

> measure the level of reduction... endotoxin challenges should not be easier to remove from the target surfaces than the endotoxin that may normally be present.

Moreover,

> Rubber compound stoppers pose another potential source of microbial and (of concern for products intended to be pyrogen-free) pyrogen contamination. They are usually cleaned by multiple cycles of washing and rinsing prior to final steam sterilization. The final rinse should be with USP water for injection. It is also important to minimize the lapsed time between washing and sterilizing because moisture on the stoppers can support microbiological growth and the generation of pyrogens. Because rubber is a poor conductor of heat, proper validation of processes to sterilize rubber stoppers is particularly important.

There should be an awareness on the part of those charged with performing depyrogenation validation that there is a distinct difference between items that may be heat-treated and those that must be washed (inactivation vs. removal, respectively). The heat treatment of bottles and vials follows the more easily reasoned path that, given appropriate time and temperature parameters, endotoxins will be destroyed. However, the wash removal of endotoxins is complicated by the tenacity with which endotoxin sticks to rubber and other porous polymers that compose such materials. Entrenched endotoxin's removal is governed by more difficult-to-assess parameters including agitation and solubility. Thus with removal, there are additional variables involved than heat and duration as in the case of incineration.

There is really no perfect way to verify the presence or recovery of low amounts of endotoxin, (i.e., 1.0 EU/stopper) given the adsorption by porous materials (Fig. 8). Common methods involve vigorous vortexing, sonication, or other means of agitation to dislodge it prior to testing. The selection of a vigorous method of dislodging endotoxins is empirical (whatever works) and various laboratories have chosen to use either intense, short-duration vortexing or prolonged but less vigorous mixing (such as shaking or sonication), or simply washing with or without added surfactants. Agalloco [98] has described a theoretical problem associated with cleaning validation studies that relate aptly to depyrogenation validation (endotoxin removal) studies by using a "tar baby" analogy:

> The cleanliness of the bath water may not necessarily relate directly to the cleanliness of the baby. If the contamination is not soluble in the cleaning agent, then the contamination will remain on the surface. If the contamination is not soluble in the final rinse, samples of the bath

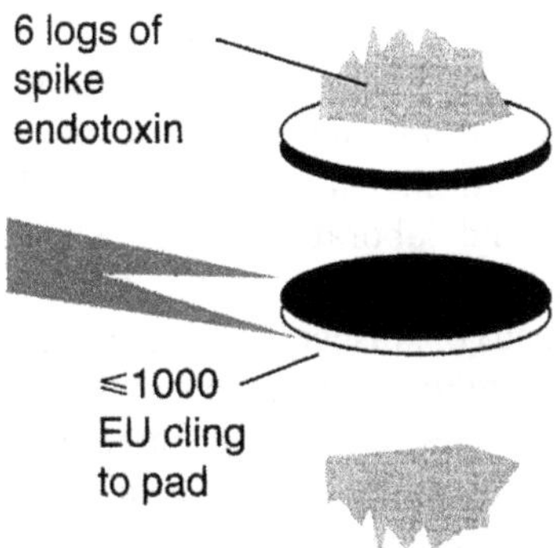

FIGURE 8 Is this validation? A mountain of applied spike is turned over (or washed) and the mountain of spike falls off. Has a >3 log reduction transpired? Increasing applied spikes to obtain better percent recovery (rather than developing better removal methods) may result in spikes that are too easily removed, thereby revealing nothing about the depyrogenation process.

water will not detect the presence of residual contamination. The conclusion will be drawn that the baby is clean, when in fact both the cleaning and evaluation methods are inadequate.

In other words, if one determines the cleanliness of the baby (stopper) by measuring the "tar" (endotoxin) remaining in the bath water (laboratory rinse method), then one has to ensure that the method used does indeed remove the "tar." There must be some validation of the method to serve as a demonstration that the method removes endotoxin from "sticky" surfaces. At least theoretically, endotoxin that clings tenaciously to a stopper (thereby escaping pyroburden detection) can be removed later by the surfactant action of a drug and become available for parenteral administration.

An added step of RSE/CSE characterization of spike solutions to be applied for depyrogenation studies can bring about a greater consistency of recovery given that the potency of the reconstituted solution of concentrated endotoxin (i.e., Difco) used can be highly variable (i.e., may vary from the label and from laboratory to laboratory). Additional characterization under laboratory conditions (as opposed to the manufacturer's assigned potency) may aid in "getting back" numerical values that are very close to the theoretical value (i.e., 48,800 EU per component of a 50,000 EU per component spike application).

It is instructive to separate manufacturing and QC laboratory division of labor in the fragmented depyrogenation validation process. Regardless of how specific companies have bridged the activities, a natural division exists between the manufacturing and QC functions in the depyrogenation validation process. The manufacturing area may have a validation group that runs the studies to document that their processes comply with cGMP requirements including depyrogenation validation. QC laboratories support these efforts by supplying expertise in the endotoxin application. Therefore, the coordination of activities involves manufacturing and laboratory support. The manufacturing group determines and documents the depyrogenation treatment process [oven (including *F* values obtained) or washer (settings, rinses)] and the laboratory supplies inoculated components, performs before-treatment and after-treatment (depyrogenation) LAL testing with accompanying controls, and documents and reports the results (as supported by a validated laboratory method). Differences exist in the intentions, activities, and requirements of laboratory validation to support pyroburden methods and depyrogenation validation processes (3 log reduction validations) although they are similar in many respects. A significant difference in the two lies in the fact that pyroburden is a release test for components to allow them to be used in marketed products in lieu of (or in addition to when obtained sterile from a vendor) a validated depyrogenation process. As such, the number to be tested should be derived from a statistical (or at least reasoned) sampling of a given lot of components based on the manufactured component lot size.

ACC (Woods Hole, MA) intends to publish a procedure to promote the use of LAL to bathe medical devices in situ.* Novitsky refers to an in-house study revealing significant LAL reactivity when implants were tested via the LAL in situ bath method vs. negative results when tested by traditional extraction in which endotoxin spike recoveries are notoriously difficult to recover. Such a method would overcome, in theory, many of the adsorption issues involved in recovering endotoxin from glass vials and rubber stoppers.

16. ENDOTOXIN REMOVAL IN PHARMACEUTICAL MANUFACTURING PROCESSES

Modern techniques used to remove endotoxins from drugs during parenteral manufacturing often involve the combination of several methods. Macromolecules cannot be removed by simple ultrafiltration given that their size may be similar to endotoxin aggregates. Two case studies will be reviewed in which endotoxin removal processes were devised for (1) a 32-kDa enzyme [superoxide dismutase (SOD)], and (2) a high M_w α-1,6 branched α-1,4 glucan (amylopectin) derived from corn or potato starch and used as an encapsulation matrix for pharmaceutical products.

---

* Ref. 126.

An endotoxin removal process to meet a proposed specification level of <0.25 EU/mg protein was performed at Sigma Chemical, St. Louis, MO (referred to as "Case Study 1") [99]. Held et al. designed the initial purification of the protein to achieve >99% purity using "extraction, heat treatment, clarification, and ammonium sulfate fractionation..." followed by three chromographic steps which removed the majority of endotoxins. Subsequently, the product yielded endotoxin values between 0.16 and 0.72 EU/mg, which provided no consistency in meeting the necessary specification (nmt 0.25 EU/mg). The authors employed a "polishing step" to perform the remaining threefold reduction of endotoxin with an eye on adding only a minimal additional cost to the process. They used a positively charged, 1-ft^2, 0.2-μm disposable Posidyne filter (Pall, East Hills, NY) to achieve the required endotoxin reduction without product loss. The natural negative charge of LPS above a pH of 2.0 allows the use of ion exchange as a means of binding the endotoxin to the filter matrix while the protein solution passes through.

In "Case Study 2," the same Sigma Chemical group had a more formidable task of reducing endotoxin in amylopectin from approximately 500 EU/g to <20 EU/g (<0.02 EU/mg). The low solubility and viscosity of the product prevented the filtration removal of endotoxin. They added 400 g of food-grade amylopectin to 20 L of 2 mM EDTA to reduce the aggregate size of the endotoxins. They heated the mixture to 85–90 °C and stirred the mix for an hour. After cooling to 54–56 °C, they added NaOH to a final concentration of 0.25 M and stirred for another hour to hydrolyze the endotoxin base labile bonds (i.e., lipid A-KDO). The solution was neutralized using hydrochloric acid (HCl) and cooled to room temperature. Repeated ultrafiltration with 300,000 M_w cutoff filters removed salts and endotoxin. On concentration to 10 L, the solution was diluted to 30 L with endotoxin-free water. This was followed by repeated reconcentration to 10 L followed by redilution in endotoxin-free water, for a total of nine times. The final solution was filtered through a 0.45-μm Posidyne filter (Pall), frozen, lyophilized, and stored overnight under vacuum. Thus the group combined three different well-known mechanisms to remove the endotoxin in stages: treatment with moderate heat and alkali, filtration separation by molecular weight cutoff filters, and ion exchange binding to the 0.45-μm filter. They quantitated the endotoxin removed by each of the processing steps to find that the reduction factors achieved were 20, 5, and 2, respectively. The final filtration resulted in a solution of <1 EU/g. The authors advise: "even water with endotoxin levels that are below the detection limit can become a major contributor to endotoxins when large volumes are used for repeated cycles of dilution and concentration of a product." Historical methods of obtaining multiple log reductions in parenteral processing have involved chromatography and adsorption. Particularly problematic is the removal of endotoxin tightly bound to biologicals drug compounds (proteins, polysaccharides, or DNA) [100,101].

17. THE FUTURE AND ENDOTOXIN TESTING

Two important reoccurring themes that may help form a view of the future direction of parenteral contamination testing are as follows:

1. Endotoxin is the major microbial cell residue, but it is not the only important cellular artifact.
2. Endotoxin is the most potent of such artifacts and induces a wide range of deleterious host effects at the cellular and systemic levels, but it is not the only one or the only potent one.

Two general questions form the broad outline for this section: (1) What are some likely paths to future prospective tests for endotoxin? (2) Might such prospective tests be expanded to include nonendotoxin parenteral contaminants? Pyrogen testing originated with a fairly insensitive but broadly inclusive method (rabbit pyrogen) to the exquisitely sensitive but narrow (specific) LAL method. Characteristics to be desired for a new assay may not only test for bacterial endotoxin but also for other potentially deleterious host-active microbial substances. A futuristic test would be more inclusive than LAL (reminiscent of the pyrogen test) and as sensitive and specific as LAL. Given the recent advances in molecular biology, the successor to the LAL test may be an LAL test using a recombinant LAL product (now available from Cambrex and soon to be from ACC [102]). The recombinant test merely maintains the status quo of LAL testing without the need to bleed horseshoe crabs.

There are three likely roads that lie ahead: (1) the expansion of the current LAL path (including the use of recombinant LAL), or (2) the supplementation and perhaps eventual replacement of LAL testing with the whole blood test,* or (3) an increased specificity for the detection of endotoxin† as one of several detected artifacts). The LAL assay is *almost* entirely specific for endotoxin but has been criticized for both its specificity (i.e., cannot detect GPB or viral contamination) and its lack of specificity (some preparations are sensitive to β-glucans). The road toward greater specificity and broader application to other microbial artifacts has been explored in that several methods are applicable to both endotoxin and nonendotoxin pyrogens (i.e., mononuclear cell assays and the use of gas chromatography mass spectrometry (GC-MS) for the detection of multiple markers) (Fig. 9).

---

*Although a broad assay, the pyrogen test is hardly sensitive enough to be all-inclusive.
†GC-MS detection of β-hydroxymyristic acid.

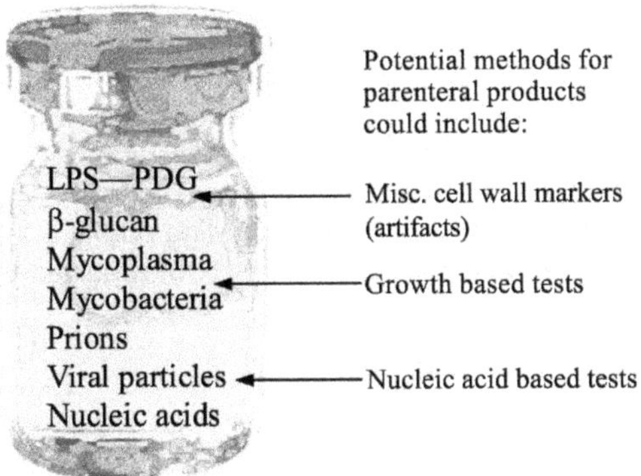

FIGURE 9 A test (including endotoxin) sensitive and specific for as many microbial markers and artifacts as possible would be desirable.

It may prove desirable to screen drug products for as many microbial contaminants as possible simultaneously with a single test [i.e., subplanting sterility, bioburden, indicator organism recovery (microbial purity), fungi (β-glucan), mycoplasma, endotoxin and other microbial by-product detection, such as enterotoxins and superantigens (many of which are not now analytically precluded)] or, more realistically, perhaps one test for living organisms and another for relevant microbial artifacts. The justification for such testing would be driven by either (1) product-specific (indication-specific) concerns of nonendotoxin artifact contamination, (2) the potency (relative biological activity) of some nonendotoxin modulins, (3) the emerging technology itself, (4) an increase in the likelihood of nonendotoxin contamination given an increase in manufacturing methods sensitive to alternative (non-GNB) contamination, or (5) necessity, in the case that LAL becomes unavailable and would therefore have to be supplanted with a new technology.

The *PDA Journal of Pharmaceutical Science and Technology* technical report no. 33 [103] describes three broad categories of microbiological testing technologies including: (1) viability-based, (2) artifact-based, and (3) nucleic acid-based technologies. Clearly, the concern for endotoxin as a contaminant lies in its occurrence as an artifact. It is the enduring potent biological activity of endotoxin as an artifact, coupled with its almost indestructible nature, that separates it from other host artifacts and modulins that are both less biologically active and less resistant to inactivation by heat, chemical, and other common pharmaceutical manufacturing treatments. Therefore, the viability-based and nucleic acid-based technologies can be viewed as much less relevant as proposed tests to any eventual replacement of LAL, although they could and do currently find utility in relevant applications such as clinical detection in blood plasma or the examination of complex media used in cell culture.

TABLE 11 Endotoxin and Nonendotoxin Assays for Microbial Contaminants

β-Glucan-insensitive LAL and endotoxin-insensitive LAL

- Factor G biosensor contained within the LAL reagent has been removed to create an endotoxin-specific LAL reagent for both gel clot and kinetic assays [105].
- The factor C pathway enzymes have been removed, resulting in reagents insensitive to endotoxin and specific to various β-glucans including curdlan, pachyman, laminaran, and lichenan. Kitagawa and coworkers [105] reported that the sensitivity toward curdlan was approximately 10^{-10} g/mL.

ELISA with monoclonal antibody against limulus peptide C (163)

GC-MS of 3-hydroxy fatty acids

- The GC-MS method quantitates endotoxin by relating (integrating) the (triangular) area in the marker fatty acid recovered (β-hydroxymyristic) from the areas obtained for standards recovered. There is a commercial effort to apply the technology to endotoxin detection. Microbial ID GC is coupled to a computer database to reference chromatograms for standard American Type Culture Collection (ATCC) organisms as well as a variety of environmental and clinical isolates. Biochemical and GC methods work side by side now in many microbial ID laboratories.
- Clinical researchers correlated meningococcal endotoxin levels (determined by GC-MS) in septic shock patients with LAL results [107]. Brandtzaeg et al. concede that the utility of the LAL assay in measuring plasma LPS activity is still debatable and, in most cases, not feasible due to the low levels of endotoxins present. Due to the high endotoxin plasma concentrations associated with patients afflicted with the deadly *Neisseria meningitidis* infection, their studies were successful. They identified 3-hydroxy lauric acid (3-OH-12:0), the neisserial lipid A marker not found in Enterobactereaceae. *N. meningitidis* LPS is potent from an endotoxin perspective due to its active production of excess outer membrane material called "blebs" [108].
- The suspected false-positive endotoxin reactions occurring in LAL assays have been confirmed using GC-MS. Maitra et al. used GC-MS to test hemodialyzer rinses containing up to 4800 ng of endotoxin equivalents per milliliter to reveal that the solutions did not contain any measurable β(OH) C_{12}, C_{14}, or C_{16} fatty acids [109]. It is incumbent on users claiming that LAL activity is not due to endotoxin (such as with β-glucans) to have an independent method to prove such a contention.
- GC-MS has been used in the clinical determination of other markers present in septic synovial fluid and septic arthritic joints via the identification of levels of GPB markers, namely muramic acid [110], and has been used to screen out background peaks to allow researchers to detect 30 ng/mL (a sensitivity increase of 1000× over prior attempts). The GC-MS method may be a valuable investigative tool utilizing multiple markers.

TABLE 11 Continued

Cultured human mononuclear cells followed by pyrogen testing (Human Leukocytic Pyrogen Test) [111], *cultured human mononuclear cells followed thymocyte proliferation assay* [112]
Silk worm larvae plasma (SLP) test detects peptidoglycan. A novel mechanism of detecting specific non-LPS microbial components including β-glucan (βG) and peptidoglycan (PG) (contained in GPB and in lesser amounts in GNB) [113–115] is available commercially [116] for experimental purposes. • In a method reminiscent of the early LAL test, the SLP test uses another primitive blood-based host defense system, namely that of the silkworm larvae (*Bombyx mori*) plasma. Melanin, a black-pigmented protein, serves as a self-defense molecule in insect hemolymph and is the end-product of a cascade reaction utilizing multiple serine proteases called the prophenoloxidase (proPO) cascade [127]. Commercialized by Wako Pure Chemical Industries, Ltd.[a] • Used as a supplementary tool in the detection of bacterial meningitis (which was also one of the first clinical applications of the early LAL test) [117]. Rapid determination of infection type is critical to the patient's treatment. • Used to show that peptidoglycan may be a pyrogen concern in dialysate contamination, as per their measurements made on 54 dialysate samples from nine facilities [118].
PCR test for specific fragments of bacterial DNA (that should not be present in parenterals): Dussurget and Roulland-Dussoix [119], *at the Institut Pasteur, amplified DNA fragments of mycoplasmas to act as probes and detected as little as 10 fg of specific mycoplasma contaminant sequences*
Recombinant Factor C test-utilizing the cascade "biosensor" factor C produced recombinantly, Cambrex has begun marketing this as an LAL substitute, albeit an alternative assay due to the fluorescence method of detection. It may find application in biologics that show interference using traditional LAL. It is glucan-nonreactive as well.

[a] Osaka, Japan.

According to the PDA report, artifact-based technologies that may prove relevant to the detection and quantification of microbial constituents include: (1) the use of fatty acid profiles (gathered by GC-MS), (2) fluorescence antibody techniques, (3) enzyme-linked immunosorbent assay (ELISA), and (4) latex agglutination (as well as the continued reliance on LAL).

A testimony to the BET test is the lack of adverse events associated with pharmaceutical or medical device contamination since the use of LAL. The difficulty of replacing LAL lies in its extreme ease of use, sensitivity, and specificity, which, in turn, is also a testament to the crab's defense system. Some non-LAL assays have served in some instances to complement the LAL and pyrogen tests, and some may hold potential as eventual alternative tests

TABLE 12 Microbial Contamination Marker Detection by GC-MS Marker Indicates the Presence of Non-GC Assays

3-OH fatty acids (lipid A)	Endotoxin (gram-negative orgs)	LAL, pyrogen SLP[b]
β-Glucans[a]	Yeast and fungi	Liquid chromatography
Ergosterol	Yeast and fungi	SLP[b]
Muramic acid	Peptidoglycan (gram-positive orgs)	Acid fast stain
Long-chain fatty acids	Mycobacteria	Broth[c] or agar culture,[c] PCR
Unique lipopeptides	*Mycoplasma* (and other mollicutes)	Broth[c] or agar culture[d], PCR

[a] Detectable by endotoxin-insensitive LAL and LC-MS.
[b] SLP = silkworm larvae plasma.
[c] Broth culture: 5% CO_2 up to 6 weeks of sediment and pH change [120].
[d] Agar culture: inverted microscopic observation—"fried egg" appearance.
Source: Refs. 120 and 121.

as they have already served as complementary or confirmatory tests to the use of LAL testing in specific applications (see Tables 11 and 12).

Some non-LAL assays such as GC-MS or polymyxin B binding may achieve a stoichiometric determination of LPS content that is not a measure of the relative biological responsiveness of a given endotoxin. Although this may seem, at first glance, to be an ideal advantage in providing a truer means of LPS quantitation, it is the biological responsiveness of the LAL test that provides the current basis for regulatory acceptability and is one that is strictly enforced (and historically is the result of much effort to achieve) through the establishment of reference standards, controls standards, LAL standardization, and the relationship of LPS activity to the threshold pyrogenic response in both humans and rabbits. In other words, the biological responsiveness of LPS as a means of quantification will not only not go away; presumably, it will have to be correlated to any truly quantitative nonbiological measure (i.e., non-LAL or nonpyrogen method) developed. Specialized immunological tests (some used in conjunction with LAL) have been developed for clinical applications such as the detection of endotoxemia and other investigational applications.

The effect of blood plasma on LAL tests has made the quantification of endotoxin in blood inconsistent (see Hurley's paper for a detailed discussion of methods of endotoxemia detection).*

* Hurley JC. Endotoxemia: methods of detection and clinical correlates. Microbiol Rev 1995; 8(2):268–292.

18. WHOLE BLOOD PYROGEN TEST

The concept of an in vitro "human pyrogen test" that utilizes whole blood [and the underlying physiological basis of the fever reaction: the activation of blood monocytes by exogenous pyrogens to produce endogenous pyrogens (cytokines)] has gained support recently with the commission of the Hartung group (University of Konstanz) by the European Commission to investigate the development of such a test with an eye toward eventual compendial inclusion [122–125]. The use of isolated monocytes/leucocytes has proved to be highly variable and, therefore, Hartung et al. have evaluated tests that employ diluted, fresh whole blood in a procedure that involves sample incubation and subsequent ELISA detection of immunoreactive monophage-secreted cytokines (IL-β, IL-6, and TNFα). The former two cytokines are largely intracellular as opposed to the latter, which is secreted into the incubated medium (blood) and, therefore, perhaps more amenable to assay. Additionally, IL-6 has been purported to be the principal endogenous precursor to fever and, therefore, the most accurate predictor of the pyrogenic response. Hartung et al. collaborated with the European Center for the Evaluation of Alternative Methods (ECVAM) beginning in 1999 to propose and perform tests needed to eventually establish such a "human pyrogen test." The test participants summarized their discussions from the ECVAM Workshop 43 (Tables 13 and 14) in ATLA/2001 and claimed a test sensitivity of 0.03– 0.1 IU/mL compared with the BET limit of detection given as 0.03 IU/mL.* The authors address the "need" for nonendotoxin pyrogen testing in several instances as shown in Table 13.

Hartung et al. state that the European Pharmacopeia Commission should examine each monograph individually to determine if replacement of the rabbit pyrogen test requirement should be done by means of LAL or IPT. One LAL supplier, Charles Rivers Laboratories (CRL, Charleston, SC), has marketed a commercial kit for investigational purposes. Some industry debate has begun on the utility of the test and some have called into question the relevance of nonendotoxin pyrogens under any circumstances. Novitsky (Associates of Cape Cod) asserts: "many microbial components once thought to be pyrogens have since been shown to be contaminated with endotoxin. A recent example is lipoteichoic acid (LTA)..." [126]. He cites a study by Gao et al. [127], which found contaminating endotoxin in commercial preparations of LTA, and another by Morath et al. [128] (that includes Hartung as a coauthor) suggesting that crude preparations of LTA are not suitable for use as indicators of immune cell activation. However, pointing to the lack of general

*Note that kinetic chromogenic assays can be as sensitive as 0.005 EU/mL.

TABLE 13 Whole Blood Assay (In Vitro Pyrogen, IPT) Claims

Need	Advantage
For nonendotoxin pyrogens	Lists 13 exogenous microbial pyrogen and two exogenous nonmicrobial pyrogen classes (the two nonmicrobial classes are drugs and devices/plastics)
Instances of nonendotoxin contamination	Cites events associated with parenterally manufactured biologicals (most referenced by the group member's own experiences including immunoglobulins, human serum albumin, hepatitis B vaccine, pertussis vaccine, influenza vaccine, tick-borne encephalitis vaccine, gentamycin (actually contaminated below the limit but given at off-label dose)
"Comparison of testability"	A range of sample types according to rabbit, LAL, or IPT test, and lists only recombinant proteins as being questionably tested via the IPT
"Special problems with biological products"	Notes that vaccines raise both pyrogen and LAL-related problems such as when vaccines derived from GNB contain endotoxin as a component, are inherently pyrogenic although LAL-nonreactive, or contain aluminum hydroxide that interferes with the LAL test, and, finally, the fact that many blood products are incompatible with LAL testing
Medical devices	Adherent pyrogens could be incubated in IPT without the need for elution, which is notably inefficient and potentially may affect biocompatibility (i.e., rejection by local inflammatory reaction)
r-DNA used for biologicals	New expression systems (GNB, GPB, fungi, mammalian, and insect cells) may be contaminated by expression organisms without LAL detection

TABLE 14 Materials That Cannot Be Tested with IPT

Drugs that interact with monocytes	IL-1, receptor antagonists, nonphysiological solutions, cytotoxic agents, r-proteins with cytokine activity (i.e., INF-γ), or cytokine detection such as rheumatic factors

agreement, Novitsky maintains that β-glucans "represent a clear case of an adulterated (i.e., contaminated) product when present in an otherwise cGMP-prepared pharmaceutical drug or device" and suggests differentiating and quantifying such contamination using ACC's glucan-specific LAL products. Elsewhere, he details ACC's current thinking on a particular nonendotoxin "pyrogen": it has been our policy to treat glucans as "bioactive" molecules and as "foreign substances" when present in pharmaceutical preparations [129]. The dismissal of LTA as a "bioactive" contaminant goes to the heart of the now-marketed CRL whole blood test that employs LTA as a positive control (for GPB) in the IPT.

In the ACC technical report, Novitsky prescribes caution in moving too quickly to IPT and details perceived shortcomings on several fronts:

IPT is not adequately characterized or validated.
There is no valid nonendotoxin pyrogen standard.
The requirement for fresh, whole human blood.
Variability associated with donor blood in that some contain endotoxin.
12–24-hr incubation for cytokine expression; assay of up to 4 hr for cytokine assay.

Changes in LAL testing probably will not occur until a driving event transpires such as the near extinction of horseshoe crabs on the Atlantic seaboard. If that happens, there will be urgency in looking to cut the use of LAL reagent. In fact, crab populations may have already declined significantly:

> Since Hall (of the University of Delaware's Sea Grant College Program) began coordinating an annual springtime census... a decade ago, the number of breeding adults on the shores of Delaware Bay—the center of the species' range and its most important spawning zone—has plummeted from 1.2 million to about 400,000. The main reasons for the decline are the loss of Atlantic beach habitat and—perhaps most significant—the crabs' value as bait for eel and conch fishermen. Though results of this year's census are not yet in, some conservationists already are worried, not just for the crab itself but also for other species, from shorebirds to humans, that depend on this living fossil for their welfare. [130]

Tangley goes on to say that the crab's populations have varied sometimes widely in the past, but have always come back. However, the year they do not come back may catch the pharmaceutical industry by surprise, either in the rise in cost of reagents or their lack of availability. Lastly, but perhaps of greatest relevance to parenteral manufacturing in the consideration of potential drivers of change in analytical testing for contamination control, is the exploding knowledge of the interrelationship of microbes, their by-products, and human disease states.

Two disease states relevant to such a discussion include: systemic fungal infection and sepsis. β-Glucan is a fungal (or cellulosic breakdown) artifact known to the bacterial endotoxin laboratory because of its LAL reactivity. Although the substance is not prohibited or excluded by testing from parenteral products and has not been found to be a common contaminant, however, because it is used as a diagnostic marker for systemic fungal infections, then it is not hard to envision that those who manufacture parenteral drugs to treat such infections may one day be expected to preclude the possibility of β-glucan contamination. A second, more complex indication and thus a more speculative proposition is the association of minute amounts of nonendotoxin contamination with the occurrence of sepsis. In a similar manner as endotoxin-containing GNB have been correlated with GNB sepsis, GPB have been implicated with GPB sepsis. Indeed, approximately 50% of the instances of sepsis are presumptively caused by GPB infections. What is not known is whether the possibility exists that minute amounts of GPB cellular artifacts introduced from medical devices, infusion solutions, or even parenteral drugs could be relevant contributing factors to this disease state. What is documented is the correlation of the historical rise of sepsis with the use of antibiotics and medical intervention.

REFERENCES

1. Hitchcock PJ, et al. Lipopolysaccharide nomenclature—past, present, and future. J Bacteriol 1986; 166:699–701.
2. Mayer H, Weckesser J. The protein component of bacterial endotoxin. In: Rietschel ET, ed. Handbook of Endotoxin. Elsevier Science Publishers: NY, 1984:339.
3. Beverage TJ. Structures of gram-negative cell walls and their derived membrane vesicles. J Bacteriol August 1999; 181(16):4725–4733.
4. Raetz C, et al. Gram negative endotoxin: an extraordinary lipid with profound effects on eukaryotic signal transduction. FASEB September 1991; 5(12):2652–2660.
5. Galanos C, Luderitz O. Lipopolysaccharide: properties of an amphipathic molecule. In: Rietschel ET, ed. Handbook of Endotoxin: Vol. 1. Chemistry of Endotoxin. Elsevier Science Publishers: NY, 1984.
6. Rietschel, Brade. Bacterial Endotoxins. Scientific America. August 1992:54–61.
7. Amro NA, et al. High-resolution atomic force microscopy studies of the *Escherichia coli* outer membrane: structural basis for permeability. Langmuir 2000, 16(6):2789–2796.
8. Kelly MT, Brenner DJ, JJF III. Enterobacteriaceae. In: Lennette EH, ed. Manual of Clinical Microbiology. Washington, DC: American Society for Microbiology, 1985:263–277.

9. Dundas S, Todd WTA. *Escherichia coli* O157 and human disease. Curr Opin Infect Dis 1998; 11:171–175.
10. Nnalue AN. All accessible epitopes in the salmonella lipopolysaccharide core are associated with branch residues. Infect Immun 1999; 67(2):998–1003. (3.3).
11. Rietschel, et al. Bacterial endotoxins: molecular relationship of structure to activity and function. FASEB February 1994; 18:217–225.
12. Raetz C. Biochemistry of endotoxins. Annu Rev Biochem 1990; 59:129–170.
13. Nowotny A. Relation of structure to function in bacterial endotoxin. Annu Rev Microbiol, 1977, 247–252.
14. Morrison DC, et al. Structure–function relationships of bacterial endotoxins, contribution to microbial sepsis. In: Opal SM, Cross AS, eds. Infectious Disease Clinics of North America. Philadelphia: Harcourt Brace and Co., 1999:313–340.
15. Reitschel, et al. Chemical structure of lipid A. Microbiology, 1977, 262.
16. Sweadner, et al. Filtration removal of endotoxin (Pyrogens) in solution in different states of aggregation. Appl Environ Microbiol October 1977; 34(4):285–382.
17. Hochstein HD. The LAL Test Versus the Rabbit Pyrogen Test for Endotoxin Detection. Pharmaceutical Technology. June 1987.
18. Bowman JP, et al. Diversity and association of psychrophilic bacteria in Antarctic sea ice. Appl Environ Microbiol 1997; 63(8):217–225.
19. Stanley JT, Gosink JJ. Poles apart: biodiversity and biogeography of sea ice bacteria. Annu Rev Microbiol 1999; 53(1):189–215.
20. Novitsky TJ. Discovery to commercialization: the blood of the horseshoe crab. Oceanus 1991; 27(1):13–18.
21. Dinarello CA, et al. New concepts in the pathogenesis of fever. Rev Infect Dis January/February 1988; 10(1):168–189.
22. Dinarello CA, Wolff SM. Molecular basis of fever in humans. Am J Med May 1982; 72:799–819.
23. Kluger MJ, et al. The adaptive value of fever. In: BAC MD, ed. Infectious Disease Clinics of North America. March 1996:1–20.
24. Kluger MJ. Fever: role of pyrogens and cryogens. Physiol Rev 1991; 71(1):93–127.
25. Horn DL, et al. Antibiotics, cytokines, and endotoxin: a complex and evolving relationship in gram-negative sepsis. Scand J Infect Dis 1996; 101(9).
26. Greisman, Hornick. Comparative pyrogenic reactivity of rabbit and man to bacterial endotoxin. Proc Soc Exp Biol Med 1969.
27. Weary ME. Pyrogens and pyrogen testing. In: Swarbrick, Boyan, eds. Encyclopedia of Pharmaceutical Technology. Marcel Dekker, Inc.: New York, 1988:179–205.
28. Braude, et al. Fever from pathogenic fungi. J Clin Invest 1960; 39, 1266–1276.
29. Wilson M, Seymour R, Henderson B. Bacterial perturbation of cytokine networks. Infect Immun 1998; 66(6):2401–2409.
30. Henderson B, Poole S, Wilson M. Bacterial modulins: a novel class of virulence

factors which cause host tissue pathology by inducing cytokine synthesis. Microbiol Rev 1996; 60(2):316–341.
31. Sriskandan S, Cohen J. Gram-positive sepsis. In: Opal SM, Cross AS, eds. Bacterial Sepsis and Septic Shock. Philadelphia: W. B. Saunders Company, 1999:397–412.
32. Idanpaan-Heikkila I, Tuomanen E. Gram positive organisms and the pathology of sepsis, chapter 7. In: Fein AM, et al., eds. Sepsis and Multiorgan Failure. Williams and Wilkins, 1997:62–73.
33. Fox A, et al. Absolute identification of muramic acid, at trace levels. Human Septic Synovial Fluids In Vivo and Absence in Aseptic Fluids. Infection and Immunity 1996; 64 (9):3911–3915.
34. van Langevelde P, et al. Antibiotic-induced release of lipoteichoic acid and peptidoglycan from *Staphylococcus aureus*: quantitative measurements and biological reactivities. Antimicrob Agents Chemother 1998; 42(12):3072–3078.
35. Hellman J, et al. Outer membrane protein A, peptidoglycan-associated lipoprotein, and murein lipoprotein are released by *Escherichia coli* bacteria into serum. Infect Immun 2000; 68:2566–2572.
36. Alberts B, et al. Chapter 10: Membrane Structure. Molecular Biology of the Cell 3rd ed. Garland Publishing, 1994.
37. Akers J, Agalloco J. Sterility and sterility assurance. J Parenter Sci Technol March/April 1997; 51(2):72–77.
38. Bruch CW. Quality assurance for medical devices. In: Avis KE, Lieberman HA, Lachman L, eds. Pharmaceutical Dosage Forms: Parenteral Medications. New York: Dekker, 1993:487–526.
39. Artiss DH. Water systems validation. In: Carlton, Agalloco, eds. Valid Aseptic Pharmaceutical Processes. New York: Dekker, 1998:207–251.
40. Nema S, Washkuhn RJ, Brendel RJ. Excipients and their use in injectable products. PDA J Pharm Sci Technol January–February 1990; 44(1).
41. European pharmacopoeia. Monograph. 3rd ed. 1997.
42. Croes RV. Maltitol solution, mannitol, sorbitol, sorbitol solution, noncrystallizing sorbitol solution—suggested revisions for harmonization of pharmacopeial specifications and procedures. Pharmacop Forum January–February 1998; 24(1).
43. U.S. Department of Health and Human Services. FDA Guideline on Validation of the Limulus Amebocyte Lysate Test as an End-Product Test for Human and Animal Parenteral Drugs, Biological Products, and Medical Devices. Dec. 1987.
44. Williams KL. Developing an endotoxin control strategy. Pharm Technol September 1998; 22(9):90–102.
45. Williams KL. Developing an endotoxin control strategy for parenteral drug substances and excipients. Pharm Tech Asia. 1998; 20–24. Spec Issue.
46. Wade A, Weller PJ, eds. Handbook of Pharmaceutical Excipients. Washington and London: American Pharmaceutical Association and the Pharmaceutical Press, 1994.
47. Cleland JL, et al. The stability of recombinant human growth hormone in poly(lactic-co-glycolic acid) (PLGA) microspheres. Pharm Res 1997; 14(4):420–425.

48. PDA. Jour Sci Tech Jan–Feb 1990; 44(1):16–18.
49. Cooper JF. Formulae for maximum valid dilution. In: Watson SW, Levin J, Novitsky TJ, eds. Endotoxins and Their Detection with the Limulus Amebocyte Lysate Test. New York: Alan R. Liss, 1982:55–64.
50. Levin J. The limulus test and bacterial endotoxins: some perspectives. In: Watson SW, Levin J, Novitsky TJ, eds. Endotoxins and Their Detection with the Limulus Amebocyte Lysate Test. New York: Alan R. Liss, 1982:7–24.
51. Novitsky TJ. Selection of the standard. LAL Update 1996; 17(1):1–4.
52. World Health Organization. Tech Rep Ser 1987; 760:29.
53. Poole S, Das REG. Report on the Collaborative Study of the Candidate Second International Standard for Endotoxin. Expert Committee on Biological Standardization, 1996.
54. Cooper JF, Levin J, Wagner HN. New rapid in vitro test for pyrogen in short-lived radiopharmaceuticals. J Nucl Med 1970; 11:310.
55. Hochstein HD. Review of the bureau of biologic's experience with Limulus amebocyte lysate and endotoxin. Endotoxins and their Detection with the Limulus Amebocyte Lysate Test. New York: Alan R. Liss, 1982:141–151.
56. Cooper JF, Hochstein DH, Seligman EB. The *limulus* test for endotoxin (pyrogen) in radiopharmaceuticals and biologicals. Bull Parental Drug Assoc 1972; 26:153–162.
57. Oettgen HF, et al. Toxicity of *E. coli* L-asparaginase in man. Cancer 1970; 25:253–278.
58. Cooper JF, Pearson SM. Detection of endotoxin in biological products by the *limulus* test. International Symposium on Pyrogenicity, Innocuity and Toxicity Test Systems for Biological Products. Budapest, 1976.
59. Bang FB. A bacterial disease of *Limulus polyphemus*. Bull Johns Hopkins Hosp 1956; 98:325–351.
60. Levin J, Bang FB. The role of endotoxin in the extracellular coagulation of *Limulus* blood. Bull Johns Hopkins Hosp 1964; 115:265–274.
61. Levin J, Bang FB. A description of cellular coagulation in the *Limulus*. Bull Johns Hopkins Hosp 1964; 337–345.
62. Nakamura S, et al. A clotting enzyme associated with the hemolymph coagulation system of horseshoe crab (*Tachypleus tridentatus*): its purification and characterization. J Biochem 1982; 92:781.
63. Nakamura T, Morita T, Iwanaga S. Lipopolysaccharide-sensitive serine-protease zymogen (factor C) found in *Limulus* hemocytes. Eur J Biochem 1986; 154:511–521.
64. Muta T, Iwanaga S. The role of hemolymph coagulation in innate immunity. Curr Opin Immunol 1996; 8:41–47.
65. Iwanaga S, et al. Hemolymph coagulation system in *Limulus*. In: Leive L, ed. Microbiology-1985; Meetings of the American Society for Microbiology. Washington, DC: American Society for Microbiology, 1985:29–32.
66. Levin J, Bang FB. Clottable protein in *Limulus*: its localization and kinetics of its coagulation by endotoxin. Thromb Diath Haemorrh 1968; 19:186–197.
67. Nachum R, Berzofsky R. Chromogenic Limulus amoebocyte lysate assay for rapid detection of gram-negative bacteriuria. J Clin Microbiol 1985; 21(5):759–763.

68. Cohen J, McConnell JS. Observations on the measurement and evaluation of endotoxemia by a quantitative limulus lysate microassay. J Infect Dis 1984; 150(6):916–924.
69. Urbaschek B. Quantification of endotoxin and sample-related interferences in human plasma and cerebrospinal fluid by using a kinetic limulus amoebocyte lysate microtiter test. Endotoxin Detection in Body Fluids. 39–43.
70. Morita T, et al. Horseshoe crab (*Tachypleus tridentatus*) clotting enzyme: a new sensitive assay method for bacterial endotoxin. Jpn J Med Sci Biol 1978; 31:178.
71. Iwanaga S, et al. Chromogenic substrates for horseshoe crab clotting enzyme—its application for the assay of bacterial endotoxins. Haemostasis 1978; 7:183.
72. Pearson FC III. Pyrogens, endotoxins, LAL testing, and depyrogenation. Advances in Parenteral Sciences. Vol. 2. New York: Marcel Dekker, Inc., 1985.
73. Tsuji K, Steindler KA, Harrison SJ. Limulus amoebocyte lysate assay for detection and quantitation of endotoxin in a small-volume parenteral product. Appl Environ Microbiol 1980; 40(3):533–538.
74. Nema S, Washkuhn RJ, Brendel RJ. Excipients and their use in injectable products. J Pharm Sci Technol 1990; 44(1).
75. Cooper J. Resolving LAL test interferences. J Pharm Sci Technol 1990; 44(1):13–15.
76. Cooper JF. Using validation to reduce LAL pH measurements. LAL Times 1997; 4(2):1–3.
77. Motise PJ. Human drug CGMP notes. PDA Lett March 1996; 12.
78. Novitsky TJ, Schmidt-Gengenbach J, Remillard JF. Factors affecting recovery of endotoxin adsorbed to container surfaces. J Pharm Sci Technol 1986; 40(6):284–286.
79. McCullough KZ. Variability in the LAL test. J Pharm Sci Technol 1990; 44(1): 19–21.
80. Roslansky PF, Dawson ME, Novitsky TJ. Plastics, endotoxins, and the Limulus amebocyte lysate test. J Pharm Sci Technol 1991; 45(2):83–87.
81. Twohy CW, et al. Comparison of Limulus amebocyte lysates from different manufacturers. J Pharm Sci Technol 1983; 37(3):93–96.
82. Feldstine P, et al. A concept in glassware depyrogenation process validation. Parenter Drug Assoc 1979; 33(3):125.
83. Banks HM. A study of hyperpyrexia reaction following intravenous therapy. Am J Clin Pathol 1934; 4:260.
84. Cherkin A. Destruction of bacterial endotoxin pyrogenicity by hydrogen peroxide. Biochemistry 1975; 12:625.
85. Novitsky TJ, et al. Depyrogenation by moist heat. PDA Monograph on Depyrogenation Technical Report No. 7. 109–116.
86. Welch H, et al. The thermostabiity of pyrogens and their removal from penicillin. J Am Pharm Assoc 1945; 34:114.
87. Robertson JH, Gleason D, Tsuji K. Dry-heat destruction of lipopolysaccharide: design and construction of dry-heat destruction apparatus. Appl Environ Microbial 1978; 36:705.
88. Tsuji K, Lewis AR. Dry-heat destruction of lipopolysaccharide: mathematical approach to process evaluation. Appl Environ Microbiol 1978; 36:710.

89. Tsuji K, Harrison SJ. Limulus amebocyte lysate—a means to monitor inactivation of lipopolysaccharide. In: Cohen E, ed. Biomedical Applications of the Horseshoe Crab (Limulidae). New York: Alan R. Liss, 1979:367–378.
90. Groves FM, Groves MJ. Dry heat sterilization and depyrogenation. In: Swarbrick J, Boylan JC, eds. Encyclopedia of Pharmaceutical Technology. New York: Marcel Dekker, Inc., 1991:447–484.
91. Akers MJ, Avis KE, Thompson B. Validation studies of the fostoria infrared tunnel sterilizer. J Parenter Drug Assoc 1980; 34:330.
92. Akers MJ, Ketron KM, Thompson BR. *F*-value requirements for the destruction of endotoxin in the validation of dry heat sterilization/depyrogenation cycles. J Parenter Drug Assoc 1982; 36:23.
93. Sweet BH, Huxsoll JF. Depyrogenation by Dry Heat. Parenteral Drug Association, Inc., PDA Technical Report No. 7. 1985. Chapter 12.
94. Berger TJ, et al. Biological indicator comparative analysis in various product formulations and closure sites. PDA J Pharm Sci Technol 2000; 54(2):101–109.
95. Ludwig JD, Avis KE. Validation of a heating cell for precisely controlled studies on the thermal destruction of endotoxin in glass. J Parenter Sci Technol January–February 1998; 42(1):9–40.
96. Ludwig JD, Avis KE. Dry heat inactivation of endotoxin on the surface of glass. J Parenter Sci Technol 1990; 44(1):4–12.
97. Sterile Drug Products Produced by Aseptic Processing, CDER, FDA, 1987.
98. Agalloco J. Points to consider in the validation of equipment cleaning procedures. J Parenter Sci Technol 1992; 46(5):163–168.
99. Held DD, et al. Endotoxin reduction in macromolecular solutions: two case studies. BioPharm March 1977; 32–37.
100. Wilson MJ, et al. Removal of tightly bound endotoxin from biological products. J Biotechnol 2001; 88:67–75.
101. Brendel-Thimmel U, Barenowski K. Pyrogen removal by adsorption. Pharm Manuf Rev, June 1991; 9–12.
102. Novitsky TJ. Letter from the President. LAL Update 1997; 15(2):1.
103. PDA. Evaluation, validation and implementation of new microbiological testing methods. PDA J Pharm Sci Technol May/June 2000; 54:1–39.
104. Tamura H, et al. A new test for endotoxin specific assay using recombined limulus coagulation enzymes. Japan J Med Sci Biol 1985; 38:256–273.
105. Kitagawa T, et al. Rapid method for preparing a β-glucan-specific sensitive fraction from Limulus (*Tachypleus tridentatus*) amebocyte. J Chromatography 1991; 567:267–273.
106. Zhang GH, et al. Sensitive quantitation of endotoxin by enzyme-linked immunosorbant assay with monoclonal antibody against *limulus* peptide C. J. Clin. Microbiol. 1994; 32(2):416–422.
107. Brandtzaeg P, et al. Meningococcal endotoxin in lethal septic shock plasma studied by gas chromatography, mass-spectrometry, ultracentrifugation, and electron microscopy. J Clin Invest 1992; 89:816–823.
108. DeVoe IW, Gilchrist JE. Release of endotoxin in the form of cell wall blebs during in vitro growth of *Neisseria meningitidis*. J Exp Med 1973; 138:1156–1167.
109. Maitra SK, Nachum R, Pearson FC. Establishment of beta-hydroxy fatty acids

as chemical marker molecules for bacterial endotoxin by gas chromatography-mass spectrometry. Appl Environ Microbial 1986; 52(3):510–514.

110. Fox A, et al. Absolute identification of muramic acid at trace levels. Human Septic Synovial Fluids In Vivo and Absence in Aseptic Fluids. Infection and Immunity 1996; 64(9):3911–3915.
111. Dinarello CA, et al. Human leukocytic pyrogen test for detection of pyrogenic material in growth hormone produced by recombinant *Escherichia coli.* J Clin Microbiol 1984; 20(3):323–329.
112. Hansen EW, Christensen JD. Comparison of cultured human mononuclear cells, limulus amebocyte lysate and rabbits in the detection of pyrogens. J Clin Pharm Ther 1990; 15:425–433.
113. Tuchiya M, et al. Detection of peptidoglycan and β-glucan with silkworm larvae plasma test. FEMS Immunol Med Microbiol 1996; 15:129–134.
114. Ashida M, Yamazaki HI. Molting and metamorphosis. In: Onishi O, Ishizaki H, eds. Biochemistry of the Phonoloxidase System in Insects: With Special Reference to Its Activation. Tokyo: Japan Scientific Society Press, 1990:239–265.
115. Tsuchiya M, et al. Reactivities of gram-negative bacteria and gram-positive bacteria with limulus amebocyte lysate and silkworm larvae plasma. J Endotoxin Res 1994; 1(suppl 1):70.
116. Wako Pure Chemical Industries, Package Insert, Osaka, 1999.
117. Kahn W. New Rapid Test for Diagnosing Bacterial Meningitis. ASM, 1996.
118. Tsuchida K, et al. Detection of peptidoglycan and endotoxin in dialysate, using silkworm larvae plasma and limulus amebocyte lysate methods. Nephron 1997; 75(4):438–443.
119. Dussurget O, Roulland-Dussoix D. Rapid, sensitive PCR-based detection of mycoplasmas in simulated samples of animal sera. Appl Environ Microbial 1994; 60(3):953–959.
120. Kenny GE. Mycoplasmas. In: Lennette EH, ed. Manual of Clinical Microbiology. Washington, DC: ASM, 1985:407–411.
121. Waris ME, et al. Diagnosis of *Mycoplasma pneumoniae* pneumonia in children. J Clin Microbiol 1998; 36:3155–3159.
122. Hartung T, Wendel A. Detection of pyrogens using human whole blood. In Vitro Toxicol 1996; 9:353–359.
123. Hartung T, Fennrich S, Wendel A. Detection of endotoxins and other pyrogens by human whole blood. J Endotoxin Res 2000; 6:184.
124. Hartung T, et al. Novel pyrogen tests based on the human fever reaction (The Report and Recommendations of ECVAM Workshop 43). ATLA 2001; 29:99–123.
125. Hartung T, Wendel A. Detection of pyrogens using human whole blood. In Vitro Toxicol 1996; 9(4).
126. Novitsky TJ. BET vs. PT non-endotoxin pyrogens. LAL Update April 2002; 20(2).
127. Gao JJ, Xue Q, Zuvanich EG, Haghi R, Morrison DC. Commercial preparations of lipoteichoic acid contain endotoxin that contributes to activation of mouse macrophages in vitro. Infect Immun 2001; 69:751–757.

128. Morath S, Geyer A, Spreitzer I, Hermann C, Hartung T. Structural decomposition and heterogeneity of commercial lipoteichoic acid preparations. Infect Immun 2002;70938–944.
129. Novitsky T. Letter From the President. LAL Update, Assoc Cape Cod June 2002; 20(3).
130. Tangley L. The decline of an ancient mariner. A Crab's Bad Tidings for Land Dwellers. U.S. News Online, 1999.

9

Proper Use and Validation of Disinfectants

Laura Valdes-Mora
Elite MicroSource Corporation, Panama City, Florida, U.S.A.

1. BASIC CONCEPTS

The use of chemicals to prevent or retard microbial contamination is believed to date back to the origins of microbiology as a science. Joseph Lister, an English surgeon, is credited with introducing in 1867 the use of phenol (carbolic acid) to decrease the probability of infections [1]. However, the need for cleanliness takes us back to biblical times and around 800 BC; sulfur dioxide was the first reported disinfectant per historical reviews by Seymour Block [2]. Interestingly, it can be concluded that the control of microorganisms (unknowingly) via chemicals preceded the birth of microbiology as a science.

Today we use a wide variety of chemicals for an array of applications to control microorganisms. In general, control mechanisms remove, inhibit, or reduce microbial populations. The field of disinfection is rich in the use of terminology. The key terms are described in this chapter.

2. DEFINITIONS

Disinfectant—Chemical agent used to destroy pathogens or inhibit their growth. A disinfectant is not effective against bacterial endospores (spores) and it is used on nonliving material. Note that fungal spores, although more resistant than vegetative cells, are not as resistant to chemicals as bacterial endospores. Bacterial endospores are the most resistant forms of microbial life. This type of survival structures is produced by members of the genera *Bacillus*, *Geobacillus*, and *Clostridium*.

Antiseptic—Chemical agent that can inhibit or destroy microorganisms. It is not effective against bacterial endospores and it is used on living tissue. Antiseptics then are the same as disinfectants except that they can be used on living entities.

Biocide—Chemical agent that kills all living microorganisms, including bacterial endospores.

Sporicide—Chemical agent that kills microorganisms including bacterial endospores. This term is a synonym to biocide.

Biostat—Chemical agent that inhibits microbial growth but does not kill microorganisms.

Germicide—Chemical agent that kills pathogenic organisms. It is not sporicidal. The term can be applied to substances used on living tissues (antiseptics) and on inanimate objects (disinfectants). The word germicide is not commonly used in the pharmaceutical industry.

Sanitization—Process by which the bioburden of an area is taken to a safe (approved) level. The term is applied to the processing of inanimate objects. The safe level is defined by the public health authorities. In the pharmaceutical industry, the term is used for processes that provide a 3-log reduction in microbial content.

Disinfection—Process that reduces or eliminates microorganisms with the exception of bacterial endospores. In pharmaceuticals, the term is applied to processes that provide a 5-log reduction in microbial content. Therefore sanitization is not a synonym of disinfection.

Decontamination—Process by which the bioburden is removed. It can refer to a mechanical process, disinfection, sanitization, or sterilization.

3. DISINFECTION SELECTION CONSIDERATIONS

There are at least 12 points to consider during the selection of a disinfectant.

3.1. Bioburden

What is the number of microorganisms present in the areas that need to be disinfected? In addition to microbial populations, the type of microorganisms present is important. Are the organisms bacteria, yeasts, molds, others?

3.2. Surface

The type of surface may react with the chemicals to be applied. Therefore the compatibility of the surface with each disinfectant must be evaluated.

3.3. End Result

What is the end result of the process? Is it stasis or cidal? If the goal is to maintain the microbial population at the current level or if reducing the populations is acceptable but not required and at the same time the target is to ensure the concentration of microorganisms does not increase significantly over time, then the end result is stasis. However, if the goal is to significantly reduce microbial populations and the target is to kill microorganisms, then the end result is cidal.

3.4. Contact Time

The disinfectants will require a certain amount of time to do their job. Here it is important to understand how long production personnel or technical personnel will wait for the disinfection process to be completed before using the area.

3.5. Organic Matter

The presence of organic matter will negatively impact the performance of some disinfectants. It is known that alcohol is affected by the presence of soil, requiring more contact time when this occurs.

3.6. Preparation Steps

The accuracy of the preparation of the disinfectant's use-dilution is crucial to the desired end result. Choose disinfectants that are easy to prepare. Consider using sterile water and aseptic techniques if you do not plan to filter sterilize the preparation. Because disinfectant concentrates and water can contain microorganisms, it is highly recommended that disinfectants be filtered sterilized before use not only in sterile operations but also in nonsterile areas including laboratory benches and laminar flow hoods. This practice will ensure that a disinfection process will take place, not an inoculation process. Contamination incidents of surface areas have been traced back to water, chemical concentrates, and/or incorrect dilutions.

3.7. Local Regulations

Ensure knowledge of the local laws and regulations regarding use of chemicals and their disposal. These binding documents dictate which chemicals are acceptable for use in your facility.

3.8. Safety

Before purchasing any disinfectant, obtain its Material Safety Data Sheet (MSDS). Review safety data and have a chemist, formulator, safety officer, or another qualified individual evaluate the compound to ensure your company has or can develop procedures to safely and responsibly handle the chemical.

3.9. Shelf Life

The concentrate of a disinfectant should have an expiration date. The date is determined by the manufacturer based on stability studies. The studies also dictate if the compound is sensitive to light, thus requiring storage in an amber container or equivalent. If the concentrate is to be diluted for use, the expiration date of the solution is unknown and it is the user's responsibility to determine it. The expiration date of the concentrate is not to be used for the solution because the properties and the stability of the compound are possibly altered by the dilution. The diluted compound may be more or less stable than its concentrate. It is recommended that the use-dilution (diluted disinfectant) be stored in a container of similar material to that of its concentrate.

3.10. Residues

Most disinfectants leave a residue on the surfaces they are applied to. Are these residues acceptable in your operation? Consider methods for removal of the residues. The most common method is to use 70% isopropyl alcohol (IPA). If more than one disinfectant will be used on the surface, the compatibility of the chemicals needs to be evaluated.

3.11. Type of Water Available

This will only be a consideration for companies that will use potable (drinking) water to prepare disinfectants. Drinking water can be hard or soft depending on its chemical makeup. Water hardness or lack of it can interfere with the action of some disinfectants. It is highly recommended that USP Purified Water or a better grade of water be used for the preparation of disinfectants.

3.12. Application Method

The mode of application of the disinfectant can affect the end result in terms of microbial populations. This is because some methods mechanically remove microorganisms.

4. CLASSIFICATION OF DISINFECTANTS

Disinfectants are classified according to their chemical composition. One can further describe them based on their sporicidal properties, if any (Table 1).

TABLE 1 Classes of Disinfectants

Class	Disinfectant	Sporicidal
Alcohol	70% Isopropyl 70% Ethanol	No
Phenol	Yes	No
Quaternary ammonium compounds	Yes	No
Sodium hypochloride	Yes	Weak
Peracetic acid	Yes	Yes
Hydrogen peroxide	3% Solution Yes	No 30% Solution
Glutaraldehyde	Yes	Yes
Formaldehyde	Yes	Yes
Chlorine dioxide	Yes	Yes
Peracetic acid/hydrogen peroxide	Yes	Yes

Each type of disinfectant will target one or more structures in a microbial cell. The targets are shown in Table 2. Detailed information on the dynamics of disinfectants can be obtained from the literature [3].

4.1. Alcohols

Alcohols are the most widely used disinfectants. Ethanol and isopropyl alcohol are the most commonly used at a 70% concentration. Alcohols denature proteins, solubilize lipids, and dehydrate cells. Alcohols require water for their activity; therefore absolute alcohol (100%) is not antimicrobial. They are fast acting (as low as 10 sec) when concentrations from 50% to 70% of

TABLE 2 Target Sites of Various Classes of Disinfectants

	QAC	Formal-dehyde	Alcohol	Phenol	H_2O_2	Sodium hypochlorite	Glutar-aldehyde
Cell wall		✓		✓		✓	✓
Enzymes		✓			✓	✓	✓
ATP		✓					
Membranes	✓	✓	✓	✓			
Coagulation	✓	✓		✓			✓
Nucleic acids		✓					
Amino groups		✓				✓	✓
Ribosomes		✓			✓		

ethanol (CH_3CH_2OH) or 40% to 80% isopropyl alcohol $(CH_3)_2CHOH$ are used in the absence of organic matter [4].

4.2. Phenols

Phenol was the first disinfectant used by Joseph Lister and to date the efficacy of disinfectants can still be compared to that of phenol using the classical Phenol Coefficient Test. Phenolic compounds can be bactericidal, bacteriostatic, fungicidal, and/or fungistatic. Phenols are affected by organic matter and alkalinities. Phenolics are not compatible with Quaternary Ammonium Compounds and Iodophors. Phenolics are also affected by detergents and by dilution.

4.3. Quaternary Ammonium Compounds

Quaternary ammonium compounds (QACs) can be bactericidal, bacteriostatic, and/or fungistatic. These compounds are cationic and as such are affected by low pH. They are not compatible with phenolics, detergents, or anionic compounds. Their activity is greatly reduced in the presence of organic matter.

4.4. Sodium Hypochlorite

Bleach can be a wide-spectrum antimicrobial and sporicidal compound. Although not much has been published, except for the work of Denny et al. [5], there are unpublished data in many pharmaceutical microbiology laboratories that indicate that sodium hypochlorite (NaOCl) is a borderline or weak sporicidal (L. Valdes-Mora, personal observation). However, there are companies that have data showing excellent kill of bacterial endospores by this chemical (A. Cundell, personal communication, 2003). Sodium hypochlorite solutions work best at an acidic pH, which also makes them unstable. Bleach is not compatible with hydrogen peroxide, detergents, or organic matter. It is well known to be corrosive.

4.5. Peracetic Acid

Peracetic acid (CH_3COOOH) is a wide-spectrum antimicrobial and sporicidal compound. It has a certain degree of toxicity and can corrode metals. Peracetic acid is not compatible with organic matter.

4.6. Hydrogen Peroxide

Hydrogen peroxide (H_2O_2) is a strong bactericidal and fungicidal agent. It is used as an antiseptic at a 3% concentration. Concentrations of at least 25%

but typically 30% are used as sterilants because of their good sporicidal action. Hydrogen peroxide is not compatible with detergents.

4.7. Glutaraldehyde

Glutaraldehyde [$CHO(CH_2)_3CHO$] is an excellent antimicrobial agent affecting bacteria, fungi, and viruses. It is also sporicidal. Typically 2% concentrations are used. In just a few minutes, it works very well as a disinfectant but it requires up to a 10-hr exposure to work as a sporicidal. Contact time here is an issue. Glutaraldehyde releases strong fumes; therefore a respirator or gas mask should be worn during preparation and use of this compound.

4.8. Formaldehyde

Formaldehyde (HCHO) can be used as a liquid or a gas. The vapors of formaldehyde are most commonly used. Formaldehyde is a wide-spectrum antimicrobial and sporicidal compound. Formaldehyde is not compatible with temperatures lower than 22°C or humidity outside a 60–80% range. Use of formaldehyde requires extensive safety training because of the high toxicity of the compound.

4.9. Chlorine Dioxide

Chlorine dioxide (ClO_2) is used as a gas in special sterilization cycles. Chlorine dioxide is a wide-spectrum antimicrobial and sporicidal compound. This yellow green gas is water-soluble, noncarcinogenic, and nonflammable at use concentrations. Its sporicidal activity takes place at low concentrations and at room temperature. Chlorine dioxide is not compatible with copper, uncoated aluminum, and neoprene.

4.10. Peracetic Acid/Hydrogen Peroxide

These two compounds together are excellent antimicrobials and sporicidals. This liquid can be used as a disinfectant using a 15-min contact time or as a sterilant using a 3-hr contact time at room temperature. This combination solution is not as toxic or corrosive as other liquid sterilants. Several combinations are commercially available and typically contain a higher concentration of hydrogen peroxide than peracetic acid. Examples are 7.35% hydrogen peroxide + 0.23% peracetic acid or 1.0% hydrogen peroxide + 0.08% peracetic acid.

5. ROTATION VS. NO ROTATION

Rotation and no rotation practices vary from company to company and even from site to site in the United States. Rotation practices are primarily driven by regulatory citations from the Food and Drug Administration (FDA), commonly known as 483s. Rotation is recommended by scientists who have been incorrectly taught that disinfectants are similar to antibiotics; that is, that microorganisms develop resistance over time. For antibiotics this is true and it is mediated by plasmids (extrachromosomal DNA). To date, there is no scientific evidence that this occurs with disinfectants. In the pharmaceutical industry, we rotate disinfectants because of pressure from the FDA. A paper by Denny and Marsik [6] provides information on the use of disinfectants at the production sites.

6. INCOMING TESTS AND IN-HOUSE QUALIFICATIONS

Disinfectants should be considered raw materials. As such, there should be Standard Operating Procedures (SOPs) in place describing Quarantine Procedures and Release Procedures for Disinfectants. Possible incoming tests to conduct are as follows:

1. Identification Test—This is a chemistry test. Depending on the chemical compound, the test may consist of infrared absorption, chromatography, ultraviolet absorption, formation of a precipitate, etc.
2. Verification of Concentration—This is another chemistry test. Concentration verifications are typically conducted using chromatographic methods from the conventional paper chromatography to thin-layer chromatography to high pressure liquid chromatography (HPLC).
3. pH Determination—This is also a chemistry test easily conducted with a pH meter.
4. Challenge Tests—These microbial tests are described in this chapter.
5. Bioburden or Sterility Test—Perform sterility test if the concentrate is labeled "sterile," if not, perform a bioburden determination.

In lieu of the tests described above, you may elect to audit the manufacturer and accept the certificate of analysis (COA).

In-house qualifications should consist of the following sections: Association of Official Analytical Chemists (AOAC) Tests using American Type Culture Collection (ATCC) Strains and Environmental Isolates, Determination of Application Mode, Determination of Contact Time, Bioburden Method Validation, and Expiration Dating.

7. ASSOCIATION OF OFFICIAL ANALYTICAL CHEMISTS TESTS

Disinfectants must pass the tests described in the book by the Association of Official Analytical Chemists (AOAC) to be approved by the Environmental Protection Agency (EPA) in the United States. The EPA regulates the safety, use, and disposal of disinfectants. The efficacy of disinfectants is determined using the procedures described in the AOAC.

The book Official Methods of Analysis of AOAC International was in its 17th Edition in July 1, 2000. This edition was used to summarize the tests described in this section [7]. The AOAC has seven tests that can be used for disinfectants. Choosing the test is primarily based on the label claims of the disinfectant.

7.1. Spray Products Test

This test uses Method 961.02 "Germicide Spray Products as Disinfectants." The test is designed for pressurized and nonpressurized sprays. Test organisms are:

Trichophyton mentagrophytes	ATCC 9533
Salmonella choleraesuis	ATCC 10708
Staphylococcus aureus	ATCC 6538
Pseudomonas aeruginosa	ATCC 15442

Note that *T. mentagrophytes* is used if fungicidal activity is to be assessed. The Spray Products Test consists of inoculating 10 slides and spraying them for the specified time and distance. Slides are held for the contact time, excess disinfectant is drained, and slides are transferred to broth. Disinfectants will show kill in 10 out of 10 trials.

7.2. Phenol Coefficient Method

This is the classical test to evaluate disinfectant efficacy. This test is for disinfectants that are miscible in water and can be performed by two methods. Method 955.11 "Testing Disinfectants against *Salmonella typhi*" or Method 955.12 "Testing Disinfectants against *Staphylococcus aureus*." The Phenol Coefficient Test is not typically performed in the pharmaceutical industry.

7.3. Test for Tuberculocidal Activity

This test is described under Method 965.12 "Tuberculocidal Activity of Disinfectants." The method uses *Mycobacterium smegmatis*. During the test, three dilutions are prepared and 10 carriers are used per dilution. The applicability of this test is for hospitals because disinfectants used in medical facilities must be tuberculocidal.

7.4. Test for Fungicidal Activity

This test uses Method 955.17 "Fungicidal Activity of Disinfectants." Test uses *T. mentagrophytes* ATCC 9533 after a 10–15-day incubation. The Fungicidal Test employs a dilution series and test times of 5, 10, and 15 min. The highest dilution that kills spores within 10 min is considered the highest dilution that can disinfect.

7.5. Use-Dilution Test

This test uses Method 955.14 "Testing Disinfectants against *Salmonella choleraesuis*." The test uses *S. choleraesuis* ATCC 10708 and employs stainless-steel cylinders. These cylinders can be referred to as penicylinders or carriers. The cylinders are 8 ± 1 mm outer diameter (od), 6 ± 1 mm inner diameter (id), and 10 ± 1 mm length type 304 stainless steel (ss18-8). Cylinders can be obtained from S&L Aerospace Metals, 58-2957 Drive, Maspeth, New York.

Before the test, there is a prescreening process for the cylinders using alkyldimethyl ammonium chloride and *S. aureus* ATCC 6538. Any cylinders giving a positive result in this test are discarded. In addition, any cylinders that are visibly damaged are to be discarded.

The procedure describes preparation of carriers involving rinses and autoclaving. After these steps, 20 carriers are transferred into the test culture, after a 15-min exposure, carriers are removed and placed in a 37°C incubator. Carriers are allowed to dry for 40 min. To perform the test itself, 10 tubes of the use-dilution of the selected disinfectant are prepared. The inoculated cylinders are transferred one at a time into each of the use-dilution tubes. This is performed at 1-min intervals of each other and it is a very critical step. All carriers are transferred from the test solution to the subculture broths using the same sequence. Sample preparations are incubated for 48 hr at 37°C. Results are reported as growth (+) or no growth (–). Positive tubes are confirmed via Gram stains to ensure that there is no contamination. If there is no growth in 10 out of 10 carriers, the use-dilution is acceptable. If growth is detected in any of the 10 carriers, the use-dilution is incorrect and not safe for use. Note that this document recommends preparation of a higher concentration (lower dilution) of the disinfectant. The overall goal of the test is to determine the maximum dilution of germicide that kills the test organism on carriers in a 10-min interval.

There are other use-dilution methods in AOAC. Method 955.15 uses *S. aureus* ATCC 6538. This test is conducted in places where bacteria are of great significance. The method indicates that killing 59 out of 60 replicates gives a confidence level of 95%. Method 964.02 uses *P. aeruginosa* ATCC 15442. The same method described for *Salmonella* is used because the *Pseudomonas* document only covers the preparation of the organism.

The dilution test can be a faulty test. It has been known to have false positives and is difficult to reproduce. It is speculated that the problems are caused by the cylinders because they can be reused. The continuous processing of the cylinders seems to have an impact on the final outcome of the test. The carrier test method is seen as a superior procedure; here carriers are only used once.

7.6. Carrier Test Method

This method is also referred to as the Hard Surface Carrier Test. In the AOAC, it is Method 991.47 "Testing Disinfectants against *Salmonella choleraesuis*." This method uses disposable borosilicate glass carriers (Fig. 1), which are of 10 ± 1 mm length, 6 ± 1 mm inner diameter (id), and 8 ± 1 mm outer diameter (od). They can be obtained from Bellco Glass, 340 Edrudo Road, Vineland, New Jersey 08360, website http://www.bellcoglass.com. AOAC cites order no. 2090-S0012 or equivalent. However, the cylinders typically sold are no. 2091-00808, which are 8 mm (od) × 8 mm in length.

As in the previous method, any defective carriers are to be discarded. However, none of the carriers are to be reused. The method has a section on how to clean, disinfect, and autoclave the carriers. Cultures are prepared by making bacterial lawns and harvesting using Dacron swabs. The microbial

FIGURE 1 Glass cylinders. (Courtesy of Bellco Glass, Inc.)

suspension is filtered and sonicated. This suspension can only be used the day it is prepared.

Cultures yielding 5×10^9–1×10^{10} CFU/mL are used. Twenty-four carriers are transferred to 24 mL of culture (1 mL per carrier). After a 15-min contact time, the carriers are transferred in an upright position onto a filter paper in a petri dish. If >1 carrier falls over, the procedure is to be repeated. Carriers are dried in an incubator at 37°C for 40 min (Fig. 2).

The 12 cylinders are used as follows: 10 for the test, 1 extra, and 1 for population determination. The average count/carrier must be 0.5–2.0×10^6 CFU/dried carrier. The disinfectant is to be prepared using sterile water unless otherwise stated on the label.

Twenty tubes are prepared (Fig. 3) with 10-mL aliquots of the disinfectant. Tubes are placed in a water bath at 20 ± 0.5°C. Carriers are transferred to the disinfectant tube one at a time every 30 sec. At exactly 10 min, carriers are removed every 30 sec. Excess disinfectant is also removed and carriers are transferred to Letheen broth (or another medium with appropriate neutralizer). Carriers are incubated at 37°C for 48–54 hr. The Criterion

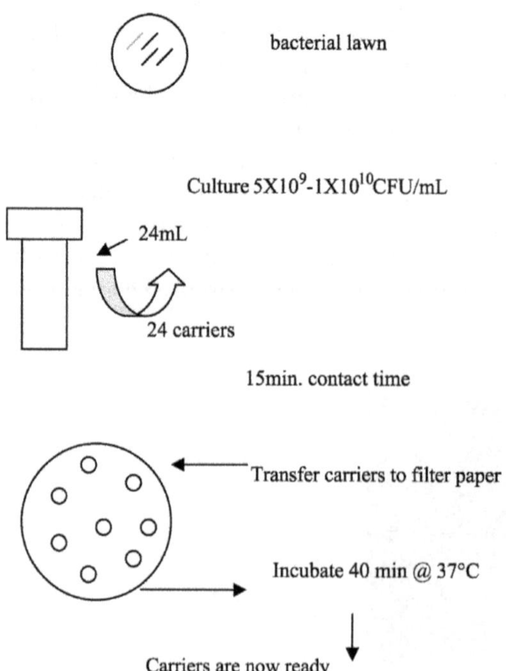

FIGURE 2 Preparation of cylinders for the carrier test method.

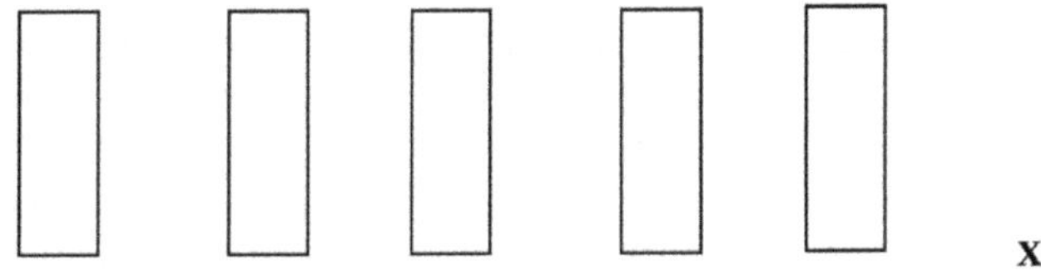

20 tubes w/ 10mL each of the Disinfectant's Use-Dilution

Place in water bath @ 20°C $\pm$ 0.5°C

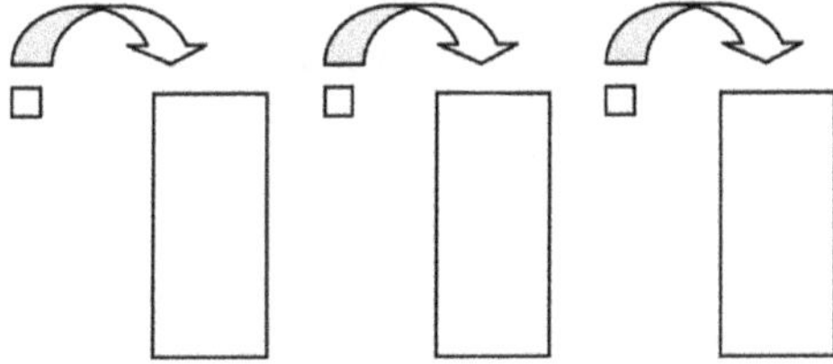

Expose 10 min. Remove carriers every 30 seconds

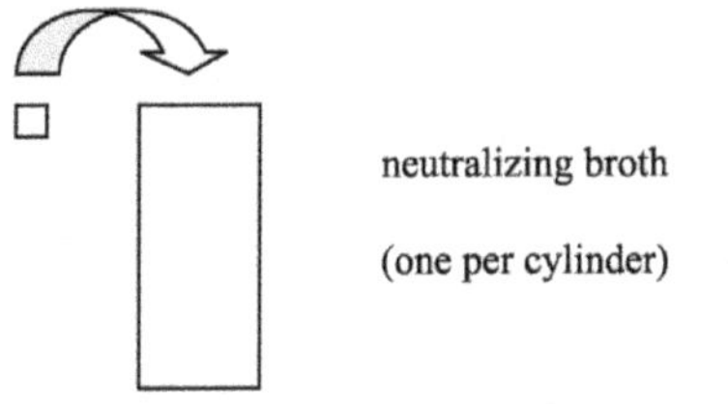

Incubate @ 37°C for 48-54 hours

FIGURE 3 Carrier test method diagram.

of Acceptance is 2 positives out of 60. The same test can be performed with *S. aureus* ATCC 6538 using AOAC Method 991.48 and/or *P. aeruginosa* ATCC 15442 using AOAC Method 991.42.

7.7. Sporicidal Activity Method

This test uses Method 966.04 "Sporicidal Activity of Disinfectants." The method uses *Bacillus subtilis* ATCC 19659 or *Clostridium sporogenes* ATCC 3584. The document indicates that it is applicable for use with other spore formers. The methodology was developed in 1966 and uses suture loops that are to be prepared by the practitioner using surgical silk suture followed by an extraction procedure using chloroform ($CHCl_3$).

Penicylinders are used in this test. These are made of porcelain having 8 ± 1 mm outer diameter (od), 6 ± 1 mm inner diameter (id), and 10 ± 1 mm length. These cylinders are available from ALSIMAG Tech Ceramic in Laurens, South Carolina. *Bacillus* species are grown in Soil Extract Nutrient Broth and *Clostridium* species are grown in Soil Extract Meat Egg Medium.

Suture loops and cylinders are sterilized. Ten suture loops or penicylinders are transferred to three tubes containing 10 mL of a 72-hr culture of spores and exposed for 15 min. Thirty-five loops or cylinders are to be prepared and placed on filter paper for drying in a vacuum desiccator containing calcium chloride ($CaCl_2$) for 24 hr.

Ten milliliters of the dilution of a liquid sporicide is placed into six tubes. The contents is brought to 20°C using a water bath. Five cylinders are placed into each of six tubes using a 2-min interval. After the contact period is completed, all carriers are removed at 2-min intervals and transferred to a subculture medium that contains a neutralizer. Each carrier is again transferred to thioglycollate medium and incubated for 21 days at 37°C. Results are reported as growth (+) or no growth (–). The acceptance criterion is 59 out of 60 (95% confidence level). This test is to be conducted using both genera, using 30 replicates of each of the 2 carriers specified for a total of 120 carriers per trial.

This AOAC method states: "For sporicidal claims, no more than 2 failures can be tolerated in this 120 carrier trial. For sterilizing claims, no failures can be tolerated." Scientists who have performed this procedure as described in AOAC indicate that it is extremely laborious. In addition, to ensure reduced variability in the test, they choose soil that contains no pesticides by typically having a soil plot dedicated for use in this test. In pharmaceuticals, microbiology practitioners find this test not to be highly applicable because it uses porcelain cylinders, and porcelain is not a common production surface. Because of this and other details, it is recommended that the tests be customized for use in pharmaceutical applications.

There are a few important details to remember from the AOAC Tests. All AOAC Tests are standardized by conducting them at 20 ± 0.5°C. Higher temperatures will increase disinfectant activity, therefore producing better kill. The pH of the disinfectants will also play a role in their activity.

8. CHALLENGE MICROORGANISMS

The disinfectants to be used in pharmaceutical facilities should be challenged in the laboratory using microorganisms from the American Type Culture Collection (ATCC) and also environmental isolates. The goal is to select a group of organisms that represent the entire spectrum of the kingdoms Monera and Fungi.

Good candidates:

1. Gram-positive coccus.
2. Gram-positive, spore-forming rod.
3. Two Gram-negative rods (one fermentative and one nonfermentative).
4. Yeast.
5. Mold.

8.1. American Type Culture Collection Strains

Why should we use organisms from a culture repository? Because we know these strains based on the following key points:

1. In the microbiology laboratory, we use these organisms as standards to verify that tests will produce expected results. Example of this is the inoculation of a positive control for eosin methylene blue agar with *Escherichia coli*. Colonies of this organism have specific morphological characteristics on this medium; when the expected colonies are not found, the microbiologist knows there are problems with the test.
2. We have knowledge of their properties.
3. We can find information in literature regarding capabilities of these organisms.
4. We can predict which disinfectants can affect them. Organisms typically used in disinfectant studies by pharmaceutical industries are as follows:

Staphylococcus aureus	ATCC 6538
Escherichia coli	ATCC 8739
Pseudomonas aeruginosa	ATCC 15442
Candida albicans	ATCC 10231
Aspergillus niger	ATCC 16404
Bacillus subtilis	ATCC 6633

In the laboratory, we usually have *P. aeruginosa* ATCC 9027; however, for this test, as stated above, strain 15442 is considered a better choice as it is a hardier, more resistant strain (L. Clontz, personal communication, 2000).

The purpose of using the ATCC strains is to verify the manufacturer's claims of each disinfectant. Let us remember that the manufacturer had to prove with data the effectiveness of the disinfectant; therefore the new laboratory tests will be considered a verification.

8.2. Choosing Environmental Isolates

The most common environmental isolates as well as the ones considered objectionable should be evaluated against the chosen disinfectant(s). At least one of the disinfectants that end up in the "qualified" group should significantly control one of the environmental strains. As a whole, the chosen disinfectants should kill or dramatically reduce the population of all environmental strains. Therefore the purpose of using environmental isolates is to demonstrate that the chosen disinfectants are effective, in theory, for the specific facility under evaluation. Other qualification tests are still required to deem a disinfectant approved for use. The qualification tests include surface evaluations (in vitro tests), expiry date determinations, and in situ studies.

9. CUSTOMIZED IN VITRO TESTS

The purpose of the customized in vitro tests is to mimic the disinfection process to determine its effectiveness. To conduct the customized in vitro tests, other tests such as contact time determinations, evaluation of mode of application, and bioburden testing are prescreened and the proper methods are selected at that time.

Always remember that to eliminate microorganisms, it is always best to choose a sterilization method. In places where absence of microorganisms is not required, then disinfection processes are implemented. The customized in vitro tests, sometimes referred to as surface evaluations, are laboratory tests conducted using samples of the materials that will be routinely disinfected. At this point, we do not inoculate equipment or facility surfaces, instead samples of the various surface materials to be disinfected are used. These samples are commonly known as coupons.

9.1. Choosing Coupons

The first task of the surface evaluation is to determine all types of surfaces to be treated with each disinfectant. Examples of surfaces are plastic, metal, glass, stainless steel (316L, 318, etc.), painted surfaces, tile, aluminum, rubber, wood, flooring, latex, epoxy, vinyl, and others. Of the materials mentioned, note that wood can occasionally be found in older laboratory areas. Wood is difficult to decontaminate because of its porous surface. It is highly recommended that any wood areas be discarded and remodeled using materials that are non-porous and suitable for a simple disinfection process such as a wipe down.

Have coupons prepared that are approximately 2 × 2 in. or larger, preferably not to exceed 4 × 4 in. A handle of the same material placed on each coupon facilitates laboratory work. A lollipop design has been used successfully for many years (Fig. 4). However, other geometric shapes such as squares, rectangles, and triangles have also been used. Based on the type of material, the placing of a handle on the coupon may not be feasible. Coupons for this type of work are typically made in-house as they are not commercially available.

9.2. Contact Time

Contact time is a preliminary test. Contact time is the time that elapses between the application of the disinfectant and its removal. This time is critical to the effectiveness of the chemical agent, as it is the time during which the disinfectant does its job. The most effective contact time for each disinfectant based on the mode of application is experimentally determined. During the experimental design, the following choices can be made:

- Use the recommended contact time (according to the manufacturer's instructions).
- One or two times below the recommended time.
- One or two times above the recommended time.

For the in vitro tests, you will determine the best contact time for your application on a per disinfectant basis. The typical recommended contact time is 10 min because that is the AOAC Test time. Therefore following the recommendation contained here, one can choose to run the tests at 2, 4, 10, 12, and 15 min. These times represent the contact time plus two times below and two times above the recommended contact time. One may choose only three

FIGURE 4 Stainless-steel coupon–lollipop design. (Courtesy of Mary Connor, aaiPharma, Wilmington, North Carolina.)

time points such as 5, 10, and 15 min, where only one point above and one point below the recommended contact time will be used. The contact time can be evaluated side by side with the mode of application to construct a more meaningful evaluation or the tests can be conducted independently as the impact of these two variables will be concurrently determined during the surface evaluations.

9.3. Mode of Application

How the disinfectant is applied to a surface also influences the final outcome of the disinfection process. It is crucial that any item targeted for disinfection be cleaned first. Some disinfectants are affected by the presence of organic matter as previously indicated.

Disinfectants can be applied by spraying, wiping, moping, soaking, immersing, or fogging. Microorganisms can be removed by physical actions. Therefore wiping, brushing, scrubbing, etc. significantly aid in the removal and reduction of bioburden. There is no one method of application better than another, although the methods that include mechanical action may seem superior.

To determine if the reduction or elimination of microorganisms comes primarily or solely from mechanical action, the appropriate control is incorporated into the validation protocol. Typically, validation protocols used in the pharmaceutical industry do not include this determination as the goal is to evaluate the efficacy of the disinfectant along with a specific application mode and a specific contact time.

9.4. Surface Evaluations

This section is extremely valuable as here you will gather data that demonstrate the suitability of the chosen disinfectant, the mode of application, and the contact time. These factors will control (reduce or eliminate) typical microorganisms represented by the chosen ATCC strains and the environmental isolates. During this experiment, you will determine:

- If the disinfectant is appropriate for the surface.
- If the disinfectant acts as expected on the experimental organisms.
- Which contact time is appropriate, if any.
- If contact time and/or application mode need to be modified.

Examples of Protocol for Surface Evaluations

1. Prepare each of the six challenge organisms and environmental isolates.

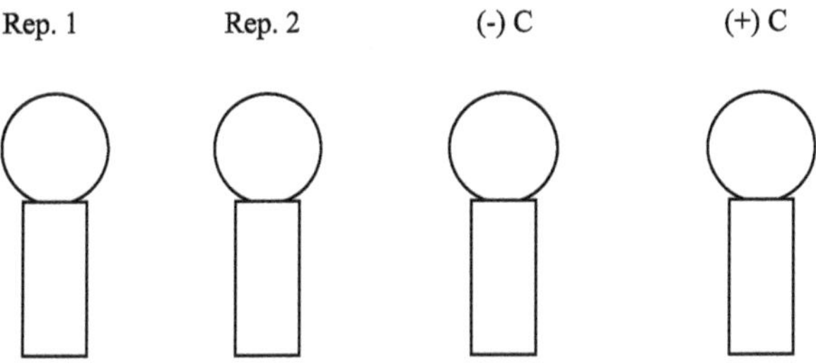

(-) C = Negative Control
(+) C = Positive Control

FIGURE 5 Diagram showing a surface evaluation setup for one challenge organism.

2. Inoculate a minimum of two coupons (sterile) with each of the challenge organisms and prepare a positive control and a negative control (Fig. 5).
3. Treat the coupons per the established or proposed disinfection procedure at each of the five experimental contact times.
4. Use a validated bioburden procedure to process each sample after completion of the contact time.

It is important to note that you must be able to neutralize the disinfectant immediately after the contact time elapses.

During this test, one will be able to gather data on the log reduction (kill) of each chemical under test. The expected results are as follows:

Classification	Log reduction
Bacteriostatic	0.5 log
Sanitizer	3 log
Disinfectant	5 log
Sporicidal	6 log (bacterial endospores)

Tables 3–7 show results obtained using various challenge organisms with various disinfectants. All the disinfectants tested show the expected 5-log reduction. Note that the performance of sodium hypochlorite (Table 7) is not any better than that of a quaternary ammonium compound (Table 6).

TABLE 3 Surface Test—Example of Results for *P. aeruginosa* (Surface: Plastic; Inoculum: 1.7×10^6 CFU; Disinfectant: Quaternary Ammonium Compound No. 1)

Contact time	Total count (CFU/plate)	Log reduction
1 min	0	6.2
5 min	0	6.2
10 min	0	6.2
30 min	0	6.2
24 hr	0	6.2

10. SAMPLING METHODS

Choosing a sampling method is part of the customization of the disinfectant program. This selection involves a great deal of details that are summarized in Table 8. The typical sampling methods are contact plates and swabbing. The chosen method is based on the size and shape of the surface to be sampled. If the surface in question is flat and it is equal or larger to the diameter of a contact plate, this is chosen as the sampling method. This is a simpler method to use. It is extremely important to remember to wipe the area after sampling to remove any medium (agar) residues that could be used as nutrients by the environmental flora. All other surfaces are typically sampled using a swab. Some manufacturing equipment are rather large and may be dangerous for anyone to sample them via any of the two mentioned methods, in this case just as in environmental monitoring, a rinse sample is analyzed.

The choice of medium is also made at this point. Most practitioners in pharmaceutical companies choose media already utilized for environmental monitoring or product testing. Although the organisms that may be recovered are suspected to have been injured by the disinfection process, special low-

TABLE 4 Surface Test—Example of Results for *P. aeruginosa* (Surface: Stainless Steel; Inoculum: 3.3×10^6 CFU; Disinfectant: Quaternary Ammonium Compound No. 2)

Contact time	Total count (CFU/plate)	Log reduction
1 min	0	6.5
5 min	6	5.7
10 min	0	6.5
30 min	0	6.5
24 hr	0	6.5

TABLE 5 Surface Test—Example of Results for *P. aeruginosa* (Surface: Plastic; Inoculum: 3.3×10^6 CFU; Disinfectant: Alcohol)

Contact time	Total count (CFU/plate)	Log reduction
1 min	32	5.0
5 min	0	6.5
10 min	0	6.5
30 min	0	6.5
24 hr	0	6.5

nutrient medium formulations have not been historically considered. Therefore a rich medium is typically used for these determinations as long as it works well during the validation. The chosen medium will be supplemented with neutralizers that will inactivate any disinfectant or residue that may be present in the sample. Lastly, the incubation time and temperature are chosen after conducting preliminary testing with several potential combinations.

If swabbing is the chosen sampling method, the type of swab to be used is of paramount importance. Swabs are primarily developed for use by clinical laboratories. One of the most prominent characteristics of a swab is that they are highly absorbent. For our clinical counterparts, this property is highly desirable as the goal is to identify the pathogen. In pharmaceutical companies, our primary goal is to quantify microorganisms present, and in many cases if a problem is found, the organism will be identified. The absorbency of the swabs does present a very significant problem to us as quantification of the organisms present will be negatively impacted.

The type of swab also impacts the recovery of microorganisms. Cotton swabs contain fatty acids that can inhibit some microorganisms. Although

TABLE 6 Surface Test—Example of Results for *B. subtilis* (Surface: Plastic; Inoculum: 8.5×10^6 CFU; Disinfectant: Quaternary Ammonium Compound No. 2)

Contact time	Total count (CFU/plate[a])	Log reduction
1 min	500	4.9
5 min	500	4.9
10 min	500	4.9
30 min	500	4.9
24 hr	TNTC	N/A

[a] Estimate (plates were crowded).

TABLE 7 Surface Test—Examples of Results for *B. subtilis* (Surface: Plastic; Inoculum: 8.5×10^6 CFU; Disinfectant: Sodium Hypochlorite)

Contact time	Total count (CFU/plate[a])	Log reduction
1 min	500	4.9
5 min	500	4.9
10 min	500	4.9
30 min	500	4.9
24 hr	500	4.9

[a] Estimate (plates were crowded).

dacron or rayon swabs may be used, calcium alginate swabs seem superior for quantitative analyses. Their superiority stems from the fact that their heads dissolve in the presence of 1% sodium hexametaphosphate or 1% sodium citrate. Other chemicals may also be used. Of these two compounds mentioned, hexametaphosphate requires less vortexing. Typically a preparation containing 9 mL Ringer's Solution and 1 mL of sodium hexametaphosphate (sterile) is used.

The swabbing technique requires further fine tuning. An important aspect of it is its standardization. The goal is to have all personnel trained to rotate swabs on a surface in a similar fashion and to cover with the swab the same area (square inches). To assist those taking samples, templates can be obtained from manufacturers of swabs. The best templates are either autoclavable or are supplied sterile.

Transportation or holding of the samples requires consideration. If more than 60 min will elapse between the time a sample is taken and the time

TABLE 8 Ten Points to Consider for the Selection of a Sampling Method

1. Choose a method based on the type of surface to be sampled.
2. Consider selecting a rich medium.
3. Typically, the growth medium contains one or more neutralizers.
4. The type of swab selected is crucial for a better recovery.
5. The swabbing technique must be standardized.
6. Transport medium (buffer) for the swabs must be carefully selected.
7. Holding time for swabs before processing needs to be established.
8. The need for refrigeration of samples collected should be determined and tested before the implementation.
9. Incubation temperature and time should be empirically evaluated.
10. Overall objective is to customize the disinfectant program.

the sample is processed in the laboratory, protection of samples to ensure microbial populations do not grow or die needs to be incorporated in the procedures. This usually entails refrigeration of the samples. However, note that also the length of time the samples can be refrigerated without affecting the results, needs to be experimentally evaluated using various time points. Determination of the appropriateness of the 60 min stated above needs to be empirically determined as it may be affected by the type of sample in question.

Another detail of the swabbing method is that the buffer in which the swab is placed can also affect the microbial populations and should be carefully selected and evaluated for the specific situation. A buffer used to keep the swabs is known as the transport medium.

Swabs must be moistened immediately before use and they must remain moistened through the sampling and holding time. Otherwise, microbial populations will be easily affected and underestimated as microorganisms will die by desiccation. The transport medium provides moisture, a balanced pH, and the proper osmotic pressure for optimum microbial survival.

11. BIOBURDEN DETERMINATIONS

A bioburden determination is an assessment of the number of bacteria, yeasts, and molds present in the sampled area. Before performing these determinations, the variables of the method need to be studied to select the most appropriate combination for the disinfectant and microbial flora under study. One can start by creating a matrix.

The matrix experiment is performed in the absence of the disinfectants and must include your environmental isolates. To choose the optimum conditions, design a matrix; test two to three combinations of medium, incubation time, and temperature. Select the method that provides the highest recovery time in the least amount of time. Highest recovery refers to significantly higher counts that differ by more than 0.5 log, as we use the concept when performing the Antimicrobial Effectiveness Test per USP Chapter 51 [8].

For this matrix, the practitioner chooses media, incubation times, incubation temperatures, and a transport medium (if working with swabs). The choices of media can be Soybean Casein Digest (SCD), Microbial Content Test Agar (MCTA*), D/E Neutralizing Agar or Letheen Agar. Of the mentioned media, D/E Neutralizing Agar should be highly considered. This medium is especially formulated to contain five types of inactivators used to neutralize disinfectants including quaternary ammonium compounds, phenols, iodines, mercurials, chlorine preparations, formaldehyde, and glutar-

* MCTA is the same as SCD with Lecithin and Polysorbate 80.

TABLE 9 Example of Combination to Study

Medium	Temperature (°C)	Time (days)
D/E Agar	30–35	2
D/E Agar	20–25	7
MCTA	30–35 then 20–25	7

aldehyde. Other media such as Microbial Content Test Agar (MCTA) and Letheen Agar contain neutralizers, the inactivator lecithin is included in these two. For incubation temperatures and times, the choices are 30–35°C for 2–3 days, 20–25° C for 5–7 days, 30–35°C for 2 days then 20–25° C for 5 days (total of 7 days of incubation at two temperatures), 20–25°C for 5 days then 30–35°C for 2 days (total of 7 days of incubation at two temperatures), or other combinations. Transport media for swabs can be Phosphate Buffer, Butterfield's Buffer, Commercial Transport Buffer, or Ringer's Solution. Table 9 shows an example of a matrix study set up. The total possible combinations are not shown. In the example, the matrix uses two types of media, three incubation temperatures, and two lengths of incubation. This is a 2 × 3 × 2 matrix, which results in 12 possible combinations.

Once the combinations are tested, the optimum one is chosen and it is the method to undergo validation. This matrix study can also be conducted in the presence of disinfectants to evaluate recovery of possible injured organisms. However, because of the presence of disinfectants or their residues, each method should be validated before use, resulting in a more laborious study.

12. METHOD VALIDATION

The method validation will demonstrate that the sample preparation and processing are suitable to recover microorganisms present even if these are in low concentrations. To do this successfully, one must inactivate any antimicrobial activity that may be present. This means that the disinfectant must be neutralized.

Each method chosen will be validated for each of the chosen disinfectants. During the validation, the proper neutralizer (Table 10) is evaluated and chosen. Five to six organisms are used to challenge the disinfectant and a low level inoculum for each organism (<100 CFU/mL) is prepared. The low-level inoculum preparations are commercially available for most of the typical challenge organisms.

On any of the methods used, any holding (waiting) times typical of sampling and processing should be included in the validation. It is crucial that

TABLE 10 Commonly Used Neutralizers for Disinfectants

Disinfectant	Neutralizer(s)
Iodine	Lecithin, sodium thiosulfate
Glutaraldehyde	Sodium sulfite, bisulfate
Chlorine	Sodium thiosulfate
Phenol	Tween (polysorbate), dilution
QAC	Lecithin, polysorbate
Hydrogen peroxide	Catalase
Chlorhexidine	Lecithin and polysorbate
Aldehydes	Glycine, dilution
Alcohols	Dilution
Formaldehyde	Ammonium carbonate

holding times be included in the method validations. Holding time is defined as the time that elapses between sampling and final processing of the sample in the laboratory. The typical challenge organisms for disinfectant studies are as follows:

Staphylococcus aureus	ATCC 6538
Escherichia coli	ATCC 8739
Pseudomonas aeruginosa *	ATCC 15442
Candida albicans	ATCC 10231
Aspergillus niger	ATCC 16404
Bacillus subtilis	ATCC 6633

As previously mentioned, note that *P. aeruginosa* ATCC 15442 is considered a hardier, more resistant strain than ATCC 9027.

12.1. Example of Method Validation for Contact Plates

1. Expose the medium to the disinfectant. Allow proper contact time and do an imprint on the contact plate. The goal is to pick up residues of the disinfectant.
2. Inoculate each organism independently of the others, using low-level inocula.
3. Run a minimum of two replicates per organism.
4. Include positive controls. These are contact plates inoculated but not exposed to the disinfectant.

5. Include medium controls. These are uninoculated and unexposed contact plates.
6. Include negative controls. These are contact plates uninoculated but exposed to the disinfectant.
7. Incubate all plates at the chosen temperature for the selected time. Consider including two different incubation times, at this stage, as the organisms may be injured by the disinfectants and may require longer times for recovery.
8. Retrieve plates and compare all test organisms to their controls.
9. Acceptance criterion: Recover ≥70% of each of the challenge organisms. Use this criterion if you would like to follow USP Chapter 1227, Validation of Microbial Recovery from Pharmacopeial Articles. However, note that recovery from contact plates as well as from the other two sampling methods present enormous challenges. Some companies have chosen 50% recovery as criterion while other companies consider any recovery acceptable. Recoveries of about 10–30% have been recorded at some pharmaceutical companies but data have not been published (laboratory analysts, personal communications 2003). The enormous challenges stem from the fact that the environments are typically dry; therefore many organisms will find their demise. The same applies to the fact that during testing, the suspension of organisms used as inoculum dries out, killing the organisms by desiccation. Other microorganisms will be killed by the disinfectant while others can be critically injured, in which case, they could be present but the organisms will not form a colony forming unit (CFU).
10. If all or any of the challenge organisms do not meet the acceptance criteria, the method needs to be modified. Modifications will entail finding a better neutralizer or a better combination of neutralizers and/or changing the medium formulation, incubation temperature, or length of incubation. When excessive problems are encountered with recovery and all feasible avenues of tweaking the method are exhausted, revising the acceptance criterion may be the last option and should be discussed with regulatory agencies before proceeding with it.

Note that any coupons and disinfectants used for these tests need to be sterile to prevent interference in the study.

12.2. Example of Method Validation for Swabs

1. Apply disinfectant to a sterile surface.
2. Use a test swab to mimic swabbing the surface.

3. Inoculate the swab with the challenge organism. Inoculate each organism independently of the others.
4. Run a minimum of two replicates per organism.
5. Include positive controls (swab inoculated but not exposed to the disinfectant).
6. Include medium controls (these are uninoculated, unexposed plates).
7. Include negative controls (these are processed swabs exposed to the disinfectant but uninoculated).
8. Process all swabs as applicable for the type of swab.
9. Incubate all plates at the chosen temperature for the selected time.
10. Retrieve plates and compare all test organisms to their controls.
11. Acceptance criterion: Recover ≥70% of each of the challenge organisms. Use this criterion if you would like to follow USP Chapter 1227. See note on this under contact plates section.
12. If all or any of the challenge organisms do not meet the acceptance criterion, the method needs to be modified. Examples of possible modifications are to change type of swabs, add neutralizers to transport medium, and/or decrease transport time. Typically the transport medium contains neutralizers; therefore, in many cases, it is not necessary to have neutralizers present in the final plating medium. However, exceptions to this may be found.

Note that any swabs and disinfectants used for these tests need to be sterile to prevent interference in the study.

12.3. Example of Method Validation for Rinse Samples

1. Rinse a test area (sterile) after application of disinfectant. Collect the typical rinse amount that production personnel will give you.
2. Divide the rinse into aliquots for each challenge organism.
3. Inoculate the aliquots with the challenge organisms. Inoculate each organism independently of the others.
4. Run a minimum of two replicates per organism.
5. Include positive controls (water inoculated not exposed to disinfectant).
6. Include medium controls (these are uninoculated, unexposed plates).
7. Include negative controls (water exposed to disinfectant but uninoculated).
8. Process rinse samples as planned (pour plates or membrane filtration). If using membrane filtration, it is important that the filter compatibility with the disinfectant be verified. The manufacturer of the filter should be able to provide information.

9. Incubate all plates at the chosen temperature for the selected time.
10. Retrieve plates and compare all test organisms to their controls.
11. Acceptance criterion: Recover ≥70% of each of the challenge organisms. See note regarding acceptance criteria in the contact plates section.
12. If all or any of the challenge organisms do not meet the acceptance criterion, the method needs to be modified.

Note that it is important that any water and disinfectants used for these tests be sterile to prevent interference in the study.

13. DETERMINATION OF EXPIRATION DATES

The expiration date for the use-dilution is to be empirically determined. The manufacturer of the disinfectant concentrate will determine its expiration date. However, once it is diluted for use at any facility, it is a different solution and it requires the determination of its stability. Even if the use-dilution is going to be prepared, used the same day, and any left-over solution discarded, it is crucial that there be data showing if the disinfectant is acceptable a few minutes or hours after preparation.

For this determination, prepare the use-dilution as indicated by manufacturer and as it was used during the validation work. Perform the appropriate AOAC Tests based on the properties of the compound. Typically, a carrier test is used. Always add a few extra carriers (two or three) to do plate counts to determine log reductions. Select additional time points to test the disinfectant. Store the disinfectant as indicated by manufacturer or per established or proposed SOP. Choose three or more time points at which to test the efficacy of the disinfectant. At each time point, conduct the AOAC Test. The time at which failures are obtained on the AOAC Tests with any of the organisms indicates that the expiration date was exceeded. Choose the prior time point as the expiration date. The company may elect to prepare its disinfectants every week while some companies prefer to prepare their disinfectants every day, regardless of the expiration dating data obtained by the laboratory.

A method variation that can also be used for expiration dating determinations is an adaptation of USP Chapter 51, Antimicrobial Effectiveness Test. For this approach, take a 10-, 20-, or 25-mL aliquot of the use-dilution of the disinfectant. Inoculate the preparation using 1×10^5–1×10^6 CFU/mL of the appropriate challenge organism. Use at least two replicates per organism. Choose at least five different challenge organisms and perform challenges on a per organism basis. Let the inoculated preparation stand for the selected (validated) contact time. After the contact time elapses, take an

TABLE 11 Expiration Dating Test for 500-ppm Sodium Hypochlorite (Carrier Test Results—60 Days Time Point)

Organism	Control (CFU/carrier)	Positives out of 60 carriers
Staphylococcus aureus	1.2×10^7	35
Pseudomonas aeruginosa	5.2×10^6	60
Escherichia coli	1.4×10^6	60
Candida albicans	1.7×10^6	59
Aspergillus niger	2.2×10^6	60
Bacillus subtilis	1.2×10^6	60
Bacillus (environmental)	1.1×10^7	60

aliquot and proceed to perform an enumeration procedure by rapidly neutralizing the preparation and plating the sample based on the validated test method. Consider membrane filtration for this test if the validation allows it.

Examples of expiration dating are as follows: Solution A with a projected expiry date of 90 days will have test points of 0, 7, 15, 30, 60, 90, and 120 days. Solution B with a projected expiry date of 30 days will have test points of 0, 7, 14, 21, 30, and 45 days. As seen in the example on Table 11, 60 days is not an appropriate expiry date for a sodium hypochlorite solution. The results of a freshly prepared solution are shown on Table 12.

The examples given above include the testing of at least one additional test point beyond the projected expiry date. This recommendation is useful as it may allow finding a longer expiration dating than predicted or requested. Based on the results, a longer expiry date can be used or one can still choose the targeted expiry date based on specific circumstances presented at your site during the brainstorming for all these studies.

TABLE 12 Example of Carrier Test Results for 500-ppm Sodium Hypochlorite

Organism	Control CFU/carrier	Positives out of 60 carriers
Staphylococcus aureus	1.2×10^7	0
Pseudomonas aeruginosa	5.2×10^6	1
Escherichia coli	1.4×10^6	0
Candida albicans	1.7×10^6	0
Aspergillus niger	2.2×10^6	25
Bacillus subtilis	1.2×10^6	60
Bacillus (environmental)	1.1×10^7	45

14. IN SITU TESTS

After completing the evaluation of the efficacy of the disinfectants under laboratory conditions, the customization of the disinfectant program concludes with tests performed in the real areas and on the equipment to be routinely disinfected. These tests are referred to as the in situ testing. The key to this test is that microorganisms will not be introduced to the areas or pieces of equipment. The in situ tests consist of performing bioburden determinations before and after disinfection procedures.

Example of an in situ study:

1. Conduct bioburden determinations before cleaning and disinfecting.
2. Proceed to clean and disinfect using the previously validated procedures.
3. Conduct bioburden determinations after the contact time has elapsed.
4. Results should show lower microbial counts after cleaning and disinfection.

These results are the final demonstration that the disinfectant, the application mode, and the contact time were properly chosen for each area or piece of equipment. These "before and after" tests are typically conducted during the validation and not routinely thereafter, unless the procedure is suspected to have become ineffective. A signal of this is the isolation of new environmental organisms.

The sections of a disinfectant program can be summarized as follows:

1. Have SOP for incoming disinfectants.
2. Conduct AOAC Tests.
3. Have validated bioburden method(s).
4. Conduct surface tests.
5. Determine expiry dates for use-dilutions of disinfectants.
6. Perform in situ tests.

Should one construct a solid disinfectant program, the company should be able to avoid FDA citations (FD-483s) such as:

"Sanitizer efficacy studies have not been conducted for the current disinfectants."

"Disinfectant effectiveness for production bacteria strains not validated."

"Lack of sanitizer effectiveness studies."

"Disinfectant evaluation did not include worst case of open solution held to expiry."

"Disinfectant evaluation did not include all types of equipment/articles entering the sterile core."

REFERENCES

1. Atlas RM. Microbiology Fundamentals and Applications. Macmillan, NY: Publishing Company, 1984.
2. Block SS, ed. Disinfection, Sterilization, and Preservation. 5th ed. Philadelphia: Lippincott Williams & Wilkins, 2001.
3. Hugo WB, Russell AD. Pharmaceutical Microbiology. 5th ed. Cambridge: Blackwell Sciences, 1992.
4. Pelczar MJ, Chan ECS, Reid RD, eds. Microbiology. 4th ed. New York: McGraw-Hill Book Company, 1977.
5. Denny VF, Kopis EM, Marsik FJ. Elements for a successful disinfection program in the pharmaceutical environment. PDA J Pharm Sci Technol 1999; 53:115–124.
6. Denny VF, Marsik FJ. Current practices in the use of disinfectants within the pharmaceutical industry. PDA J Pharm Sci Technol 1997; 51:227–228.
7. Horwitz W, ed. Official Methods of Analysis of AOAC International. 17th ed. Gaithersburg: AOAC International, 2000.
8. United States Pharmacopeia. 27th ed. Rockville: United States Pharmacopeial Convention, Inc., 2004.

10

Antimicrobial Effectiveness Test and Preservatives in Pharmaceutical Products

Luis Jimenez
Genomic Profiling Systems, Inc., Bedford, Massachusetts, U.S.A.

1. INTRODUCTION

When nonsterile and sterile pharmaceuticals are formulated, chemical preservatives are added to protect the products from microbial contamination and spoilage. An adulterated pharmaceutical product represents a serious health threat to consumers by the loss of the drug potency, drug efficacy, presence of high numbers of microorganisms, and microbial pathogens.

Preservatives are needed when pharmaceutical products do not have strong antimicrobial activity. However, they must not be used as a replacement for good manufacturing practices (GMP). Manufacturing of pharmaceutical products must minimize the possibility of microbial survival and growth. Furthermore, some of the processes control the environmental conditions in the facility and the quality of the materials. Before formulation, raw materials and water are screened for the presence of bacteria, yeast, and mold [1,2].

The efficacy of a preservative system is enhanced or inhibited by the different chemical ingredients in a formula and final package. Preservatives can be absorbed or inactivated by organic compounds in the formula or the packaging material. The stability of the preservative systems must be also ascertained over time by incubating samples for long periods, e.g., 3 mo, 6 mo, or 1 yr. After incubation, testing is performed to determine whether the efficacy of the preservative system changes over time. Testing is based on the inoculation of different types of bacteria, yeast, and mold into a given formulation. After incubation, the numbers of microorganisms reduced or inhibited over time is determined to evaluate the efficacy of the product to control different types of microbial populations [3,4].

The tests used to evaluate the efficacy of preservative present in different pharmaceutical formulations are based on compendial requirements by the United States (USP), European (EP), and Japanese (JP) Pharmacopeia [3,4,5]. The methods describe the test microorganisms, growth media, incubation conditions, method validation, and pass and fail criteria.

2. TYPES OF PRESERVATIVES

Preservatives are toxic chemicals. Therefore before using a preservative in a pharmaceutical product, several toxicological tests are performed. To protect patients receiving pharmaceutical dosages, preservative concentrations are kept to a level that cannot be toxic to consumers [6]. When a drug is formulated, other ingredients in the formula might add to the antimicrobial activity of the product that might simultaneously control and eliminate microbial growth and viability [7]. If a preservative inhibits the growth of a given bacterial or fungal species, it is called bacteriostatic and fungistatic. However, bactericidal and fungicidal activity reflects the reduction in the numbers of bacteria and mold as a result of microbial death.

A synergistic antimicrobial effect is also possible when more than one preservative type is used [8]. However, several types of preservatives are not compatible and should not be used in the same formulation. This incompatibility results in the inactivation of the antimicrobial activity of the preservative system allowing microorganisms to survive and proliferate in the pharmaceutical formulation. In some cases, the intrinsic nature of the drug is strongly antimicrobial by itself and does not require the addition of preservatives. For instance, some antibiotic solutions demonstrate intrinsic antimicrobial activity against microorganisms.

There are different types of preservatives. The general categories are based on the chemical structure of the different chemical compounds. Table 1 shows the different categories of preservatives. These general categories are:

- Alcohols
- Aldehydes

TABLE 1 List of Commonly Used Preservatives in Pharmaceutical Formulations

Alcohols
Benzyl, chlorbutol, phenylethanol, bronopol
Aldehydes
Formaldehyde, glutaraldehyde
Biguanides
Chlorhexidine, PHMB
Halogens
Chlorine, hypochlorite, chloroform, iodine
Heavy metals
Mercurials
Hydrogen peroxide and peracid compounds
Phenols
Surface active agents (surfactants)
Anionic
Cationic
Ampholytic

- Biguanides
- Halogens
- Heavy metals
- Hydrogen peroxide and peracid compounds
- Phenols
- Surface active agents (surfactants)

3. PRESERVATIVE EFFICACY TEST METHODS

3.1. Inoculum Preparation

According to the USP, all stock cultures of the test microorganisms must not be subculture more than five times from the original commercial culture container [4]. Stock cultures can be developed by adding sterile glycerol to the growth media. The limit for incubation of these stock cultures is $-50\,^{\circ}$C or in liquid nitrogen. Once the cultures are resuscitated from the stock culture, they are grown in soybean casein digest broth (SCDB) or soybean casein digest agar (SCDA), for bacteria, and Sabouraud dextrose agar (SDA) or Sabouraud dextrose broth (SDB) for yeast and mold. Temperatures are $32.5 \pm 2.5\,^{\circ}$C and $22.5 \pm 2.5\,^{\circ}$C for bacteria, and yeast and mold, respectively. Incubation times are 18–24 hr for bacteria, 44–52 hr for yeast, and 6–10 days for mold. The USP requires the use of sterile saline (TS) to harvest the bacterial and yeast growth from the agar plates. Several washes are performed to obtain a microbial count of approximately 1×10^8 colony-forming units (CFU) per milliliter. To harvest the mold, polysorbate 80 (0.05%) is added to the saline to obtain a similar count as with bacteria and yeast.

When the cells are grown in liquid media, harvesting is performed by centrifugation, washing, and resuspension of the cells in sterile saline to obtain a similar count as previously described [9]. The CFU of the different inocula is determined by plate counts on SCDA and SDA. Once the microbial suspensions are prepared, they are refrigerated for 24 hr (bacteria and yeast) or up to 7 days (mold). The JP also follows the same procedure for the inoculum preparation [5].

The EP recommends growing the bacteria on SCDA and the mold on SDA and incubating the cultures for 18–24 hr at 30–35°C, yeast for 48 hr at 20–25°C, and mold for 7 days at 20–25°C [3]. In contrast to the USP, there is no limitation in the number of passages from the original container. To harvest the bacterial and yeast growth, a 0.9% sodium chloride solution with 0.1% peptone is used. The procedure for harvesting the mold is similar to the USP. Final microbial densities are approximately 1×10^8 colony-forming units (CFU) per milliliter.

3.2. Inoculation of Pharmaceutical Articles

The three different pharmacopeias, United States (USP), European (EP), and Japanese (JP), describe the types of microorganisms used in the antimicrobial effectiveness test (AET). Table 2 shows the different types of bacteria, e.g., gram negative and gram positive, yeast, and mold. Gram-negative bacteria such as *Escherichia coli* and *Pseudomonas aeruginosa* are used for the challenge studies. *Staphylococcus aureus* is the only gram-positive bacterial species

TABLE 2 Microorganisms Used for Preservative Effectiveness Test Per Unites States (USP), European (EP), and Japanese Pharmacopeia (JP)

	Pharmacopeia		
Microorganism	USP	JP	EP
Candida albicans ATCC 10231	Yes	Yes	Yes
Aspergillus niger ATCC 16404	Yes	Yes	Yes
Escherichia coli ATCC 8739	Yes	Yes	No[a]
Pseudomonas aeruginosa ATCC 9027	Yes	Yes	Yes
Staphyloccocus aureus ATCC 6538	Yes	Yes	Yes
Zygosaccharomyces rouxii	No	Yes[b]	Yes[b]

[a] Only for oral pharmaceuticals.
[b] Oral preparations containing high concentration of sugars.

used. For mold and yeast, *Aspergillus niger* and *Candida albicans* are the representative species, respectively.

There are major similarities in the types of microorganisms recommended by the three major pharmacopeias. The only difference is the exclusion of *E. coli* by the EP while the USP and JP use this microorganism for the regular challenge test. *E. coli* is only recommended by the EP for oral pharmaceuticals. Furthermore, *Zygosaccharomyces rouxii* is also recommended for challenge studies in pharmaceutical formulations with sugars. Both the EP and JP state that common microbial contaminants can be used for challenging studies. However, the USP does not mention any other types of microorganisms.

Per USP, EP, and JP regulations, individual containers with the pharmaceutical formulations are inoculated with the suspensions of the microorganisms listed in Table 2. The final numbers of viable microorganisms in the product are 10^5–10^6 cells/g or mL of the preparation. However, for antacids made with aqueous bases, the final number must be within 10^3 and 10^4 cells/g or mL of product. The volume of the inoculum is not exceeding 1% of the volume of the product. The samples are mixed thoroughly to ensure a homogenous solution. The inoculated samples are incubated at 20–25°C. Testing is usually performed in the final container to determine the compatibility of the packaging material to the chemical ingredients in the formulation [10,11].

After incubating the samples for the different times described in Table 3, samples are withdrawn from the different containers. The sample volume is 1 g or mL. This sample is added to a diluent and further diluted to determine the numbers of viable microorganisms by plate count or membrane filtration.

When products are not soluble in water, heating of the samples might increase the solubility of the samples. Surfactants are also added to optimize the dispersion of the formulations and increase the miscibility between the liquid media and formulations containing ointments and oils [12].

3.3. Sampling of Inoculated Test Materials

After samples are incubated at 25°C, aliquots are taken at different time intervals. The time intervals recommended by the different pharmacopeias are shown in Table 3. Of all the three protocols, the EP requires a more intensive testing program. All samples are analyzed at time zero while the USP and JP do not. This means that as soon as the samples are inoculated with the microorganisms, aliquots are withdrawn and plated to determine the CFU/mL or g. After this, two different categories are described. There are target (EP-A) and acceptable (EP-B) level criteria. Efficacy results at the EP-B level are acceptable if there are dramatic reasons for EP-A levels not to be fulfilled.

TABLE 3 Criteria for Evaluating Preservative Effectiveness of Parenteral and Ophthalmic Pharmaceutical Formulations

	Log_{10} reduction					
	6 hr	24 hr	7 days	14 days	21 days	28 days
Bacteria						
USP			1	3		NI
EP-A	2	3				NR
EP-B		1	3			NI
JP			1	3	3	NI
Yeast						
USP			NI	NI	NI	NI
EP-A			2			NI
EP-B				2		NI
JP			NI	NI	NI	NI
Mold						
USP			NI	NI	NI	NI
EP-A			2			NI
EP-B				2		NI
JP			NI	NI	NI	NI

NI = No growth increase.
NR = No recovery on plates.
Source: Ref. [3–5].

The faster time interval for determining bacterial reduction is set up by the EP. After a 6-hr incubation time, the numbers of CFU/g or mL are determined and converted to $\log_{10}$ values. The EP-A continues the sample monitoring after a 24-hr and 28-day incubation. However, criteria EP-B extends further testing at 7 and 28 days. Its not until 7-day incubation time that the USP and JP determine the numbers of CFU followed by 14 and 28 days. The JP adds an additional time interval by sampling at 21 days.

Yeast and mold are sampled first after 7 days by all three test protocols. Further testing proceeds to 14 and 28 days by the USP and EP-A and EP-B. As with the bacterial test, the JP adds a day 21 analysis.

When samples are withdrawn from the incubated samples, serial dilution are performed in neutralizing agents (Table 4) and plated on different types of microbiological media (Table 5). These media support the growth of the different types of microorganisms inoculated into the products. For instance, Dey/Engley (D/E) Agar is a universal media containing neutralizers and other agents that inhibit the antimicrobial activity of a wide variety of preservatives [13]. On the other hand, media such as SCDA with neutralizers do not provide a broad neutralization efficacy.

TABLE 4 Preservatives and Their Neutralizing Agents

Preservative	Neutralizing agents
Hypochlorites	Sodium thiosulfate D/E broth
Phenolics	Polysorbate 80 Letheen broth with lecithin D/E broth
Aldehydes	Sodium sulfite Glycine D/E broth
Mercury compounds	Sodium thioglycollate Cysteine D/E broth
Quaternary ammonium compounds	Letheen broth with lecithin D/E broth

3.4. Criteria for Passing and Failing Preservative Efficacy

What are the criteria to determine the efficacy of preservative systems in pharmaceutical products? The criteria are based on the type of product that is analyzed. For instance, the USP describes four different categories of products (Table 6). Category 1 comprises injections, parenterals, emulsions, otic, sterile nasal, and ophthalmic products made with aqueous bases. Category 2 includes topical products with aqueous bases, nonsterile nasal, and emul-

TABLE 5 Different Types of Growth Media Used in Preservative Efficacy Studies

Media	Microorganisms detected
Soybean casein digest agar (SCDA)	Bacteria
D/E agar	Bacteria
Sabouraud dextrose agar	Yeast, mold
Potato dextrose agar	Mold
Letheen agar	Bacteria
Thioglycollate agar	Bacteria
SCDA-containing lecithin and polysorbate 80 or 20	Bacteria
Eugon agar	Bacteria, yeast, mold
Nutrient agar	Bacteria

TABLE 6 Product Categories per United States Pharmacopeia [4]

Category	Product description
1	Injections, other parenteral including emulsions, otic products, sterile nasal products, and ophthalmic products made with aqueous bases.
2	Topically used products made with aqueous bases, nonsterile nasal products and emulsions, including those applied to mucous membranes.
3	Oral products other than antacids, made with aqueous bases.
4	Antacids formulated with an aqueous base.

Source: Reference 4.

sions. Products applied to mucous membranes are also part of this group. Category 3 is comprised of oral products made with aqueous bases. These categories and groups are based on the route of administration of the products. The risk of having a fatal infection is higher when microorganisms are present in category 1 than in category 2 because products in category 1 are supposed to be sterile and are used in critical areas of the body where microbial infection can be fatal, e.g., lungs, blood, eyes.

Products cover by category 4 are antacids produced with an aqueous base. The criteria for yeast and mold are basically similar for categories 2, 3, and 4 products. After inoculation of the yeast and mold into the samples, no increase from the initial inoculum at 14 and 28 days is required. No increase is defined as not more than 0.5 $\log_{10}$ (log) higher than the previous values obtained. Category 1 recommends a no increase at 7, 14, and 28 days after inoculation.

When it comes to bacteria, category 1 products requires not less than 1-log reduction at 7 days, 3-log reduction at 14 days, with no increase from the day 14 counts at 28 days. Category 2 shows a longer time for the first reduction to be observed with not less than 2-log reduction at 14 days and no increase from the day 14 counts at 28 days. Category 3 criteria are not less than 1-log reduction from the initial count at 14 days, and no increase from the day 14 counts at 28 days. The criteria for category 4 products require no increase from the initial counts at 14 and 28 days.

The EP divides the different pharmaceutical samples into four categories [3]. These are:

- Parenteral and ophthalmic
- Oral
- Topical
- Ear

For parenteral and ophthalmic preparations, they provide two different types of criteria. The A criteria (target) provides the recommended efficacy while the B criteria (acceptable) can also be used in cases when adverse reaction to a formulation is reported. When compared to USP category 1 products, both criteria are more stringent. The target reduction (A criteria) for bacteria is 2 log from the initial count at 6 hr, 3 log at 24 hr, and no recovery after 28 days. Fungi reduction must be 2 log from the initial count and no increase after 28 days. The acceptable criteria is less stringent with 1-log reduction at 24 hr, 3 log at 7 days, and no increase at 28 days.

However, criteria for oral, topical, and ear pharmaceuticals do not provide target and acceptable criteria. For bacteria, oral products require a 3-log reduction from the initial count at 14 days and no increase at 28 days. Topical formulations criteria are 3-log reduction at 48 hr with no increase in the counts at 7, 14, and 28 days. Ear products are more stringent than oral and topical products. Bacterial criteria require 2-log reduction in the counts from the initial count at 6 hr, 3 log at 24 hr, with no recovery at 28 days.

For mold, oral preparations are required to demonstrate a 1-log reduction from the initial count at 14 days and no increase at 28 days while ear formulations are supposed to show a 2-log reduction at 7 days and no increase at 28 days. The criteria for topical products are based on a 2-log reduction at 14 days and no recovery at 28 days.

The JP divides the products into two categories [5]: Categories 1 and 2. Products in category 1 are subdivided into four groups. These groups are 1A, 1B, 1C, and 1D. Groups 1A, 1B, and 1C are similar to the first three categories described in the USP. Group 1D comprises antacid formulations including solid forms with aqueous bases. Category 2 describes all products under category 1 that are formulated with nonaqueous bases. However, the criteria for mold and yeast is the same for all categories. The numbers of microorganisms obtained after testing must be the same or less than the inoculum at 14 and 28 days of testing. Group 1D and category 2 product recommendations for bacteria are the same or less than the inoculum count at 14 and 28 days. The required reductions for bacteria for groups 1A, 1B, and 1C are 0.1%, 1%, and 10% of the inoculum counts or less at 14 days with no increase in numbers at 28 days.

Why do products fall into different categories? These categories reflect the nature of the product and route of application. However, other factors such as water activity, growth potential, use conditions, packaging, and container closure configurations are also considered when categories are defined. Regarding packaging configuration, if a product is a multiple-dose type, continuous withdrawing from containers increases the challenge to its microbiological quality [14]. These types of containers are used for products such as tablets, pills, and creams.

The more times the consumer opens and closes the container, the more chances for microorganisms to be introduced into the product by consumers or aerosols. Furthermore, microorganisms are part of the human skin. The frequent use of the products by consumers increases the chances of microbial insult. If any microorganism is introduced, preservatives must inhibit microbial growth. If a microbial challenge is introduced, the preservative will significantly reduce the numbers of microorganisms. The use of single-dose containers prevents the contamination of the products during use because these types of containers are used only once. For instance, a multidose container for injections and other sterile products must be avoided because of the high risk of microbial contamination. In this case, microbial contamination can be fatal. Single-dose containers are also favored for pharmaceutical products applied to the eye and mucous membranes.

Unlike aqueous-based formulations, anhydrous products do not provide ideal situations for microbial growth and survival [15]. Because of the low moisture content in a formula, a growing population of microorganisms will not develop as long as proper conditions of storage and handling are followed. Products with high oil, alcohol, and solids, e.g., tablets and capsules, do not contain high water concentration. They provide hostile conditions for microorganism to grow. The conditions affecting microbial growth in pharmaceutical formulations are:

- Moisture
- Oxygen
- pH
- Temperature
- Nutrients
- Water activity (a_w)
- Viscosity
- Oil/water ratio
- Percent of solids

Of all the factors, moisture content and pH can be controlled by the pharmaceutical formulation laboratory. Moisture and pH are critical factors in a product enhancing or inhibiting the antimicrobial activity of the formulation. Some preservatives are extremely susceptible to pH changes. Optimal antimicrobial activity is observed at pH values greater than 7 while others exhibit strong antimicrobial activity at pH values below 5.

Moisture content is defined as water activity (a_w). Water activity is the amount of water available in a product for microbes to grow. Because available water is critical for microorganisms to survive, the lower the water concentration, the more hostile the product is to microbial survival and growth. Some of these products do not require the addition of a preservative.

An example of this is dry tablets and capsules that do not contain any water. Nevertheless, in some cases, low concentrations of preservatives are added to prevent the accidental introduction of water and microorganisms by the consumer during usage. When products contain a high water concentration, preservation becomes difficult. In these cases, the addition of one or two preservatives enhances the antimicrobial activity of the formulation.

Microorganisms respond to low water activity by developing different survival strategies. For instance, *Bacillus* spp. form spores while some gram-negative bacteria undergo a starvation-survival stage. This is because water activity requirements vary within different microbial species and genera. Available water is a more critical requirement for bacteria than yeast and mold. Bacteria require higher levels of water to survive and grow. This makes bacteria more sensitive to changes in the water activity in a given formulation while molds are more resistant.

3.5. Neutralization Validation

To validate the recovery of microorganisms from artificially inoculated product samples, neutralization validation of the conditions before testing must be performed [16]. Validating the microbial recovery from spiked product samples demonstrate the efficacy and reproducibility of a given method. There are several methods to neutralize the antimicrobial activity of a preservative system. These methods are:

- Use of an inactivating agent
- Dilution of the product
- Filtration of the product

These methods are performed in triplicates using different batches of the product to demonstrate the recovery of microorganisms from inoculated samples. If an inactivator is used, it is necessary to demonstrate it does not inhibit microbial growth. This is called neutralizer toxicity studies. A neutralizer agent can be added to the diluent, plate media, or both (Tables 4 and 5). The agent must be shown to inactivate any antimicrobial activity of the formula by recovering all the spiked microorganisms. This is called neutralizer efficacy studies. A low inoculum of less than 100 CFU of the test organisms is introduced into a sample of the product diluted in the neutralizing agent. A similar inoculum is also introduced into a sample of sterile saline control. If the recovery numbers are the same, the neutralization is considered effective. Similar results are considered to be within 0.5-log difference between the two treatments. Table 7 shows the validation of PBS with 1% Tween 20 as the neutralizer agent. Results indicate the lack of neutralization efficacy to recover *C. albicans* and *A. niger* using this agent. The $\log_{10}$

TABLE 7 Validation of Preservative Challenge Diluents per United States Pharmacopeia

Microorganism	Colony-forming units (CFU)			
	Control		Diluent	
Pseudomonas aeruginosa	90	89	78	80
Average	89.5		79	
Log 10	1.95		1.89	
Escherichia coli	87	78	76	76
Average	82.5		76	
Log 10	1.91		1.88	
Staphylococcus aureus	87	78	54	43
Average	82.5		48.5	
Log 10	1.92		1.69	
Aspergillus niger	90	91	13	14
Average	90.5		13.5	
Log 10	1.96		1.13	
Candida albicans	93	87	10	10
Average	90		10	
Log 10	1.95		1.00	

Diluent = phosphate-buffered saline with 1% Tween 20; plating media = SCDA; product = pharmaceutical emulsion; dilution = 1/100.

values of the control and diluent counts are higher than 0.5 log. Therefore the neutralizing agent cannot be used for preservative testing. However, when the concentration of Tween 20 is increased to 4%, similar recoveries are obtained demonstrating that the neutralizing agent is capable of neutralizing the preservative system in the formula (Table 8).

Toxicity testing is performed by comparing the microbial growth of samples spiked into the neutralizing agent without the product and a rich nutrient broth such as SCD broth. The numbers of microorganisms recovered on the media between treatments must be within 0.5-log difference. Table 9 shows the results for the toxicity testing of D/E broth and SCDB. Samples were plated on SCDA agar. There were no differences between the numbers of CFU on SCDA plates from samples diluted in D/E broth and SCDB, which indicated that D/E broth is not toxic to the cells.

Lecithin, polysorbate 20 or 80, Dey/Engley (D/E) broth, Letheen broth, etc. are some examples of the different types of neutralizing diluents used during preservative efficacy testing. For instance, formulations containing mercurial compounds can be neutralized using thioglycollate broth as the neutralizer diluent while hypochloride preservatives (halogens) are neutral-

TABLE 8 Validation of Preservative Challenge Diluents per United States Pharmacopeia

Microorganism	Colony-forming units (CFU)			
	Control		Diluent	
Pseudomonas aeruginosa	90	89	78	80
Average	89.5		79	
Log 10	1.95		1.89	
Escherichia coli	87	78	76	76
Average	82.5		76	
Log 10	1.91		1.88	
Staphylococcus aureus	87	78	54	43
Average	82.5		48.5	
Log 10	1.92		1.69	
Aspergillus niger	90	89	78	80
Average	89.5		79	
Log 10	1.95		1.89	
Candida albicans	90	91	76	76
Average	90.5		76	
Log 10	1.96		1.88	

Diluent = phosphate-buffered saline with 4% Tween 20; plating media = SCDA; product = pharmaceutical emulsion; dilution = 1/100.

ized with sodium thiosulfate. The concentration of the neutralizer agent must be determined for optimization of microbial recovery.

In some cases, diluting the preservative can effectively neutralize its antimicrobial activity. For instance, a 1/100 dilution of a preservative might recover all spiked microorganisms more efficiently than a 1/10 dilution. The lower the concentration of the preservative in the sample, the higher the numbers of microorganisms detected because the antimicrobial activity is directly correlated to the agent's concentration.

If dilution is not effective for neutralization purposes, membrane filtration is also tested. Filtration is based on the retention of the microorganisms on a 47-mm, 0.45-μm filter with the antimicrobial ingredient passing through the filter assembly [3–5]. Once the sample is filtered, several rinses are performed with different types of rinsing fluids such as saline, phosphate buffer, etc. The rinses remove any residual antimicrobial activity from the filters. After rinsing, the filters are placed onto growth media to determine the numbers of microorganisms over time.

Once the neutralization studies of the diluents and media are completed, the full preservative testing is performed. Table 10 shows the test results for

TABLE 9 Neutralization Toxicity Studies

Microorganism	Colony-forming units (CFU)			
	Diluent 1		Diluent 2	
Pseudomonas aeruginosa	90	89	78	80
Average	89.5		79	
Log 10	1.95		1.89	
Escherichia coli	87	78	76	76
Average	82.5		76	
Log 10	1.91		1.88	
Staphylococcus aureus	87	78	54	43
Average	82.5		48.5	
Log 10	1.92		1.69	
Aspergillus niger	90	89	78	80
Average	89.5		79	
Log 10	1.95		1.89	
Candida albicans	90	91	76	76
Average	90.5		76	
Log 10	1.96		1.88	

Diluent 1 = D/E broth; diluent 2 = SCDB; plating media = SCDA; dilution = 1/10.

TABLE 10 Preservative Challenge Test Results per United States Pharmacopeia

Microorganism	Day 7		Day 14		Day 28	
Inoculum counts = 3.0×10^6						
Pseudomonas aeruginosa	90	90	80	80	0	0
Average =	90	80	0			
Log_{10}	1.95	1.90	0			

Log reduction = $\log_{10}$ value of inoculum − $\log_{10}$ value of given time point
= 6.48 − day 7 value (1.95)
= 4.53

Microorganism	Day 7		Day 14		Day 28	
Inoculum counts = 5.0×10^6						
Escherichia coli	87	87	76	76	0	0
Average =	87		76	0		
Log_{10}	1.94		1.88	0		

Log reduction = $\log_{10}$ value of inoculum − $\log_{10}$ value of given time point
= 6.69 − day 7 value (1.94)
= 4.75

Product = pharmaceutical liquid; plating media = D/E agar; diluent = Letheen broth; dilution = 1/10.

P. aeruginosa and *E. coli* based on USP testing protocols. After 7 days, more than a 4-log reduction in the initial counts were obtained for both bacteria, indicating that the formulation is passing the challenge. The tested product (oral pharmaceutical) belongs to category 3, which only requires not less than 1-log reduction from the initial count at day 14.

4. ALTERNATIVE TESTING METHODS

Several pharmaceutical companies rely on the use of alternative test protocols during the research and development of the preservative systems used in the final product. However, these protocols do not replace the compendial tests.

4.1. Double Challenge

After product inoculation with all the microorganisms, the samples are challenged with the same inoculum at 14 or 28 days (double challenge). A second microbial challenge represents a significant test to determine the efficacy of the preservative system. This type of test is commonly performed on eye and ear pharmaceuticals. The double challenge is followed by an additional incubation of 28 days for a total of 56-day test. However, in some cases, incubation is not extended. Therefore after the double challenge, samples are completed at 28 days.

4.2. Mixed Inoculum

Mixed cultures of microorganisms are inoculated into the products to determine the resistant of different mixtures to the preservative chosen and robustness of the formulation to mixed culture contamination. For instance, a mixed bacterial inoculum containing *S. aureus*, *E. coli*, and *P. aeruginosa* is spiked into the individual samples. Furthermore, yeast and mold are also combined and spiked resulting in two mixed challenges: One of bacteria and a second of yeast and mold.

4.3. Use of Environmental Isolates

Additional microorganisms are incorporated into the test. Environmental monitoring and product testing during manufacturing reveal the presence of different types of bacteria and mold in the plant (environmental microorganisms). The addition of these microorganisms to the test provides an estimation of the resistance of environmental isolates to the preservative system used in a given formulation. Studies have demonstrated that environmental isolates are more resistant to preservative systems than the five standard microorganisms used for compendial testing [17]. This is because these isolates are continuously exposed to antimicrobial products such sanitizers, disinfectants, and preservatives.

Microbial resistance to preservatives is based on the reduced permeability to the chemical agent, production of enzymes to degrade the preservative, growth rate (fast growing cells are less sensitive), biofilm formation, and spore formation. Microbes can adapt to different types of preservatives changing the sensitivity of a microbial population against specific preservative systems. The environmental isolates can be added individually or in different mixed cultures.

5. CONCLUSION

The addition of preservatives to pharmaceutical formulations enhances the antimicrobial activity of a finished product. These substances prevent the microbial contamination of a given formula during manufacturing, storage, and consumer use. Test methods are developed and validated to determine the efficacy of preservative effectiveness. Samples are inoculated with representative microbial species to demonstrate inhibitory activity and microbial death resulting in either a reduction in the numbers of microorganisms inoculated and/or no increase after a given period of time. Optimization of antimicrobial activity by a preservative system combines the chemical composition of the preservative with conditions such as pH, packaging configuration, water activity, temperature, viscosity, and percent of solids.

REFERENCES

1. United States Pharmacopeial Convention. Microbial limit test. US Pharmacopoeia. Vol. 25. Rockville, MD: United States Pharmacopeial Convention, 2002:1873–1878.
2. European Pharmacopoeial Convention. Microbiological examination of non-sterile products. European Pharmacopoeia. 3rd ed. Strasbourg, France: Council of Europe, 2001:70–78.
3. European Pharmacopoeial Convention. Efficacy of antimicrobial preservation. European Pharmacopoeia. 3rd ed. Strasbourg, France: Council of Europe, 2001:293–294.
4. United States Pharmacopeial Convention. Antimicrobial effectiveness testing. US Pharmacopoeia. Vol. 26. Rockville, MD: United States Pharmacopeial Convention, 2003:2002–2004.
5. The Japanese Pharmacopoeia. Preservative Effectiveness Tests. Vol. 14. Tokyo, Japan: The Society of Japanese Pharmacopoeia, 2002:1321–1323.
6. Beasley R, Fishwick D, Miles JF, Hendeles L. Preservatives in nebulizer solutions: Risks without benefit. Pharmacotherapy 1998; 18:130–139.

7. Kellog DS. Preservative testing as applied to quality control systems. Bull Parenter Drug 1972; 26:216–220.
8. Hodges NA, Denyer SP, Hanlon GW, Reynolds JP. Preservative efficacy tests in formulated nasal products: Reproducibility and factors affecting preservative activity. J Pharm Pharmacol 1996; 48:1237–1242.
9. Gilbert P, Brown MRW, Costerton JW. Inocula for antimicrobial sensitivity testing: A critical review. J Antimicrob Chemother 1987; 20:147–154.
10. Brannan DK, Dille JC. Type of closure prevents microbial contamination of cosmetics during consumer use. Appl Environ Microbiol 1990; 56:1476–1479.
11. Farrington JK, Martz EL, Wells SJ, Ennis CC, Holder J, Levchuk JW, Avis KE, Hoffman PS, Hitchins AD, Madden JM. Ability of laboratory methods to predict in-use efficacy of antimicrobial preservatives in experimental cosmetic. Appl Environ Microbiol 1994; 60:4553–4558.
12. Abshire RL, Schelech BA. A method for evaluating the effectiveness of preservative systems in water-immiscible ointments. J Parenter Sci Technol 1982; 36:216–221.
13. Dey BF, Engley EB. Methodology for recovery of chemically treated *Staphylococcus aureus* with neutralizing medium. Appl Environ Microbiol 1983; 45:1533–1537.
14. Houlsby RD. An alternate approach for preservative testing of ophthalmic multiple-dose products. J Parenter Drug Assoc 1980; 34:272–276.
15. Grigo J. Microorganisms in drugs and cosmetics—Occurrence, harms and consequences in hygienic manufacturing. Zentralbl Bakteriol 1976; 162:233–287.
16. United States Pharmacopeial Convention. Validation of microbial recovery from pharmacopeial articles. US Pharmacopoeia. Vol. 25. Rockville, MD: United States Pharmacopeial Convention, 2002:2259–2261.
17. Zani F, Minutello A, Maggi L, Santi P, Mazza P. Evaluation of preservative effectiveness in pharmaceutical products: The use of a wild strain of *Pseudomonas cepacia*. J Appl Microbiol 1997; 43:208–212.

Index

Acinetobacter spp., 1, 33, 56
Acridine orange direct counts (AODC), 6
Adenosine triphosphate (ATP), 1, 2, 6
  total adenylate, 6
Aerobic microorganisms, 91
Aerobic respiration, 2
Agrobacterium spp., 8
Air sampling, 109
  methods of, 111
Alcaligenes spp., 9
Alcohol, 2, 62, 98, 105, 106, 253–256, 292
American Association of Medical Instrumentation (AAMI), 55
American National Standard, 139
Anaerobic microorganisms, 91
Anaerobic respiration, 2
Antiseptic
  definition, 252
Arthrobacter spp., 4
Aseptic processing, 77–79, 88–90, 95, 99, 104, 105, 115
Aspergillus niger, 29, 157, 163, 287, 293
Aspergillus spp., 8
Association of Official Analytical Chemists (AOAC), 258, 259
  carrier test method, 261–263
  fungicidal test, 260
  phenol coefficient method, 259
  sporocidal activity method, 263, 264
  spray products test, 259
  tuberculocidal test, 260
  use dilution test, 260, 261
ATP bioluminescence, 10, 152
  assays, 153
  for clean rooms, 158
  enrichment broths, 155
  for products, 154
  validation, 154, 155
  for water, 153

Bacillus coagulans, 135
Bacillus pumilus, 86
Bacillus sphaericus, 68, 153
Bacillus spp., 8, 9, 19, 122, 150, 227, 264, 293
Bacillus stearothermophilus, 134, 135, 143, 144
Bacillus subtilis, 26, 29, 86, 153, 263
Bacteria. *See* Gram negative bacteria, gram positive bacteria
Bacterial photosynthesis, 2
Bacteriostasis, 91
Bacterial endotoxin test (BET), 199–203, 214, 228
Bactericidal, 284
Bacteriostatic, 284
Baird Parker agar, 37
BIER vessel, 137, 139
Biocide
  definition, 252
Biological indicators (BIs), 82, 86, 133
  applications, 133
  bacteria, 134
  BIER vessels, description of, 139
  calibration, 133
  configurations, 134
  D values, 86, 133, 140
  D values, discrepancies in, 137
  end users responsibilities, 136
  equipment validation, 138
  FDA recalls, 135
  manufacturers, responsibilities, 135
  media qualification, 140
  media supplements, 141
  method validation, 140
  organisms, 134
  performance standards, 133, 136
  performance tests, 136
  population control, discrepancies in, 137
  qualification of personnel, 141
  recovery, 144
  resistance parameters, 135
  validation master plan, 138
Bismuth sulfite agar (BSA), 35
Blood agar, 38
Bovine serum albumin (BSA), 107
Bradyrhizobium spp., 9, 70, 167
Brevundimonas diminuta, 60, 84
Brilliant green agar (BGA), 35
Brilliant green lactose bile broth, 108
Burkholderia cepacia, 9, 19, 30, 33, 38, 45, 64, 68, 79, 121, 153
Burkholderia picketti (Ralstonia picketti), 9, 19
Burkholderia spp., 1

Campylobacter jejuni, 53
Candida albicans, 29, 38, 157, 163, 287, 293
Cetrimide agar, 35, 108
ChemChrome B (CB) dye, 69, 159
Chromogenic substrates, 53
Citrobacter, 17, 51
Cladosporium spp., 8
Cleaning, 106
  acceptance criteria, 107
  microbiological cultures, 107
  validation, 106
Clean room, 95, 105
  air flow, 105, 106, 110
  calibration, 106
  classification, 106, 110
  cleaning, 106
  definition, 96
  EP requirements, 97
  humidity, 105
  particulates, 105, 109
  surfaces, 105
  USP classification, 96
  validation studies, 106
Chlorine dioxide, 257
Clostridium botulinum, 193
Clostridium broth, 37
Clostridium perfringes, 9, 37
Clostridium sporogenes, 134, 263
Clostridium spp., 8, 19, 33, 37, 38, 122, 150, 264
Code of federal regulations (CFR), 16
Coliforms, 26, 38, 46–47, 50–51, 53, 108

Colilert, 108
  detection, 108
Columbia agar, 37
Comamonas acidovorans, 64
Comamonas spp., 9
Compressed gases, 109
  sampling of, 109
  types of, 109
Corynebacterium spp., 8
Cryptosporidium, 52, 67
Current good manufacturing practices (cGMP), 128, 130, 204
  lack of compliance with, 151
Cytophaga spp., 9

DAPI (4, 6-diamidino-2-phenyl-indole), 6, 159
Decontamination
  definition, 252
Denaturing gradient gel electrophoresis (DGGE), 70, 167
Deoxyribonucleic acid (DNA), 10, 149, 150, 161, 163–167, 234, 258
  direct extraction of, 7
  extraction from sample enrichments, 125, 166, 168
  fingerprinting, 125, 170
  hybridization, 170
  microchips, 170, 171
  primers, 163, 167
  ribosomal analysis, 9, 70
  sequencing, 5, 125, 167, 168, 170
Depyrogenation
  definition, 224
  removal, 224
Desulfovibrio desulfuricans, 171
Dey/Engley (D/E) agar, 120
Dey/Engley (D/E) broth, 121, 294
Direct microbial counts, 6
Direct viable counts, 159
  dyes, viability, 69, 159
Disinfectant, 106
  application, mode of, 106, 268
  bioburden determinations, 273
  challenge microorganisms, 264–266
[Disinfectant]
  classification, 254–257
  contact time, 267
  coupons, 266
  customized in vitro tests, 266
  definitions, 251–252
  efficacy testing, 107
  expiration dates, determination of, 278
  in situ tests, 280
  in vitro tests, 266
  methods, sampling, 270
  preparation, 253, 254
  rotation, 107, 258
  selection, 252–254
  tests, 258–263
  validation, method, 106, 274
Disinfection
  definition, 252
  surfaces, 254, 266, 270
  water systems, 67
Dry heat sterilization, 78, 83, 228
Durham tube, 37

Enzyme-linked immunoassay (ELISA), 172
Endo agar, 37
Endospores, 251, 252, 256, 269
Endotoxin, 83, 85, 183
  bacterial test, 199–203, 214
  basic structure, 186
  biological activity, 183
  concentration, 209
  contamination control, 193
  control standard endotoxin (CSE), 200
  control strategy, 194–199
  definition, 184
  depyrogenation, 224
  destruction of, 224
  FDA guidelines for validation of LAL test, 195, 222
  importance in parenterals, 188
  limit concentration, 200
  lysate sensitivity (λ), 200, 210

[Endotoxin]
mammalian response, 189
maximum valid dilution, 200, 223
method development and validation, 213–218
minimum valid concentration, 218, 223
nomenclature, 184
potential sources, 194
pyrogens, 184
removal, 233
safely tolerated level, 195–197
specifications, 222–224
structure, 186–188
test, 199–205, 208–213
validation, 213–222, 224–232
whole blood test, 239–242
Enterobacter spp., 9, 33, 51, 121
Enterobacter gergoviae, 30, 34
Enterobacteriaceae, 17, 37, 126, 186, 187, 189
Enterobacteriaceae enrichment broth, 37
Enterococcus spp., 9
Environmental fluctuations, 2, 150
Environmental monitoring, 103
air monitoring, 109
air pattern, 110
air sampling methods, 111
air velocity, 110
alert and action levels, 117
alert and action levels and corrective actions, 118
alert and action levels for air, 117
alert and action levels for surfaces, 118
in aseptic processing, 105
common deficiencies, 105
compressed gases, 109
corrective actions, 116
data systems for laboratory information management (LIMS), 129
gowning requirement certification program, 114
[Environmental monitoring]
gowning requirements, 113
isolates, characterization of, 121–125
microbiological methods, 118–121
nonsterile products, 104
personnel sampling, 113
plan, 103
practices in nonsterile areas, 104
product testing program, 126
quality control testing laboratory, 128
sampling sites frequency, 115
sampling sites, 114
sterile products, 103
surfaces sampling methods, 112
surfaces, 111
water, 107
Environmental Protection Agency (EPA), 53, 66, 259
acceptable risk for infectious disease, 53
Enzyme
ATP somase, 158
beta-d-galactosidase, 53
beta-glucuronidase, 53
complex, luciferase-luciferin, 152
esterase activity, 159
Eosine methylene blue media, 37
Epidermophyton spp., 9
Escherichia coli, 8, 9, 17, 25–27, 29, 33, 35, 37, 47, 52, 121, 122, 126, 150, 157, 171, 186, 187, 189, 226, 265, 286, 297
Escherichia spp., 8, 17, 19, 51
European community (EC), 110
European Pharmacopeia (EP). *See also* Pharmacopoeia Europa, 16, 126, 284
antimicrobial effectiveness test, 284, 286–288, 290
inoculation of articles, 286–287
inoculum preparation, 286
microbial limits, 16, 19, 26, 27, 29, 30, 32–35, 37–39

[European Pharmacopeia (EP)]
sterility testing, 91, 97
test requirements for nonsterile products, 27

F_0 value, 87, 226
Fatty acids, 186, 187, 189
Fermentation, 2
Fetal bovine serum, 193
Fetal calf serum, 107
Flavin-adenine dinucleotide hydrogen (FADH), 2
Flavobacterium spp., 9
Flexibacter spp., 9
Fluid selenite cysteine medium (FSCM), 34, 35
Fluid tetrathionate medium (FTM), 34, 35, 94
Fluorescein isothiocyanate (FITC), 6
Flow cytometry, 159
antibiotic testing, 161
bioburden, 161
biological indicators, 144, 162
water testing, 159, 160
Fluid thioglycollate media (FTM), 91, 92, 98, 127
Food and Drug Administration (FDA), 45, 79, 258
Formaldehyde, 106, 257
Fungi, 92
Fungicidal, 284
Fungistasis, 91, 94, 97
Fungistatic, 284
Fusarium spp., 8

Genetic identification, 167
Gene probes, 5
Geobacillus stearothermophilus, 134, 143
Geobacter chapellei, 171
Germicide
definition, 252
Glutaraldehyde, 257
Glycolysis, 2
Good manufacturing practices (GMP), 16, 77, 95, 105, 117, 283
non compliance with, 79
Gram negative bacteria, 19, 27, 33, 37, 45, 92, 121, 122, 128, 150, 185–193, 206, 207, 236, 238, 240, 242
Gram positive bacteria, 4, 19, 92, 122, 150, 168, 190, 286
Gram stain, 121, 125, 128
Growth Direct^{TM}, 174
biological indicators, 144
detection of microcolonies, 175

Heterotrophic plate count (HPC), 49
High-efficiency particulate air (HEPA) filters, 96, 99, 105, 110
Hoechst 33258, 6
Humidity, ventilation, air conditioning units (HVAC), 8, 105
Hydrogen peroxide, 226, 256

Identification of environmental isolates, 150
Immunoassays, 172
procedure, ELISA, 172
Immunocompromised people, 25
Impedance, 162
detection time (T_d), 162
INT (2-[p-iodophenyl]-3-[p-nitro-phenyl]-5-phenyl tetrazolium chloride), 6

Japanese pharmacopeia (JP), 16, 284
antimicrobial effectiveness test, 286–288, 291
inoculum preparation, 286
microbial limits, 16, 19, 29, 32–35, 37, 39
sterility testing, 91
test requirements for nonsterile products, 27

Klebsiella spp., 9, 17, 33, 51

Lactose broth, 29, 35, 37
Lactose monohydrate sulfite medium, 37
Lauryl tryptose broth, 50, 108
Lecithin agar, 120
Letheen agar, 30, 120
Letheen broth, 29, 30, 294
Limulus amebocyte lysate (LAL), 184, 188
  chromogenic, 211–213
  coagulation, 207–208
  concentration, 209
  discovery, 205–207
  endpoint, 210–213
  FDA guidelines for validation of, 195
  gel clot test, 210
  kinetic, 210–213
  origin, 204–205
  sampling, 204
  test interferences, 218–222
  test, 208–213
  turbidimetric, 211, 212
Lipid A, 86, 187, 190, 191, 200, 233, 237, 239, 243
Lipopolysaccharide
  amphiphilic nature, 191
  definition, 184
  structure, 186
Listeria, 171
Low nutrient media, 151

MacConkey agar, 35
MacConkey broth, 35
Mannitol salt agar, 37
Membrane filters, 30
  ability of water-borne bacteria to pass through, 60, 122, 151
  challenge studies, 60
  integrity tests, 85
  materials, 85
  rating of, 84
Membrane filtration, 30, 31, 58, 93, 98, 295
m-ENDO, 108
Methylobacterium spp., 79
Microbial bioburden, 85
Microbial biomass, 7
Microbial contamination
  air, 18, 19
  distribution, 19
  EP validation protocol, 30
  frequency, 19
  nonsterile products, 17
  recalls by FDA, 19
  water, 18
Microbial content test agar (MCTA), 30
Microbial distribution, factors affecting, 39
Microbial identification, 16, 121, 122
Microbial limits, 1, 2, 6–8, 24, 26
  definition, 15
  final interpretation of the quantitative results, 33
  history and harmonization, 37
  incubation times and temperatures for bacterial plates, 30
  incubation times and temperatures for mold and yeast plates, 30
  manufacturing, 16
  pathogen indicators, 16, 17,
  recommended specifications and limits, 26
  resampling, 40
  sample dilution, 34
  sampling, 39
  testing, 38
  testing for herbal and nutritional supplements, 40
  test method validation
    qualitative test, 33
    quantitative test, 28
  test requirements for EP, JP, USP, 27, 28
Micrococcus pp., 8
Microorganisms
  clinical significance in nonsterile pharmaceuticals, 25
  growth, 150
  survival, 150

Microsporon spp., 9
Mold, 8, 9, 16, 18, 19, 26–29, 33, 37, 38, 40, 54, 79, 92, 94, 108, 120, 122, 125–127, 157, 158, 161, 163, 164, 166, 167, 175, 252, 273, 283–288, 290, 291, 293, 297
Most probable number (MPN), 32, 47, 108
Mycobacterium smegmatis, 260
Mycological agar, 30
Mycoplasma, 84

Nicotinamide-adenine dinucleotide hydrogen (NADH), 2
Nonviable air particulates, 109, 110

Objectionable microorganisms, 26, 38
Oligotrophic, 150
Out-of-specification (OOS), 150

Pathogen indicators
  EP, 19
  JP, 19
  USP, 19
Pathogen screening, 33
Parenteral Drug Association (PDA), 121, 148
Particulates, 11, 96, 105, 109, 110
Penicillium spp., 8, 9, 79
Peptidoglycan (PGN), 190
Peracetic acid, 256
Pharmaceutical products, nonsterile
  risk of infection, 116
Pharmaceutical products, sterile
  risk of infection, 116
Pharmacopoeia Europa (Ph Eur), 61, 64, 73
Phenols, 256
Plate count agar (PCA), 108, 153
Polymerase chain reaction (PCR), 5, 9, 163
  assays for pharmaceutical microbes, 165
  for water, 70
Polysorbate 20, 30, 293, 294
Polysorbate 80, 30, 286, 294
Potato dextrose agar (PDA), 30
Pour plate method, 57, 108
Preservatives, 283
  anhydrous products, 292
  alternative testing methods, 297
  aqueous based formulations, 292
  conditions affecting microbial growth, 292
  criteria for passing and failing, 289
  definition, 283
  efficacy, 283, 284
  inoculation of pharmaceutical articles, 286
  inoculum preparation, 285
  microbial resistance, 298
  neutralization validation, 293
  samples categories, 289, 291
  sampling, 287
  synergistic effect, 284
  types, 284, 285
  time intervals for sampling, 287
  water activity, 292
Process control, 16–19, 25, 41, 78, 88, 95, 99, 100, 106, 114, 121, 128, 130, 150–152, 179, 200
Propionibacterium acnes, 8
Propionibacterium spp., 8
Proteus spp., 17
Pseudomonas aeruginosa, 8, 16, 17, 19, 26, 27, 30, 34, 35, 37, 40, 45, 107, 108, 121, 126, 153, 157, 163, 173, 174, 226, 259, 261, 265, 286, 297
Pseudomonas alcaligenes, 45
Pseudomonas baleurica, 45
Pseudomonas fluorescens, 33
Pseudomonas isolation agar, 35, 108
Pseudomonas putida, 33, 45
Pseudomonas spp., 1, 8, 9, 19, 35, 45, 108, 121, 189
Pseudomonas vesicularis, 64
Pyrogens, 184

Quaternary ammonium compounds (QACs), 256

Quality control, 15–17, 30, 40, 46, 64, 69, 70, 90, 121, 123, 125, 128, 147, 149, 151–153, 162, 165, 171, 172, 177–179, 193, 194, 204

Ralstonia pickettii, 19, 45, 64, 68, 79, 153
RBD system, 144
Ribonucleic acid (RNA), 7, 170, 171
Rapid methods, 147
  ATP bioluminescence, 149
  direct viable counts, 159
  DNA microchips, 170
  flow cytometry, 159
  genetic identification, 167
  growth Direct™, 174
  immunoassays, ELISA, 172
  impedance, 162
  lack of implementation, reasons for, 147
  PCR technology, 163
  validation parameters, 148, 149
Relative light units (RLU), 157
Replicate organism detection and counting plates (RODAC), 112, 119, 121
R2A media, 54, 55, 108, 120, 123, 151, 153
  composition of, 57

Sabouraud dextrose agar (SDA), 30, 37, 38, 87, 108, 120
Safe drinking water act, 66
Salmonella choleraesuis, 35, 107, 259, 260, 261
Salmonella spp., 9, 16, 17, 19, 25–27, 29, 33, 34, 37, 51, 126, 187, 261
Salmonella typhi, 53, 189, 226, 259
Salmonella typhimurium, 8, 122, 150, 157, 174
Sanitization
  definition, 252
Sanitizers, 106
Serratia marcescens, 84, 226
Serratia spp., 9, 79
Scan RDI, 69, 144
Shigella spp., 9, 17, 51
Sodium hypochlorite, 106, 256
Sodium thiosulfate, 158
Soybean casein digest agar, 30, 38, 54, 87, 108, 120, 123, 294
Soybean casein digest broth, 29, 35, 37, 89, 91, 92, 94, 98, 127, 294
Sphingomonas spp., 9
Sphingomonas paucimobilis, 64
Spores, 2, 3, 18, 78, 86, 106, 122, 123, 134, 137, 141, 150, 151, 161, 226, 251, 260, 264, 293. *See also* Endospores
Sporocide
  definition, 252
Standard methods, 149
  problems with, 150
Standard methods for the examination of dairy products, 54
Standard methods for the examination of water and wastewater, 54
Standard operating procedures (SOP), 16, 79, 258
Staphylococcus aureus, 8, 16, 19, 25, 26, 29, 34, 37, 107, 126, 157, 158, 163, 173, 174, 259, 260, 261, 286, 297
Staphylococcus capitis, 8
Staphylococcus epidermidis, 8, 9, 121
Staphylococcus hominis, 8, 121
Staphylococcus spp., 8, 19, 68, 154
Stenotrophomonas maltophilia, 9, 68, 79, 154
Stenotrophomonas spp., 1, 9, 70, 167
Sterile pharmaceutical products, 77
  critical areas, 89
  manufacturing, 103
  microbial contamination, 79, 88
Sterility assurance level (SAL), 77, 94, 193
Sterility testing, 90, 94
  control sample, 92
  definition of, 94
  direct transfer, 92, 97

[Sterility testing]
isolators, 98, 99
membrane filtration, 93, 98
neutralizers, 93
sample volume, 92
steritest, 98
testing failures, 95
test method validation, 91
test requirements, USP, 90
Sterilization
autoclave, 78
bioburden (BB), 85, 87, 90, 95
class I recalls, 78
class II recalls, 78
conditions
cycle, 87
definition, 77
D value, 86, 87
media fills acceptance criteria, 90
methods, 81
process, 78
process control, 88
process for aseptic processing, 78, 88
process for dry heat, 78, 83
process for ethylene oxide gas, 78, 83
process for filtration, 78, 84
process for ionizing radiation, 78, 83
process for steam sterilization, 78, 82
validation, 78, 81, 85
validation for aseptic processing, 89
validation of filtration, important parameters for, 85
validation of media fills, 89
Stress induced response, 150
Streptococcus spp., 8, 9, 19
Streptococcus mutans, 8
Streptoccocus salivarius, 8

Taxeobacter spp., 9
Trichophyton mentagrophytes, 259, 260
Trichophyton spp., 9
Tryptic soy agar (TSA), 55

Ultramicrobacterium spp., 9
United States Pharmacopeia (USP), 83, 91, 110, 126, 148, 188, 193, 194, 201, 202, 214, 216, 227, 228, 230, 254, 273, 276–278, 284, 297
antimicrobial effectiveness test, 278, 284–289, 291, 297
bacterial endotoxins test, 228
bacterial indicators, 25
biological indicators, 135, 136, 140
microbial limits, 16, 19, 25, 26–28, 32–35, 37–40
pyrogen test, 188
sterility test, 90, 91, 94, 96, 97
test requirements for nonsterile products, 27
water for pharmaceutical purposes, 46, 58, 61, 63, 64, 68, 73

Vapor-phase hydrogen peroxide (VHP), 99
Viable air particulates, 109
Viable but not culturable, 4, 48, 49, 150
Vibrio, 51
Vibrio cholerae, 48, 53
Violet red bile glucose agar (VRBG), 37
Vogel-Johnson agar, 37

Water
action levels, 64
biofilms, 70
coliforms, 50
disinfection, 41
distilled, 46
environmental monitoring, 107
factors affecting microbial recovery from water, 54
fecal coliform procedure, 53
fungal counts, 54
grades of, 45, 46, 108
as a major pharmaceutical ingredient, 45
membrane filtration, 108
microbial contamination, 45
microbial diversity, 70

[Water]
  new testing methods, 68
  Pharmacopoeia Europa, 61
  potable, 45, 49, 108
  purified, 46, 61, 108
  quality control program, 65
  as a raw material, 45, 107
  recalls, 45
  regulations, 65
  thermophilic bacteria, 71
  total organic carbon (TOC), 71
  types of, 61, 108
  uses in pharmaceuticals, 107
  USP, 46, 61
Water activity, 39
Water for injection, 46, 61, 108, 194
Water system validation, 61

Xanthomonas spp., 9, 70, 167
Xanthomonas maltophilia, 64
Xylose lysine deoxycholate agar (XLDA), 35

Yeast, 19, 26, 27, 54, 57, 92, 94, 108, 122, 126–127, 157–158, 161, 163–164, 66–167, 175, 252, 273, 283–287, 289–291, 293, 297

Zygosaccharomyces rouxii, 287
Z-value, 87

Milton Keynes UK
Ingram Content Group UK Ltd.
UKHW020019071024
449327UK00032B/2855

9 780367 393946